Scholastic PHONICS™ A

Teacher's Guide

John Shefelbine
California State University at Sacramento
Story Editor

Published by Scholastic Inc.
Printed in the U.S.A.
ISBN 0590-76471-3

1 2 3 4 5 6 7 8 9 10 14 03 02 01 00 99 98 97

Credits

Teacher's Guide

Illustration Cover: ©Brian Dugan for Scholastic Inc.
Illustrations: ©Julie Durrell for Scholastic Inc.: pp 5, 7, 57, 240, 268. ©Vince Andriani for Scholastic Inc.: p8, 11, 19, 29, 33, 39, 43, 55, 59, 63, 87, 91, 95, 110, 111, 115, 150, 159, 198, 209, 211, 217, 224, 230, 233, 237, 241, 250, 254, 267, 269. ©Rick Brown for Scholastic Inc.: pp 13, 23, 54, 121, 187. ©Claude Martinot for Scholastic Inc.: pp14, 22, 28, 46, 77, 78, 79, 81, 85, 93, 99, 101, 109, 125, 131, 183, 203, 218, 221, 234, 260. ©Chris Reed for Scholastic Inc.: p 18. ©Susan Miller for Scholastic Inc.: pp 27, 137,165, 197, 223, 271. ©Laura Blanken for Scholastic Inc.: pp 37, 134. ©Nate Evans for Scholastic Inc.: pp 89, 118, 210, 255. ©Yvette Banek for Scholastic Inc.: pp102, 162. ©Holy Kowitt for Scholastic Inc.: pp 17, 21, 149, 249, 251, 253. ©Mary Jo Koeck for Scholastic Inc.: pp 173, 184. ©Kees de Kiefte for Scholastic Inc.: pp 175. Rusty Fletcher for Scholsatic Inc.: pp. 18-26

Photo Cover: ©John Curry for Scholastic Inc.
Photography: p. 9: © Ana Esperanza Nance for Scholastic Inc. p. 15: © David Lawrence for Scholastic Inc. p. 35: Images ©1997 PhotoDisc, Inc. p. 38: © Clara von Aich for Scholastic Inc. p. 45: © David Lawrence for Scholastic Inc. p. 47: Images ©1997 PhotoDisc, Inc. p. 49: © Bill Barley for Scholastic Inc. p. 51: Images ©1997 PhotoDisc, Inc. p. 53: Images ©1997 PhotoDisc, Inc. p. 61: Images ©1997 PhotoDisc, Inc. P. 65: Images ©1997 PhotoDisc, Inc. p. 69: © David Lawrence for Scholastic Inc. p. 70: © Ana Esperanza Nance for Scholastic Inc. p. 75: © David Lawrence for Scholastic Inc. p. 83: © Stanley Bach for Scholastic Inc. p. 97: © Bob Daemmrich/Tony Stone Images p. 113: Images ©1997 PhotoDisc, Inc. p. 117: © Tom Brakefield/The Stock Market. p. 123: © Clara von Aich for Scholastic Inc. p. 129: © David S. Waitz for Scholastic Inc. p. 130: Images ©1997 PhotoDisc, Inc. p. 133: © David Lawrence for Scholastic Inc. p. 139: © David Lawrence for Scholastic Inc. p. 145: © Ana Esperanza Nance for Scholastic Inc. p. 151: © David E. Franck for Scholastic Inc. p. 153: © Francis Clark Westfield for Scholastic Inc. p. 157: © Martin Simon for Scholastic Inc. p. 161: © W.M. Waterfall/The Stock Market. p. 167: © David E. Franck for Scholastic Inc. p. 171: Images ©1997 PhotoDisc, Inc. p. 177: © Cole Riggs for Scholastic Inc. p. 180: Images ©1997 PhotoDisc, Inc. p. 181: © John Lei for Scholastic Inc. p. 189: © Francis Clark Westfield for Scholastic Inc. p. 191: © Stephen Carr for Scholastic Inc. p. 193: Images ©1997 PhotoDisc., Inc. p. 194: © John Lei for Scholastic Inc. p. 195: © Lillian Gee for Scholastic Inc. p. 199: Images ©1997 PhotoDisc, Inc. p. 201: © Juan M. Renjifo/Animals Animals. p. 204: Images ©1997 Photo Disc, Inc. p. 205: © Cole Riggs for Scholastic Inc. p. 207: © David S. Waitz for Scholastic Inc. p. 213: © Mickey Gibson/Animals Animals. p. 215: © David E. Franck for Scholastic Inc. p. 225: © Stephen Dalton /Animals Animals. p. 227: © David E. Franck for Scholastic Inc. p. 231: © Bie Bostrom for Scholastic Inc. p. 235: © Ana Esperanza Nance for Scholastic Inc. p. 243: © Stouffer Enterprises/Animals Animals. p. 245: © Francis Clark Westfield for Scholastic Inc. p. 247: © Lillian Gee for Scholastic Inc. p. 259: Images ©1997 PhotoDisc, Inc. p. 263: © Ana Esperanza Nance for Scholastic Inc.

For reduced student page credits refer to Student Edition.

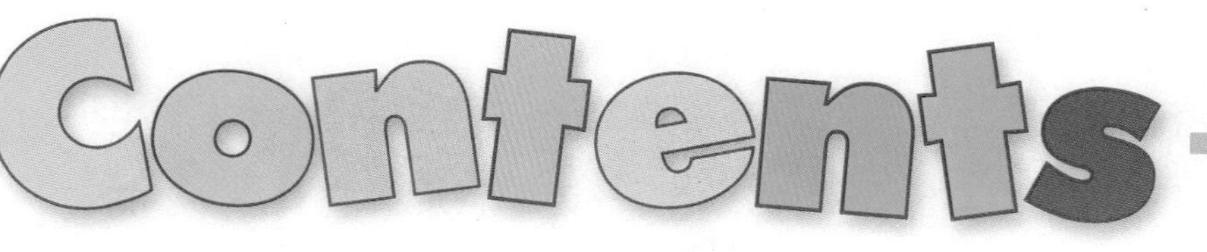

Contents

Unit 1

leaf

Unit 2

Contents

Unit 3

Unit 4

horse

clock

Unit 5

top

Unit 6

Creating Successful

Teachers want a program that's up-to-date, research-based, and able to meet the needs of the learners in their classrooms. Completely new, pedagogically sound, and attractively designed, ***Scholastic Phonics:***

is research-based.

Decisions about what to teach, how to teach it, what order to teach it in, and how to pace it are all based on research.

offers more opportunities for children to write and to read.

Children write on every page of this activity book, and decodable stories and poems appear at the end of every unit. You no longer have to wait half a year for children to be able to read or write meaningful, connected text.

top

has a front-loaded scope and sequence that gets children reading and writing faster.

Children learn high-frequency words and key sound-spelling relationships at the outset. That means they can read words in context and immediately get the satisfaction and reinforcement of making sense of the printed page.

presents a complete phonemic awareness program.

From rhyming, to blending, to segmenting, to manipulating sounds, children work with sounds to discover, learn, and review words and word parts.

offers built-in review through blending and word-building activities.

Children learn a sound-spelling relationship and put it to work within the context of a word. Not only do they consistently and continually

cat

blend sounds to make new words, but they also engage in word-building activities—such as working with word ladders and word tiles—as well as phonogram activities that reinforce the knowledge of sound-spelling relationships.

hat

contains activities for every child.

Suggestions are included for every learner—tactile, auditory, visual, or kinesthetic. In addition, scores of ideas for ESL, reteaching, and home-school connections are provided.

contains Grades 1 and 2 stories edited by Scholastic program author, John Shefelbine.

Literacy Place program author John Shefelbine, professor of reading education at California State University at Sacramento and nationally recognized authority on phonics and skill instruction, knows that a good approach to phonics instruction contains "skills in the context of a story, poem, or rhyme."

contains a complete handwriting program for Kindergarten and Grade 1.

Working with one letter at a time, children trace and write every consonant and vowel in both small and capital forms. An illustrated key word on each page helps children focus on the associations between sound, letter, and meaning. In Grade 1, children also trace a complete sentence in which small and capital forms of the letter appear.

One Program That Helps Kids Learn to Read

The Pupil Book

BUILD WORDS

Look at each picture. Use the letter tiles to write each picture name.

o b f x

1. box
2. fox

k l o ck i

3. lock
4. lick
5. kick

126 Build Words

BLEND WORDS

bride trip grade

brick

Grade 1

- Common, recognizable words used.

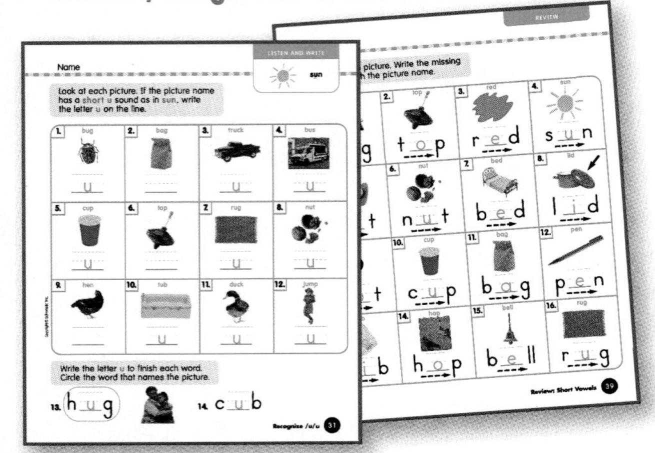
Name

LISTEN AND WRITE

Look at each picture. If the picture name has a short u sound as in sun, write the letter u on the line.

Write the letter u to finish each word. Circle the word that names the picture.

13. hug
14. cub

Recognize /u/u 31

REVIEW

top red sun nut bed lid cup bag pen hop bell rug

Review: Short Vowels 39

Grade 2

- Auditory discrimination leads to blending.

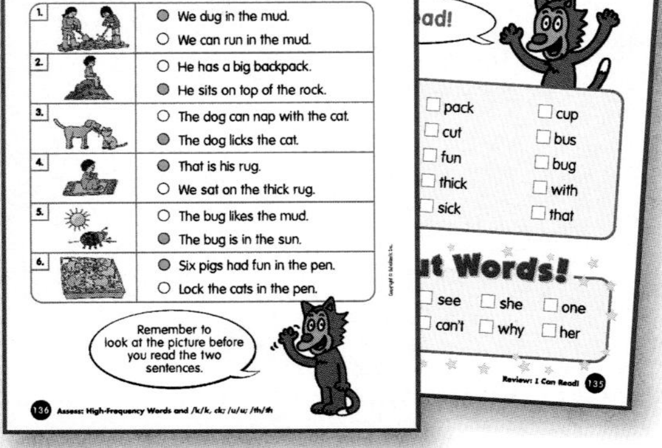
TEST YOURSELF

Fill in the bubble next to the sentence that tells about each picture.

1. We dug in the mud. / We can run in the mud.
2. He has a big backpack. / He sits on top of the rock.
3. The dog can nap with the cat. / The dog licks the cat.
4. That is his rug. / We sat on the thick rug.
5. The bug likes the mud. / The bug is in the sun.
6. Six pigs had fun in the pen. / Lock the cats in the pen.

Remember to look at the picture before you read the two sentences.

136 Assess: High-Frequency Words and /k/k, ck; /u/u; /th/th

Name

REVIEW

pack cup cut bus fun bug thick with sick that

see she one can't why her

Review: I Can Read! 135

Grade 1

- Assessment pages monitor children's progress.

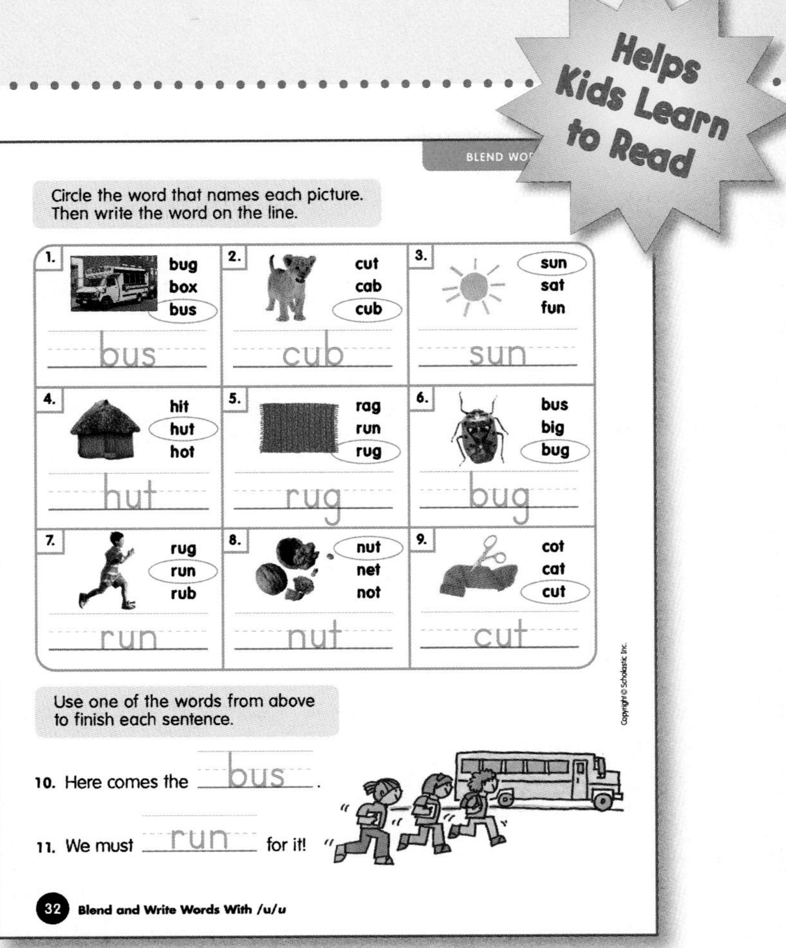
BLEND WORDS

Circle the word that names each picture. Then write the word on the line.

1. bug box bus — bus
2. cut cab cub — cub
3. sun sat fun — sun
4. hit hut hot — hut
5. rag run rug — rug
6. bus big bug — bug
7. rug run rub — run
8. nut net not — nut
9. cot cat cut — cut

Use one of the words from above to finish each sentence.

10. Here comes the bus.

11. We must run for it!

32 Blend and Write Words With /u/u

Copyright © Scholastic Inc.

Grade 2

- Initial and final consonants taught simultaneously.
- Writing, reading, and meaning work hand in hand.
- Clear, engaging photographs.
- As children blend, they write the whole word, not just a word part.
- Writing on every page.

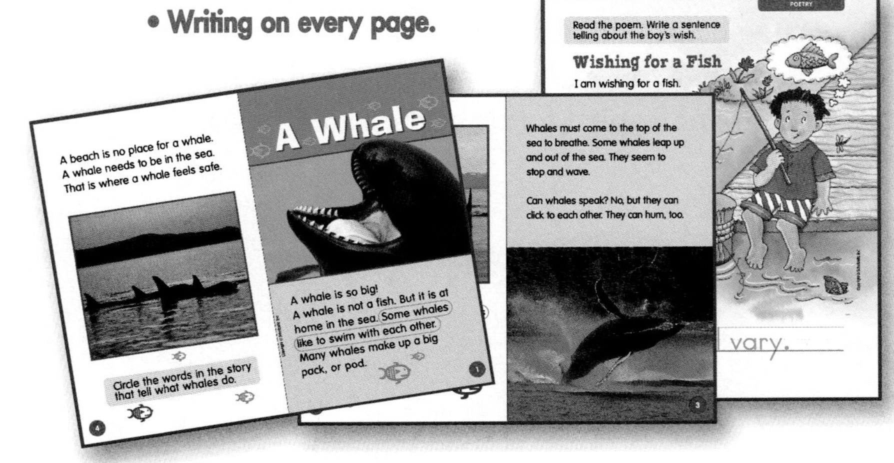
POETRY

Read the poem. Write a sentence telling about the boy's wish.

Wishing for a Fish

I am wishing for a fish.

A Whale

A beach is no place for a whale. A whale needs to be in the sea. That is where a whale feels safe.

Circle the words in the story that tell what whales do.

A whale is so big! A whale is not a fish. But it is at home in the sea. Some whales like to swim with each other. Many whales make up a big pack, or pod.

Whales must come to the top of the sea to breathe. Some whales leap up and out of the sea. They seem to stop and wave.

Can whales speak? No, but they can click to each other. They can hum, too.

Grade 1

- More practice reading decodable stories and poems.

Does It All!

The Teacher's Guide

Easy Step-by-Step Lesson Plan

- Explicit, systematic phonics instruction.
- QuickCheck informs instruction.
- Varied strategies for developing phonemic awareness are provided throughout the program.
- Modeling and think alouds employ a connected form of blending.
- Connections to useful and meaningful contexts in literature appear throughout the program.
- Hands-on activities provide reinforcement and reteaching.
- Suggestions for using the Letter Cards in the back of the workbook.

- The purpose of each activity page is stated.
- Familiar rhymes and rhythms are referenced and capitalized upon throughout the program.
- Suggestions help children acquiring English learn the regularities and irregularities of the language.
- Curriculum connections for cross-curricular activities.

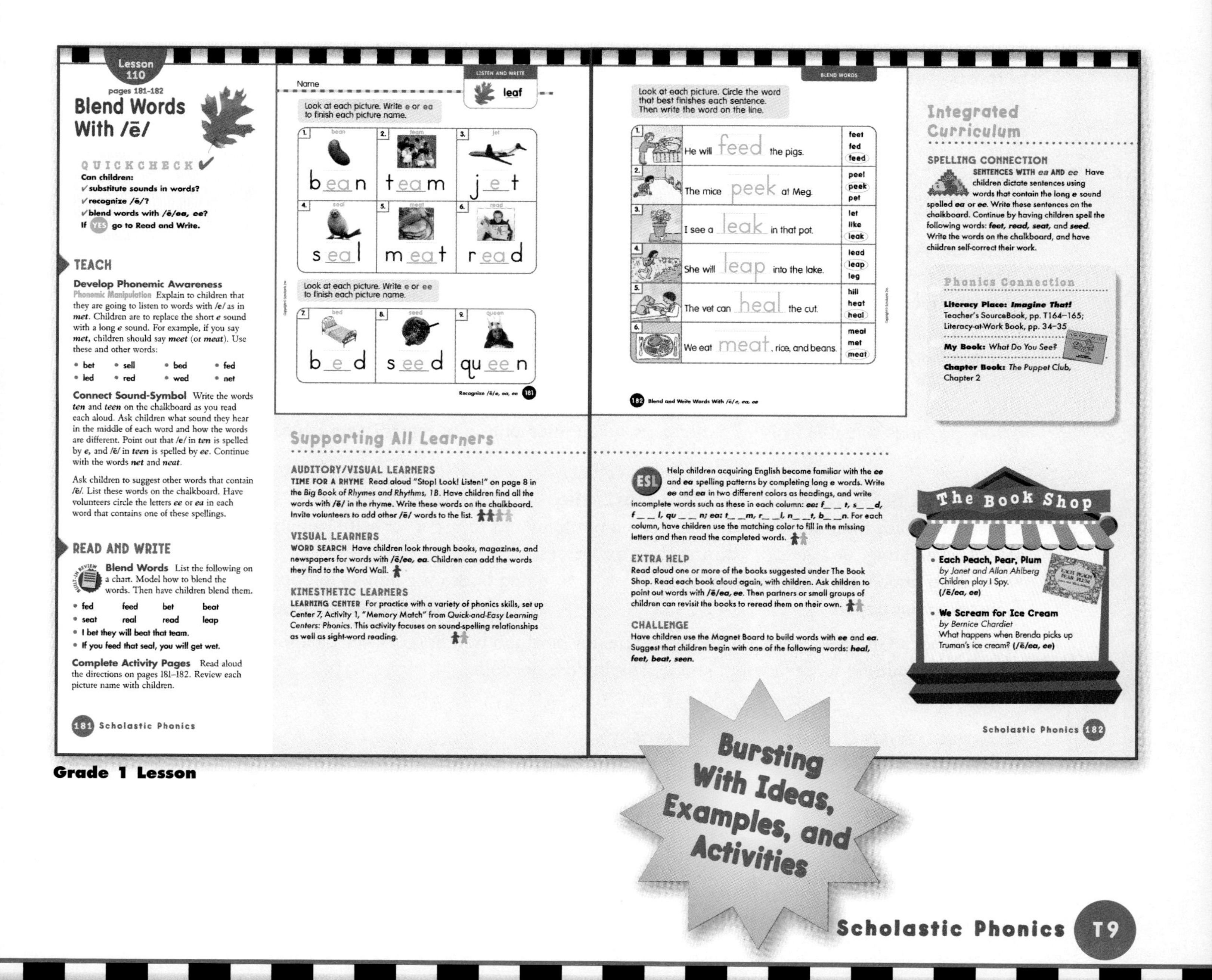

Lesson 110

pages 181-182

Blend Words With /ē/

QUICKCHECK ✓

Can children:
✓ substitute sounds in words?
✓ recognize /ē/?
✓ blend words with /ē/ea, ee?
If YES go to Read and Write.

TEACH

Develop Phonemic Awareness
Phonemic Manipulation Explain to children that they are going to listen to words with /e/ as in *met*. Children are to replace the short *e* sound with a long *e* sound. For example, if you say *met*, children should say *meet* (or *meat*). Use these and other words:

- bet • sell • bed • fed
- led • red • wed • net

Connect Sound-Symbol Write the words *ten* and *teen* on the chalkboard as you read each aloud. Ask children what sound they hear in the middle of each word and how the words are different. Point out that /e/ in *ten* is spelled by *e*, and /ē/ in *teen* is spelled by *ee*. Continue with the words *net* and *neat*.

Ask children to suggest other words that contain /ē/. List these words on the chalkboard. Have volunteers circle the letters *ee* or *ea* in each word that contains one of these spellings.

READ AND WRITE

Blend Words List the following on a chart. Model how to blend the words. Then have children blend them.

- fed feed bet beat
- seat real read leap
- I bet they will beat that team.
- If you feed that seal, you will get wet.

Complete Activity Pages Read aloud the directions on pages 181–182. Review each picture name with children.

181 Scholastic Phonics

LISTEN AND WRITE

Name ______ leaf

Look at each picture. Write e or ea to finish each picture name.

1. b ea n 2. t ea m 3. j e t
4. s ea l 5. m ea t 6. r ea d

Look at each picture. Write e or ee to finish each picture name.

7. b e d 8. s ee d 9. qu ee n

Recognize /ē/e, ea, ee 181

BLEND WORDS

Look at each picture. Circle the word that best finishes each sentence. Then write the word on the line.

1. He will feed the pigs. (feet / fed / feed)
2. The mice peek at Meg. (peel / peek / pet)
3. I see a leak in that pot. (let / like / leak)
4. She will leap into the lake. (lead / leap / leg)
5. The vet can heal the cut. (hill / heat / heal)
6. We eat meat, rice, and beans. (meal / met / meat)

182 Blend and Write Words With /ē/e, ea, ee

Supporting All Learners

AUDITORY/VISUAL LEARNERS
TIME FOR A RHYME Read aloud "Stop! Look! Listen!" on page 8 in the *Big Book of Rhymes and Rhythms, 1B*. Have children find all the words with /ē/ in the rhyme. Write these words on the chalkboard. Invite volunteers to add other /ē/ words to the list.

VISUAL LEARNERS
WORD SEARCH Have children look through books, magazines, and newspapers for words with /ē/ee, ea. Children can add the words they find to the Word Wall.

KINESTHETIC LEARNERS
LEARNING CENTER For practice with a variety of phonics skills, set up Center 7, Activity 1, "Memory Match" from *Quick-and-Easy Learning Centers: Phonics*. This activity focuses on sound-spelling relationships as well as sight-word reading.

ESL Help children acquiring English become familiar with the *ee* and *ea* spelling patterns by completing long *e* words. Write *ee* and *ea* in two different colors as headings, and write incomplete words such as these in each column: ***ee:* f__ __t, s__ __d, f__ __l, qu__ __n; *ea:* t__ __m, r__ __l, n__ __t, b__ __n**. For each column, have children use the matching color to fill in the missing letters and then read the completed words.

EXTRA HELP
Read aloud one or more of the books suggested under The Book Shop. Read each book aloud again, with children. Ask children to point out words with /ē/ea, ee. Then partners or small groups of children can revisit the books to reread them on their own.

CHALLENGE
Have children use the Magnet Board to build words with *ee* and *ea*. Suggest that children begin with one of the following words: ***heal, feet, beat, seen.***

Integrated Curriculum

SPELLING CONNECTION
SENTENCES WITH *ea* AND *ee* Have children dictate sentences using words that contain the long *e* sound spelled *ea* or *ee*. Write these sentences on the chalkboard. Continue by having children spell the following words: ***feet, read, seat,*** and ***seed***. Write the words on the chalkboard, and have children self-correct their work.

Phonics Connection

Literacy Place: *Imagine That!*
Teacher's SourceBook, pp. T164–165; Literacy-at-Work Book, pp. 34–35

My Book: *What Do You See?*

Chapter Book: *The Puppet Club*, Chapter 2

The Book Shop

- **Each Peach, Pear, Plum**
  by Janet and Allan Ahlberg
  Children play I Spy. **(/ē/ea, ee)**
- **We Scream for Ice Cream**
  by Bernice Chardiet
  What happens when Brenda picks up Truman's ice cream? **(/ē/ea, ee)**

Scholastic Phonics 182

Grade 1 Lesson

Bursting With Ideas, Examples, and Activities

As every teacher knows, assessment is an ongoing process, and a variety of strategies make up a successful assessment program. That's why many opportunities for assessment are built into every unit of both the Pupil Book and Teacher's Guide of *Scholastic Phonics.*

FORMAL AND INFORMAL ASSESSMENT, UNIT BY UNIT

I Can Read! Activity Pages On each I Can Read! page, children read new decodable words based on sounds they have learned in the unit. They also read high-frequency words that include words in the unit as well as words already learned. Although the I Can Read! page is designed to be used informally by children working with partners, it can also be used as a self-assessment by individuals.

Unit Assessment Pages These standardized test format pages are designed for formal assessment purposes, but may also be used for practice and reinforcement by children. You can use the product to generalize about phonics and reading progress or to diagnose ability with specific words, sounds, and skills.

ONGOING ASSESSMENT

Teacher Assessment and Anecdotal Records Various routes to assessment are suggested throughout the Teacher's Guide, including suggestions for making observations and using anecdotal records.

- **Observation** Firsthand observations help you record important information with freshness and immediacy. Note children's attitudes, level of interest, motivation, and self-confidence.
- **Anecdotal Records** Because the purpose of anecdotal records is to record learning as it is occurring, they help you focus on each child's process of learning rather than the product. This, too, provides valuable insight, particularly when intervention or remediation is needed.

Children's Self-Assessment: The Portfolio The foundation of children's self-assessment is their portfolio. Encourage children to keep a portfolio and to use it for self-evaluation. Children are self-evaluating, or assessing, during these processes:

- **selecting papers for their portfolio**
- **weeding out papers**
- **going back over papers to gauge progress**
- **using their portfolios to discuss progress**

The Teacher's Guide suggests beginning points for portfolio building, as well as appropriate junctures for adding to, reviewing, and discussing the portfolio.

Comprehensive, Formal Assessment

- **See pages T18–27 for formal assessments, which also employ a standardized test format and span several units.**
- **See pages T32–33 for a comprehensive skills checklist. You can use this checklist or make other observational notes to record baseline and developing phonemic awareness and phonics skills.**

Because of its auditory nature, phonics is a natural for the use of technology. In fact, one of the easiest and most productive ways to integrate technology with early learning is through the teaching of phonics.

WHY USE TECHNOLOGY?

Technology accommodates non-linguistic learners. Sounds, graphics, and screens that pulse all pique interest and provide alternatives to exclusively print formats.

Technology provides rapid or instant feedback. Programs let children know immediately whether an answer is right or wrong. Feedback can be played back as many times as needed.

Technology is a boon to ESL students. Besides instant feedback, other advantages for children acquiring English include the opportunity to learn at their own pace and in privacy.

USING TECHNOLOGY IN PHONICS INSTRUCTION

Here are just a few possibilities for using technology in your classroom:

- **WiggleWorks Plus Magnet Board** Children can move letters around on a computer screen to form word families, spell words, identify word parts, and hear words read.
- **Audiotapes** Children can listen to, record, play back, and hear individual sounds as they become connected text. Both children and teachers can use tapes as records of progress.
- **Electronic Multimedia Books** Children can read text and see it illustrated—and they can hear the text read and replay it at will. They can also manipulate the text and respond to it easily, creatively, and at the point of greatest interest.
- **Word Processors** Children can easily act upon text or try out multiple possibilities when writing or revising, or when doing replacement activities such as rhyming and word building.
- **Art Programs and Libraries** Children can select or create illustrations; they can also label, categorize, and write about graphics.
- **Networks** Children can develop a sense of audience through the use of e-mail and by networking with children in other classrooms. Technology is also one more way to connect the school and home.

English as a Second Language

ESL **Researchers agree that a second language is best taught through its use. With its front-loaded scope and sequence, which introduces the most versatile vowels and consonants at the outset, *Scholastic Phonics* enables children acquiring English to learn to listen to new sounds and make new sound-spelling relationships while encouraging them to start reading and writing *right away.***

From the very beginning of the program, ***Scholastic Phonics*** provides all students with ***fully decodable*** poems and stories. These stories and poems contain only words that are made up of predictable sound-spelling relationships. This approach to learning to read and write decreases reliance on context clues, and increases reliance on sound-spelling relationships. As a result this approach puts the second-language learner on a level footing with native speakers in the classroom. Similarly, phonemic awareness activities suggested on almost every page of the Teacher's Guide help children focus on the most essential and most commonly repeated sounds and word patterns. With this firm foundation built in a meaningful context, children acquiring English can begin to use the language receptively, productively, and for real purposes.

leaf

Here are some of the other ways in which *Scholastic Phonics* meets the needs of children acquiring English.

NEEDS OF ESL STUDENTS	HOW SCHOLASTIC PHONICS MEETS THAT NEED
Illustrated Vocabulary Featuring Common Words	Hundreds of full-color photos of useful, high-frequency words appear throughout the program—on Pupil Book pages as well as on picture cards.
Individual, Partner, and Small Group Work	Suggestions for individual, partner, and small group work appear throughout the Teacher's Guide.
Multiple Opportunities to Hear Discrete Sounds	Phonics instruction in general, and the modeling in this program, enable second-language learners to hear and learn the discrete sounds that are blended to become words.
Multiple Opportunities to Hear Sounds Blended	In *Scholastic Phonics*, the Teacher's Guide provides consistent emphasis on developing phonemic awareness through blending that elongates and emphasizes discrete sounds even as it blends them.
Visual, Tactile, Auditory, and Kinesthetic Activities	Suggestions for a wide range of activities appear throughout the Teacher's Guide.
Concrete Tasks, Mapping of Concepts, Bridges Between the New and the Known	ESL activities throughout the Teacher's Guide provide suggestions for all of these and more.
Multiple Opportunities to Pronounce Sounds, Get Feedback, and Try Again	The Teacher's Guide helps teachers through modeling that elicits choral responses and repeated experiences with the same letters in varying sequences. This provides the ESL student with multiple opportunities to pronounce within the safety of group work.

Dear Family,

Your child is about to begin a phonics program. This program will help your child make connections between the sounds and words he or she has heard and words in print. As your child makes these connections, skills in both reading and writing will develop and grow.

You can help your child in many ways on this exciting journey to literacy. When you are with your child, consider engaging in these activities:

- **Point out environmental print.** This includes street signs and shop signs, signs in buildings, and signs on buses and subways. Anything that is posted, from "For Sale" to "Exit," can be read or partially decoded. At first your child may recognize only a single letter or word part, but encourage any and all attempts to sound out words or to connect the printed word with spoken word.
- **Sing songs and recite rhymes.** From "Mary Had a Little Lamb" to "Ten in the Bed," children can experience and learn the sounds and rhythms of language through the joy of singing and reciting. Even the silliest rhyme provides a lesson in how we communicate through language and the shapes our words take.
- **Choose a "letter of the day."** Look for this letter with your child. Find it in books, in junk mail, in signs, and in household notes posted on your refrigerator. As appropriate, sound out words in which you find it.
- **Engage in wordplay of any kind.** Any kind of word game will increase your child's exposure to, and delight in, words. Children love riddles and puns. They also may enjoy coming up with lists of rhyming words, making up rhymes, making lists of words or silly sentences with words that all begin with the same letter, and, later, playing games such as Hangman and Concentration with words.
- **Celebrate and encourage!** Your child will be learning, in most cases, one sound and one corresponding letter at a time. Learning each correspondence is an accomplishment! Even if your child cannot decode a complete word, remember that decoding part of it shows progress.

As we engage in these and more literacy activities in school, your insights into your child's needs, progress, and development as a reader and writer are always welcome.

Sincerely,

Estimada familia:

Su hijo está por comenzar un programa de fonética. Este programa lo ayudará a hacer conexiones entre los sonidos y las palabras que ha escuchado y las palabras impresas. En la medida que su hijo haga esas conexiones, las destrezas de lectura y escritura irán desarrollándose y creciendo.

Usted puede ayudar a su hijo de diferentes maneras en este emocionante viaje a la lectoescritura. Cuando esté con su hijo, considere participar en estas actividades:

- **Señalen el material impreso que los rodea.** Incluyendo letreros de las calles y tiendas, letreros en los edificios y letreros en los autobuses y estaciones del metro. Cualquier letrero, desde "Se vende" hasta "Salida", puede leerse o descifrarse parcialmente. Al principio, su niño podrá reconocer solamente una letra o parte de la palabra, pero anímelo en cualquier intento de pronunciar una palabra o de conectar la palabra escrita con la palabra hablada.
- **Canten canciones y reciten rimas.** Desde "Mary tenía un borreguito" hasta "Diez en la cama", los niños pueden experimentar y aprender los sonidos y ritmos del lenguaje a través de la alegría de cantar y recitar. Hasta la rima más sencilla nos enseña cómo nos comunicamos a través del lenguaje y la forma que toman nuestras palabras.
- **Escojan una "letra del día".** Busque esta letra con su niño. Encuéntrenla en libros, en la correspondencia que no es importante, en letreros y en notas puestas en la puerta del refrigerador de la casa. Cuando sea apropiado, pronuncien las palabras donde encuentren esa "letra del día".
- **Participen en cualquier tipo de juego de palabras.** Cualquier tipo de juego de palabras aumentará el contacto de su niño con las palabras y hará que las disfrute. A los niños les fascinan las adivinanzas y los juegos de palabras. También pueden disfrutar creando listas de palabras que rimen, inventando rimas, haciendo listas de palabras u oraciones divertidas con palabras que empiecen con la misma letra, y después, jugando juegos como " El Ahorcado" y "Concéntrese" (Memoria).
- **¡Celebren y anímense!** Su hijo estará aprendiendo, en la mayoría de los casos, un sonido y la letra correspondiente a la vez. ¡Aprender cómo corresponden es un logro! Aunque su hijo no pueda descifrar una palabra completa, recuerde que descifrar parte de ella muestra que hay progreso.

Al mismo tiempo que nos vamos iniciando en todas estas actividades de lectoescritura y más en la escuela, sus comentarios y observaciones acerca de las necesidades , progresos y desarrollo de su hijo como lector y escritor serán bien recibidos.

Atentamente,

List of Decodable Words (from Scholastic Phonics)

am
as
at
baby
back
bad
bags
bake
ball
bat
bead
beak
bean
beat
bed
beep
beg
Ben
bench
best
bet
bib
Bill
bit
bite
block
blouse
blow
boat
Bob
bone
book
boot
box
brace
brake
brave
brick
bride
bring
broke
broom
brush
bud
bug
bugs
bun
bus
by
cab
cake
called
came
can
cast
cat
cave
cheek
cheese
child
chill
chin
chop
clam
clap
clay
clean
click
clip
clock
cloud
coat
cob
cot
crab
cub
cube
cup
cut
cute
dad
dash
date
day
deal
deep
did
dig
dime
dip
dish
dock
dog
dog's
donkey
don't
dot
down
dress
duck
dug
dust
Ed
egg
fade
fan
fast
fat
fed
feed
feel
feet
fight
fill
find
fine
fish
fist
fit
five
fix
fog
food
football
fox
frog
fry
fun
game
gate
gave
get
go
goat
good
got
grab
grade
grape
gray
green
greet
greeted
grill
grin
grow
gum
ham
hand
hat
hate
hay
heal
heat
hen
hid
hide
hike
hill
him
hip
hit
hive
hole
home
hop
hope
hose
hot
hug
line
lip
list
lit
lock
log
lot
loud
low
luck
lunchbox
mad
made
mail
make
man
man's
mat
may
meal
mean
meat
meet
men
met
mild
mine
mix
mixed
mom
mop
mouth
mud
mule
hugs
hum
hush
hut
it's
Jack
jam
Jan
jaw
jeans
jeep
jet
job
jog
jug
keep
key
kick
king
kite
lake
last
late
lead
leaf
leak
leap
led
leg
let
lick
light
like
lime
must
name
neat
need
nest
net
night
nine
nose
not
note
nut
ox
pack
paid
Pam
pan
past
pat
paw
peek
peg
pen
pet
pick
pig
pill
pin
pine
pink
plain
plan
plane
play

plug
pole
pony
pop
pot
pup
queen
quit
quiz
rag
rain
rake
ran
rat
real
red
rest
ride
right
ring
rip
road
rob
robe
rock
rode
room
rope
rose
rot
rub
rug
run
runs
rush
rust
sack
sad
sailboat
Sam
same
sat
save
saw
seal
seam
seat
seed
seemed
seen
shake
shell
shin
ship
shook
shop
shot
shout
shut
shy
sick
side
sight
sit
six
sky
small
smash
smile
smoke
snack
snake
snap
so
sock
soon
south
speak
speed
spill
spin
spoke
spot
spun
stab
steal
steam
steep
stick
still
stone
stool
stop
sun
sunlight
tag
tags
take
tall
tan
tap
teach
team
Ted
teen
ten
test
than
that
them
then
thick
thin
thing
think
this
time
tip
toad
top
trace
track
trade
train
tray
treat
tree
trick
trim
trot
truck
try
tub
tug
up
use
van
vase
vest
vet
vine
vote
wag
wait
wall
wave
wax
weak
weed
week
wet
whale
wheat
wheel
white
why
wig
wild
will
win
wink
wish
yes
yet
zap
zip

High-Frequency Words

about
after
and
around
because
before
boy
can't
come
could
do
does
down
first
for
from
girl
have
he
her
his
I
is
like
look
many
more
my
no
now
one
or
other
out
said
school
see
she
some
the
their
them
then
there
they
to
up
very
want
was
we
were
when
where
which
who
why
with
you
your

Lesson 1

pages 5-6

Recognize the Letters of the Alphabet

QUICKCHECK ✔

Can children:

✔ **recognize rhyme?**

✔ **identify the letters of the alphabet?**

✔ **write the letters of the alphabet?**

If YES go to Read and Write.

TEACH

Develop Phonemic Awareness

Rhyme Show children three pictures, and say the names of each. Use the picture cards from the Phonemic Awareness Kit for **cat, hat,** and **yellow;** or make your own picture cards by drawing or pasting down pictures on large index cards. Ask children which picture names rhyme. Point out that the words ***cat*** and ***hat*** rhyme because they both end in ***-at***. Continue with picture cards such as these:

- pen, hen, game
- dig, green, wig
- run, hat, sun
- coat, boat, plane
- vase, fox, box
- red, can, fan
- gate, mop, top
- clock, quilt, sock

Review Letter Formation Review how to form the capital and small forms of letters ***Aa, Bb,*** and ***Cc***. Have volunteers model on the chalkboard how to write each letter. Continue with the rest of the alphabet.

READ AND WRITE

Recognize Letters Using the letter cards from the back of the book, divide the set of cards into six groups of four or five letters each. Beginning with the first group ***(Aa, Bb, Cc, Dd),*** hold up and name each letter. Then place the cards in random order on the chalkboard ledge. Have volunteers sequence them.

Complete Activity Pages Read aloud the directions on pages 5–6. Have children complete the pages independently.

Supporting All Learners

KINESTHETIC LEARNERS

GO TO SCHOOL! To practice associating the consonants with their corresponding sounds, children may enjoy playing "Go to School" on pages 9–12 in *Quick and Easy Learning Games: Phonics.*

TACTILE LEARNERS

SHAPE AND MOLD Children can form letters out of clay or dough. You can make dough by mixing four cups of flour, one cup of salt, and one and three-quarter cups of warm water.

VISUAL/AUDITORY LEARNERS

ALPHA PICK GAME Place the letter cards in a bag. Have children take turns closing their eyes and choosing a card from the bag. After children identify each letter they chose, they can display it on the chalkboard ledge. When five or more letters cards are displayed, have volunteers arrange the letters in alphabetical order. Then continue the Alpha Pick Game.

Integrated Curriculum

WRITING CONNECTION

NAME THAT LETTER! Call out each letter of the alphabet in order. After each letter you call out, have a volunteer write the capital and small form of that letter on the chalkboard. Then call out the letters in random order. Have volunteers erase each letter as it is called.

SOCIAL STUDIES CONNECTION

LETTER SEARCH Walk around your school or neighborhood with children. Ask them to point out capital and small letters of the alphabet in signs you see.

Phonics Connection

Literacy Place: *Hello!*
Teacher's SourceBook, pp. T20–21, T42–43, T66–67

Big Book of Rhymes and Rhythms, 1A: "Alphabet Song," pp. 4–5

Phonics and Word Building Kit: Sounds of Phonics Audiocassette (Personal Voice) "ABC Soup"; pocket ABC cards and pocket chart

EXTRA HELP

Give children one letter card each. Have children arrange themselves in alphabetical order. Then randomly say letter names and/or sounds. Have children each raise his or her letter card if it matches the name or sound you said.

CHALLENGE

Invite children to begin forming an Alphabet Book. On each page they should write the capital and small form of one letter. Then they can add a drawing of an object or animal whose name begins with the sound that the letter stands for. As the year progresses, have children add words and other pictures to each page.

ASSESSMENT

To begin a baseline assessment, consider using this page in combination with pages 7 and 8, as well as your own observations. You may record each child's ability to recite the alphabet, form capital and small letters, match capital and small letters, and sequence letters.

FLEXIBLE GROUPING KEY

= individual = partners/small groups = whole class

Lesson 2

page 7

Match Capital and Small Forms of Letters

QUICKCHECK ✔

Can children:

✔ **orally segment words by syllables?**

✔ **recognize the capital and small forms of letters of the alphabet?**

✔ **match the capital and small forms of letters of the alphabet?**

If YES go to Read and Write.

TEACH

Develop Phonemic Awareness

Oral Segmentation Explain to children that you are going to say a word and then clap the number of syllables, or word parts, you hear. For example, if you say the word ***turtle***, you will clap twice—***tur...tle***. Do several examples with one-, two-, and three-syllable words such as ***mouse, river***, and ***tomato***. Continue with the following words, and have children clap the syllables with you as you repeat each word:

- pencil
- book
- frog
- eraser
- can
- bat
- kitten
- ball
- triangle
- water
- school
- kangaroo

Match Capital and Small Letters

Divide the class into small groups. Give each group several pairs of capital and small letters from the packet of pocket ABC cards in the Phonics and Word Building Kit, or make your own pairs of alphabet cards. Ask groups to name each letter. Then have them match the pairs. Finally, have groups identify which letter in each pair is capital and which is small. Groups can then exchange letters.

READ AND WRITE

Complete Activity Page Read aloud the directions for page 7. Have children complete the page independently.

Supporting All Learners

KINESTHETIC/AUDITORY LEARNERS

NAME GAME Distribute the letter cards from the back of the book among small groups. Ask groups to name the object or animal shown on each card, to repeat the sound they hear at the beginning of the name of each object or animal, and to turn over each card to see the letter that represents the sound. When groups finish with one set of cards, they can trade with other groups for different sets.

CHALLENGE

Give children one capital and small letter from the pocket ABC cards in the Phonics and Word Building Kit. The letters should not match. Have children make matching sets by trading letters with classmates. Remind children that some of them may have to trade several times before they have a matching pair of letters. When children get a matching set of letters, have them name the letter and tell its sound.

Connect the dots in ABC order.

Supporting All Learners

KINESTHETIC/VISUAL LEARNERS

LETTER SHUFFLE Shuffle the letter cards from the back of the book. Challenge pairs or small groups of children to put the cards in ABC order. Then children can check the ABC order by reciting or singing "ABC Soup" (from the Phonics and Word Building Kit: Sounds of Phonics Audiocassette) as they flip through their arranged cards.

Pair children acquiring English with native English speakers who can pronounce the name of each picture on the letter cards. Suggest the following pattern of response as children look at the letter cards one by one: The child whose primary language is English names the word that identifies the picture. His or her partner repeats the word, and both children work together to say the sound that begins each picture name.

Recognize Alphabet Sequence

QUICKCHECK ✔

Can children:

✓ orally blend word parts?

✓ sequence the letters of the alphabet?

If YES go to Read and Write.

TEACH

Develop Phonemic Awareness

Oral Blending Explain to children that you will read words in two parts. Children are to say the word as a whole. For example, say ***ta...ble,*** and ask children what the word is. Blend the word ***table*** for children. Continue with these word parts:

- teach...er
- home...work
- safe...ty
- to...day
- lunch...box
- re...cess
- pen...cil
- sev...en

Sing Alphabet Song Read aloud or sing the traditional "Alphabet Song." (The song appears on pages 4–5 in the *Big Book of Rhymes and Rhythms, 1A*.) Explain to children that the alphabet song tells all the letters in order. Everybody learns this order because it helps us find things. For example, words in dictionaries are arranged in this order.

READ AND WRITE

Recognize Letters Say each letter of the alphabet. As you do so, hold up a letter card to show it, or write the letter on the chalkboard. Invite every child whose first name begins with that letter to stand. If you wish, children may arrange themselves in alphabetical order by first name.

Complete Activity Page Read aloud the directions on page 8. Children can complete the page independently.

Lesson 4

pages 9-10

Recognize Rhyme

QUICKCHECK ✔

Can children:

✓ recognize rhyme?

If YES go to Read and Write.

TEACH

Develop Phonemic Awareness

Rhyme Write the rhyme "To Market, To Market" on chart paper. Track the print as you read it aloud. Reread the rhyme several times, having children clap the rhythm. Then have children point out the rhyming words in the poem.

To Market, To Market

To market, to market, to buy a fat pig.
Home again, home again, jiggety jig.
To market, to market, to buy a fat hog.
Home again, home again, jiggety jog.

Introduce Rhyme Point out that *hog* and *jog* rhyme because they sound the same at the end. They both end in /og/. Ask children to suggest other words that rhyme with *hog* and *jog*. Model how to make a rhyme.

THINK ALOUD

I know that the words *hog* and *jog* rhyme because they both end in /og/. I can make another word that rhymes with *hog* and *jog*. This word begins with /f/. It's *fog*. What other words rhyme with *fog*?

State aloud one of the following words, or display a picture of the object or animal: ***bat, can, fish, fox, goat, hen, nose, sun, top.*** Have children suggest words that rhyme with each word or picture name. List these rhyming words. Use the words to create rhyming poems or stories with the children.

READ AND WRITE

Complete Activity Pages Read aloud the directions on pages 9–10. Review each picture name with children.

RHYME

Name

Say the name of the first picture in each row. Then color the pictures whose names rhyme.

Supporting All Learners

KINESTHETIC LEARNERS

PICTURE SORT Display the following picture cards from the Phonemic Awareness Kit: **bat, box, can, cat, dog, fan, frog, fox, run, sun.** Have children sort the pictures into pairs whose names rhyme, or as you hold up a picture and say its name, have children identify the picture whose name rhymes with it.

TACTILE LEARNERS

RHYME TIME Have children build rhyming words that they talked about in this lesson on the Magnet Board or with the magnetic letters. See the Wiggleworks Plus My Books Teaching Plan for information on this and other Magnet Board activities.

VISUAL/AUDITORY LEARNERS

COLLECTING RHYMES Have children fold four sheets of construction paper into cards. Children can then write the words ***fog, pig, man,*** and ***bun*** on the cover of these cards, one word per card. Then have children write words that rhyme with each word on the inside of each card.

CHALLENGE

Write the following groups of three words. Tell children that two of the words in each group rhyme. Ask children to circle the word in each group that doesn't rhyme. Children can then write their own groups of three words, two of which rhyme.

• **rest**	**nest**	**cook**
• **bun**	**ox**	**fun**
• **dig**	**tub**	**pig**
• **dress**	**mice**	**rice**

EXTRA HELP

Model how you recognize that the words ***hot*** and ***pot*** in "Pease Porridge Hot" rhyme. Frame the rhyming words as you reread the poem. Then ask children to name other words that rhyme with ***hot*** and ***pot***.

Integrated Curriculum

SPELLING CONNECTION

LETTER SWITCH Help children write the words ***bat, can, fox, goat, hen, sun, top, hog,*** and ***pig***. Then ask children to change the first letter in each word to write a new word. Have children compare their new words.

SOCIAL STUDIES CONNECTION

WORD SEARCH Children can look in their school and neighborhood to find words that rhyme. Have children create a list of the words they find. Then children can suggest more words to rhyme with the words on the list.

Phonics Connection

Phonics and Word Building Kit: Sounds of Phonics Audiocassette (Personal Voice), "Willoughby Wallaby Woo," "Jack and Jill," and "Two Little Blackbirds"; magnetic letters

Big Book of Rhymes and Rhythms, 1A: "Pease Porridge Hot," p. 21

Phonemic Awareness Kit: rhyme exercises

Lesson 5

pages 11-12

Recognize and Write /m/*m*

QUICKCHECK ✔

Can children:

- ✔ **orally segment word parts?**
- ✔ **write capital and small *Mm*?**
- ✔ **recognize /m/?**
- ✔ **identify the letter that stands for /m/?**

If YES go to Read and Write.

TEACH

Develop Phonemic Awareness

Oral Segmentation Explain to children that you are going to say a word that has two parts. Children are going to listen to the word and then say it in two parts. For example, say the word ***morning***. Then slowly say ***mor...ning***. Continue with the following words:

- mailbox
- mountain
- mitten
- monkey
- marker
- mushroom

Write the Letter Explain to children that the letter ***m*** stands for **/m/,** the sound at the beginning of ***mop***. Write ***Mm*** on the chalkboard. Point out the capital and small forms of the letter. Model how to write the letter.

Have children write both forms of the letter in the air with their fingers. Have volunteers practice writing the letter on the chalkboard.

READ AND WRITE

Connect Sound-Symbol Write the word ***mat*** on the chalkboard, and have a volunteer circle the letter ***m.*** Remind children that the letter ***m*** stands for **/m/**. Ask children to suggest other words that begin with **/m/**. List these words on the chalkboard, and have volunteers circle the letter ***m*** in each one.

Write the following on the chalkboard: ***_at, _ad, _op***. Have volunteers add the letter ***m*** to each. Model how to blend each word. Then have children say each word and write it.

Complete Activity Pages Read aloud the directions on pages 11–12. Review each picture name with children.

HANDWRITING

Name

Mop begins with the m sound.

Mm

M M M

m m m

Max has a mop.

Write the Letter *Mm* 11

Supporting All Learners

VISUAL LEARNERS

***Mm* SEARCH** Invite children to find objects in the school or classroom whose names contain **/m/**. Children can draw a picture of each object. Have children write or dictate the name of the object under the picture.

KINESTHETIC/VISUAL LEARNERS

ALL ABOUT LETTERS Give children five or more pairs of capital and small letters from the pocket ABC cards in the Phonics and Word Building Kit. Include the capital and small ***Mm***. Have children find ***M*** and ***m,*** identify the capital and small forms, and then write the letters.

TACTILE LEARNERS

M* IS FOR MACARONI** Have children make ***Mm posters by writing large forms of ***M*** and ***m*** on sheets of paper. Children can write words with ***Mm*** on the posters and draw pictures of things with names that contain **/m/**. Then have children paste macaroni on each ***Mm*** on the poster. Children can trace the macaroni ***m*** shapes with their fingers.

ESL Display objects (or pictures of objects) whose names begin with **/m/**. These may include the following: **map, mug, money, mask, mittens, mouse, monkey, moon,** and **mustard**. Prompt children acquiring English to take turns identifying each object, emphasizing **/m/** as they say the word. Then give each child a self-sticking note with the letter ***m*** written on it. Have children trace the ***m*** and attach the note to one of the objects as they repeat its name.

EXTRA HELP

Children who need extra help may wish to draw pictures of things with names that begin with **/m/m**. Guide children to label the items by writing the name of each.

Integrated Curriculum

WRITING CONNECTION

STORY TIME Have children generate a list of words with **/m/m**. Record these words on the chalkboard. Then have children create a story using as many of the words in the list as possible. Write the story on chart paper. Return to the story in subsequent days, rereading it with children.

MATH CONNECTION

WORD COUNT Read aloud "The Muffin Man" from the *Big Book of Rhymes and Rhythms, 1A*. Have children find and count all the words with **/m/** in the rhyme. Then have children find and count all the words with **/m/** in their story.

Phonics Connection

Literacy Place: *Hello!*
Teacher's SourceBook, pp. T120–121; Literacy-at-Work Book, pp. 9–10

My Book: *My Dog Max*

Big Book of Rhymes and Rhythms, 1A: "The Muffin Man," p. 6

Chapter Book: *I Am Sam*, Chapter 6

OBSERVATION

Select two or three children each day to observe. Keep anecdotal records of each child's progress. Provide additional support for children displaying patterns of difficulty over time.

Lesson 6

page 13

Blend Words With /m/m

QUICKCHECK ✔

Can children:

✔ orally blend word parts?

✔ identify /m/?

✔ blend words with /m/m?

If YES go to Read and Write.

TEACH

Develop Phonemic Awareness

Oral Blending State aloud the following word parts, and ask children to orally blend them. Provide corrective feedback and modeling when necessary.

/m/...at	**/m/...om**	**/m/...op**
/m/...ad	**/m/...onster**	**/m/...ilk**

Connect Sound-Symbol Review with children that the letter *m* stands for /m/ as in *mat*. Write the word *mat* on the chalkboard, and have a volunteer circle the letter *m*. Then model for children how to blend the word.

THINK ALOUD

I can put the letters *m, a,* and *t* together to make the word *mat*. Let's say the word slowly as I move my finger under the letters. Listen to how I string together the sound that each letter stands for to make the word: *mmmmaaaat, mmaat, mat.*

Continue by helping children to blend the following words: ***mom*** and ***mop***.

READ AND WRITE

Blend Words To practice using the sound taught and to provide practice with blending, list the following words on a chart. Have volunteers read each aloud. Model blending when necessary.

- **mat** **man** **map**

Complete Activity Page Read aloud the directions on page 13. Review the art with children.

Name

Color all the pictures whose names begin with m.

Write m words from the picture.

mice monkey

mail mittens

Supporting All Learners

KINESTHETIC/VISUAL LEARNERS

POSTCARD MAIL Have children draw an ***m*** and pictures of things with names that have **/m/m** on the front of paper cut into the shape of a postcard. Children can write words with **/m/m** on the back of their postcards. Then ask children to exchange postcards.

The Book Shop

- **A Funny Man**
  by Patricia Jensen
  Children follow a man all around his town. **(/m/m)**
- **My Messy Room**
  by Mary Packard
  This story is about a child who is proud of his messy room. **(/m/m)**

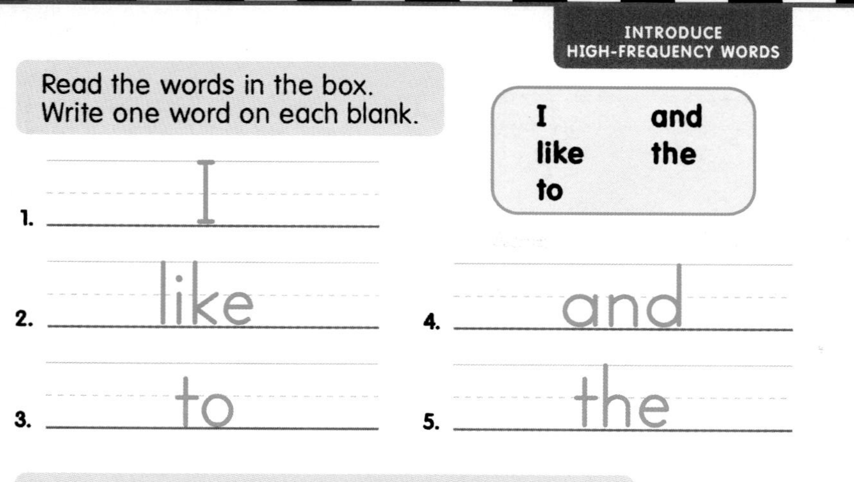
INTRODUCE HIGH-FREQUENCY WORDS

Read the words in the box.
Write one word on each blank.

I	and
like	the
to	

1. I
2. like
3. to
4. and
5. the

Use a word from the box to finish each sentence.

6. I like the rain.
7. I like to run and jump.
8. I run to the house.

Copyright © Scholastic Inc.

14 Recognize and Write High-Frequency Words

Supporting All Learners

KINESTHETIC LEARNERS

PORTFOLIO

WORD COUNT Children can count the number of high-frequency words they find in their classroom or school. Have children write each word they find on a sheet of paper. Keep a tally of the number of times they see it in books and on signs.

EXTRA HELP

For children who need additional support reading the high-frequency words, select the appropriate cards from the high-frequency word cards in the Phonics and Word Building Kit, or make your own cards using index cards. Have pairs of children practice reviewing the words by using the cards as flashcards. Children may also use the cards to help them find the high-frequency words in books and in the classroom.

Recognize High-Frequency Words

QUICKCHECK ✔

Can children:

✔ **recognize and write the high-frequency words *I, like, to, and, the?***

✔ **complete sentences using the high-frequency words?**

If YES go to Read and Write.

TEACH

Introduce the High-Frequency Words

Write the high-frequency words ***I, like, to, and, the*** in sentences on the chalkboard. Read each sentence aloud. Underline the high-frequency words, and ask children if they recognize them. You may wish to use the following sentences:

1. **I am mad.**
2. **I like the man.**
3. **Mom and I like to mop.**

Then ask volunteers to dictate sentences using the high-frequency words. You may wish to begin with the following sentence starters: ***I am*** ___ or ***Mom and I like to*** ___.

READ AND WRITE

Practice Write each high-frequency word on a note card. Read each word aloud as you display the cards. Then do the following:

- Mix the cards.
- Display one card at a time, and ask children to state each word aloud.
- Have children spell each word aloud, clapping on each letter.
- Ask children to write each word in the air as they state aloud each letter. Then have them write each word on a sheet of paper.

Complete Activity Page Read aloud the directions on page 14. Review the art with children.

Lesson 8

pages 15-16

Recognize and Write /a/a

QUICKCHECK ✔
Can children:
- ✔ **orally blend word parts?**
- ✔ **write capital and small *Aa*?**
- ✔ **recognize /a/?**
- ✔ **identify the letter that stands for /a/?**

If YES go to Read and Write.

TEACH

Develop Phonemic Awareness

Oral Blending Explain to children that they are going to put some sounds together to make words. You will say only the beginning sound of a word and then the rest of the word. Then you will ask them to tell you what the word is. For example, you might say /s/*...at*, and guide children to tell you the word ***sat***. Continue with these word parts:

/m/...ap	/r/...an	/s/...ad
/m/...at	/f/...an	/m/...an

Write the Letter Explain to children that the letter ***a*** stands for /a/ as in ***cat***. Write ***Aa*** on the chalkboard. Point out the capital and small forms of the letter. Then model for children how to write the letter.

Have children write both forms of the letter in the air with their fingers. You may also wish to have volunteers practice writing the letter on the chalkboard.

READ AND WRITE

Connect Sound-Symbol Write the word ***mat*** on the chalkboard, and have a volunteer circle the letter ***a***. Remind children that the letter ***a*** stands for /a/. Ask children to suggest other words that contain /a/. List these words on the chalkboard, and have volunteers circle the letter ***a*** in each one.

Complete Activity Pages Read aloud the directions on pages 15–16. Review each picture name with children.

HANDWRITING

Name

Apple begins with the **short** a sound.

Aa

A A A

a a a

Ann ate an apple.

Supporting All Learners

VISUAL/AUDITORY LEARNERS

READ ALOUD TIME Read aloud "Rags" on page 7 in the *Big Book of Rhymes and Rhythms, 1A*. Have children find all the words with **/a/** in the rhyme. Write these words on the chalkboard. Invite volunteers to add other **/a/** words to the list.

KINESTHETIC/VISUAL LEARNERS

WORD WALL On a large note card, write the letter ***a*** and the word ***mat*** with ***a*** underlined. Display the card on the Word Wall. Have children suggest other words with **/a/*a***. Write their words on small note cards, and add them to the wall. Remind children to look for words that contain **/a/** during their reading. Add these words to the wall.

TACTILE LEARNERS

NAME CARDS Display the ABC card for ***Aa*** from the Phonics and Word Building Kit. On the ABC card, have children whose names contain ***a*** write their names. Then have children suggest other names or words to add to the list. Display the card for future reference when reading and writing.

Integrated Curriculum

SPELLING CONNECTION

ADD AN *A*! Write the following word parts on the chalkboard: ***m_n, h_t, t_p, m_t***. Have children write each word part, then add the letter ***a*** to each. Model for children how to blend each word. Then have children say aloud each word as they blend it.

SCIENCE CONNECTION

ANIMAL MOBILE Have children draw and cut out pictures of cats, rats, and bats. Tell children to label their animals. Children can also take turns telling what they know about these animals. Then attach one end of string to the animal cut-outs and the other to a hanger to make an **/a/** animal mobile.

Phonics Connection

Literacy Place: *Hello!*
Teacher's SourceBook, pp. T122–123;
Literacy-at-Work Book, pp. 11–12

My Books: *I Like Cats* and *Look Again!*

LOOK AGAIN!

I LIKE CATS

Big Book of Rhymes and Rhythms, 1A: "Rags," p. 7

Chapter Book: *I Am Sam*, Chapter 7

ESL To help children acquiring English develop phonemic awareness of **/a/**, draw a happy-looking cat on a large sheet of paper, then pat it as you say ***Pat the cat***. Have children repeat the phrase as they come up and pat the cat. You may also sing this song with them to the tune of "Row, Row, Row Your Boat": ***Pat, pat, pat the cat/Pat and pat and pat/Pat, pat, pat the cat/Pat the happy cat.***

EXTRA HELP

Display the following picture cards from the Phonemic Awareness Kit: **bat, bus, can, cat, cup, fan, hat, log, man, nut, pen, pig, red,** and **six**. Invite children to sort the cards into two piles: one pile with pictures whose names contain **/a/,** and one pile with pictures whose names do not contain **/a/**.

Lesson 9

page 17

Blend Words With /a/a

QUICKCHECK ✔

Can children:

✔ **orally blend word parts?**

✔ **identify /a/?**

✔ **blend words with /a/a?**

If YES go to Read and Write.

TEACH

Develop Phonemic Awareness

Oral Blending State aloud the following word parts, and ask children to orally blend them. Provide corrective feedback and modeling when necessary.

/m/...at	/s/...at	/f/...an
/s/...ad	/m/...an	/r/...at

Connect Sound-Symbol Review with children that the letter *a* stands for /a/ as in *am*. Write the word *am* on the chalkboard. Have a volunteer circle the letter *a*. Then model for children how to blend the word.

I can put the letters *a* and *m* together to make the word *am*. Let's say the word slowly as I move my finger under the letters. Listen to how I string together the sound that each letter stands for to make the word: *aaaammmm, aamm, am*.

READ AND WRITE

Blend Words To practice using the sound taught and to review the previously taught sound, list the following words and sentence on a chart. Have volunteers read each aloud. Model blending when necessary.

- **at** **mat**
- **I like the mat.**

Complete Activity Page Read aloud the directions on page 17. Review the art with children.

Name

Find out what's hiding in the box. Draw a path through the maze. Follow the pictures whose names have the **short a** sound as in **hat**.

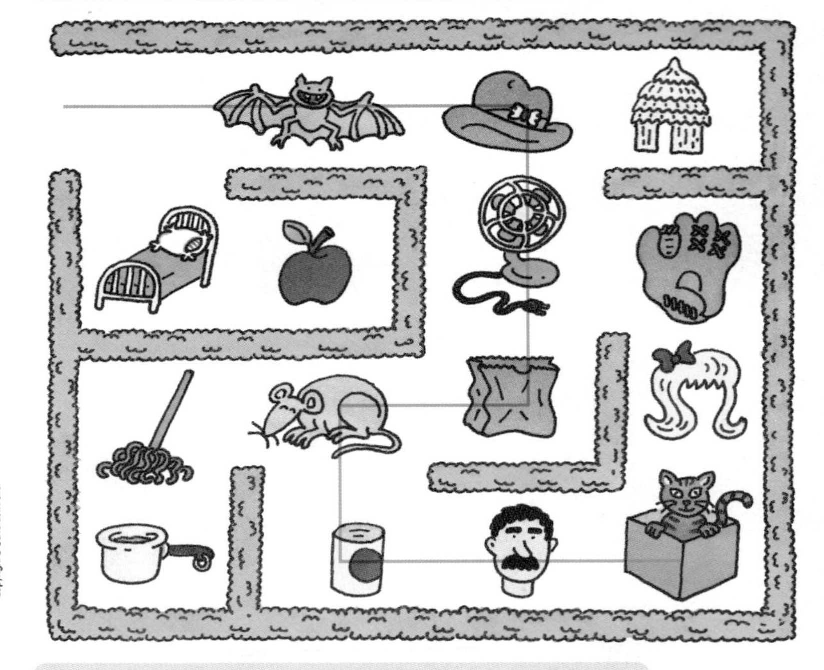

What's in the box? Trace the sentence to find out.

I am a cat.

Supporting All Learners

TACTILE/VISUAL LEARNERS

BLENDING FUN Have children use the Magnet Board or magnetic letters from the Phonics and Word Building Kit to build and blend the word **am**. Children may then compose the sentence ***I am*** ____, completing it with their first name.

The Book Shop

- **A Funny Man**
  by Patricia Jensen
  Children follow a man all around his town. **(/a/a)**

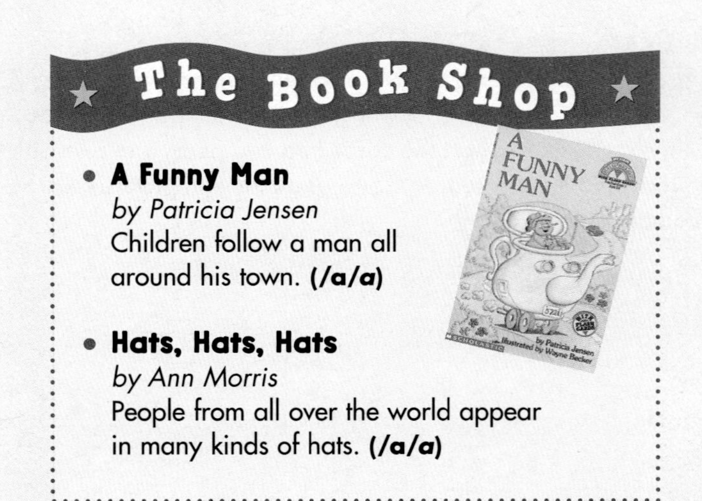

- **Hats, Hats, Hats**
  by Ann Morris
  People from all over the world appear in many kinds of hats. **(/a/a)**

Look at the pictures in each row.
Color the two pictures whose names have the same beginning sound.

Supporting All Learners

AUDITORY LEARNERS

LISTENING FUN Provide an opportunity for small groups of children to listen to any of the following songs on the Sounds of Phonics Audiocassettes in the Phonics and Word Building Kit: "Mary Had a Little Lamb," "Looby Loo" (Teamwork); "Row Your Boat" (Creative Expression); "Sing a Song of Sixpence," "Sing!," "Roses Are Red," and "Go Round and Round the Village" (Managing Information). Ask groups to identify the sounds they hear at the beginning of repeated words.

EXTRA HELP

Children can sing familiar songs such as "The Wheels on the Bus" with partners or small groups. Have children emphasize words that begin with a specific consonant sound.

CHALLENGE

PORTFOLIO

Have children find four words with /a/*a* in classroom books or in the school library. Children can list these words, then use them in sentences or a simple story.

Recognize Beginning Sounds

QUICKCHECK ✓

Can children:

✓ discriminate beginning sounds?

If YES go to Read and Write.

TEACH

Develop Phonemic Awareness

Alliteration Explain to children that you are going to read some silly sentences. Many of the words in the sentences begin with the same sound. Children will say the sound they hear again and again. For example, say ***Bob the bear likes bunches of bananas***. Repeat the words ***Bob, bear, bunches,*** and ***bananas***. Call attention to /**b**/ at the beginning of each word. Guide children to say the sound. Continue with the following sentences:

- **Sam sells socks and soap to Sarah.**
- **Take Tom the turtle to the tent.**
- **Laura likes looking at lions and lizards.**

READ AND WRITE

Practice Name the following pairs of objects or animals. Invite children to clap when they hear two objects whose names begin with the same sound.

- **book, bat**
- **fan, foot**
- **net, soap**
- **vase, van**
- **mop, kite**
- **sand, sink**
- **rug, rat**
- **tiger, fox**

Complete Activity Page Read aloud the directions for page 18. Review each picture name before children complete the page.

Lesson 11

pages 19-20

Recognize and Write /l/

QUICKCHECK ✔

Can children:

- ✔ **recognize rhyme?**
- ✔ **write capital and small *Ll*?**
- ✔ **recognize /l/?**
- ✔ **identify the letter that stands for /l/?**

If YES go to Read and Write.

TEACH

Develop Phonemic Awareness

Rhyme Display the following sets of picture cards: Set 1: **dog, log, hand;** Set 2: **king, ring, jar;** Set 3: **ten, pen, girl;** and Set 4: **sun, bun, five**. Mix the cards in each set, and have volunteers pick the two cards whose picture names rhyme. When two cards are selected, say aloud the name of each picture, and ask children to tell you another word that rhymes with the picture names.

Write the Letter Explain to children that the letter ***l*** stands for /l/, the sound they hear at the beginning of ***lion***. Write ***Ll*** on the chalkboard. Point out the capital and small forms of the letter. Then model for children how to write the letter.

Have children write both forms of the letter in the air with their fingers. You may also wish to have volunteers practice writing the letter on the chalkboard.

READ AND WRITE

Connect Sound-Symbol Write the word ***log*** on the chalkboard, and have a volunteer circle the letter ***l***. Remind children that the letter ***l*** stands for /l/. Ask children to suggest other words that begin with /l/. List these words on the chalkboard, and have volunteers circle the letter ***l*** in each one.

Complete Activity Pages Read aloud the directions on pages 19–20. Review each picture name with children.

HANDWRITING

Name

Lamp begins with the l sound.

L l

L L L

l l l

Look at the lamp.

Supporting All Learners

AUDITORY LEARNERS

LISTENING FUN Read aloud "Looby Loo" in the *Big Book of Rhymes and Rhythms, 1A*. Have children find all the words with **/l/** in the rhyme. Write these words on the chalkboard. Invite volunteers to add other **/l/** words to the list.

KINESTHETIC/VISUAL LEARNERS

LOTS OF LEAVES Cut out large leaf shapes. Ask volunteers to name words that begin with **/l/**, and write them inside the shapes. Post the shapes on a bulletin board.

Children whose first language is Japanese or Mandarin might have difficulty distinguishing between **/l/** and **/r/**. Provide picture cards of objects or animals whose names begin with **/l/** and **/r/**. Say aloud each name, and have children repeat it. Then have children sort the cards according to beginning sound. You may wish to add labels to the back of each card to make the task self-checking.

EXTRA HELP

Have pairs of children find and copy words with **/l/*l*** on cards, one word per card. Then have children write or dictate other words on cards, one word per card. Have children take turns displaying a card. If a card shows a word with **/l/*l***, children should hold up Smiley Face Response Cards from the Phonemic Awareness Kit.

Integrated Curriculum

SPELLING CONNECTION

ADD AN *L* Write the following word parts on the chalkboard: ***__eg, __et, __ot***. Have children write the word parts on pieces of paper and add the letter ***l*** to each. Next, have children write each word. Model for children how to blend each word. Then have children blend each word independently.

SOCIAL STUDIES CONNECTION

WORD HUNT You may wish to have children look in newspapers for words with **/l/*l***. Children can collect these words on an ***l*** poster or in an ***l*** book.

Phonics Connection

Literacy Place: *Hello!*
Teacher's SourceBook, pp. T164–165;
Literacy-at-Work Book, pp. 19–20

My Books: *I Like Cats* and *Look Again!*

Big Book of Rhymes and Rhythms, 1A: "Looby Loo," p. 8.

Chapter Book: *I Am Sam*, Chapter 8

ASSESSMENT

Keep a running record of each child's emerging skills. For each letter, you may note whether the child can write capital and small forms, match the forms, connect the sound and symbol, and blend words with the sound.

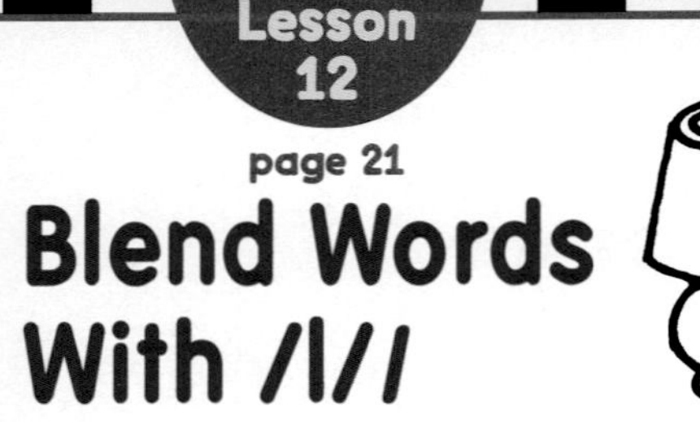

Lesson 12

page 21

Blend Words With /l/

Name

Color all the pictures whose names begin with l.

QUICKCHECK ✔

Can children:

✔ **orally blend word parts?**

✔ **identify /l/?**

✔ **blend words with /l/?**

If YES go to Read and Write.

TEACH

Develop Phonemic Awareness

Oral Blending Say the following word parts, and ask children to blend them. Provide corrective feedback and modeling when necessary.

/l/...et	**/l/...ot**	**/l/...ip**
/l/...ion	**/l/...eaf**	**/l/...amp**

Connect Sound-Symbol Review with children that the letter ***l*** stands for /l/ as in ***let***. Write the word ***let*** on the chalkboard, and have a volunteer circle the letter ***l***. Then model for children how to blend the word.

THINK ALOUD

I can put the letters *l*, *e*, and *t* together to make the word *let*. Let's say the word slowly as I move my finger under the letters. Listen to how I string together the sound that each letter stands for: *llllleeeet, lleet, let*.

Continue by helping children to blend the following words: ***lot*** and ***lip***.

READ AND WRITE

BUILT-IN REVIEW

Blend Words To practice using the sound taught and to review previously taught sounds, list the following words and sentence on a chart. Have volunteers read each aloud.

- **Al** **like**
- **I like Al.**

Complete Activity Page Read aloud the directions on page 21. Review the art with children.

Supporting All Learners

KINESTHETIC/VISUAL LEARNERS

WORD TRAIN Make a word train on a bulletin board. Each time children learn a new letter, add a car to the train for that letter. Provide space in each car for volunteers to write words that begin with, end with, or contain the letter.

The Book Shop

- **Lambs for Dinner**
  by Betsy and Guilio Maestro
  Do lambs make good dinner guests? Children will delight in this easy-to-read story. **(/l/)**

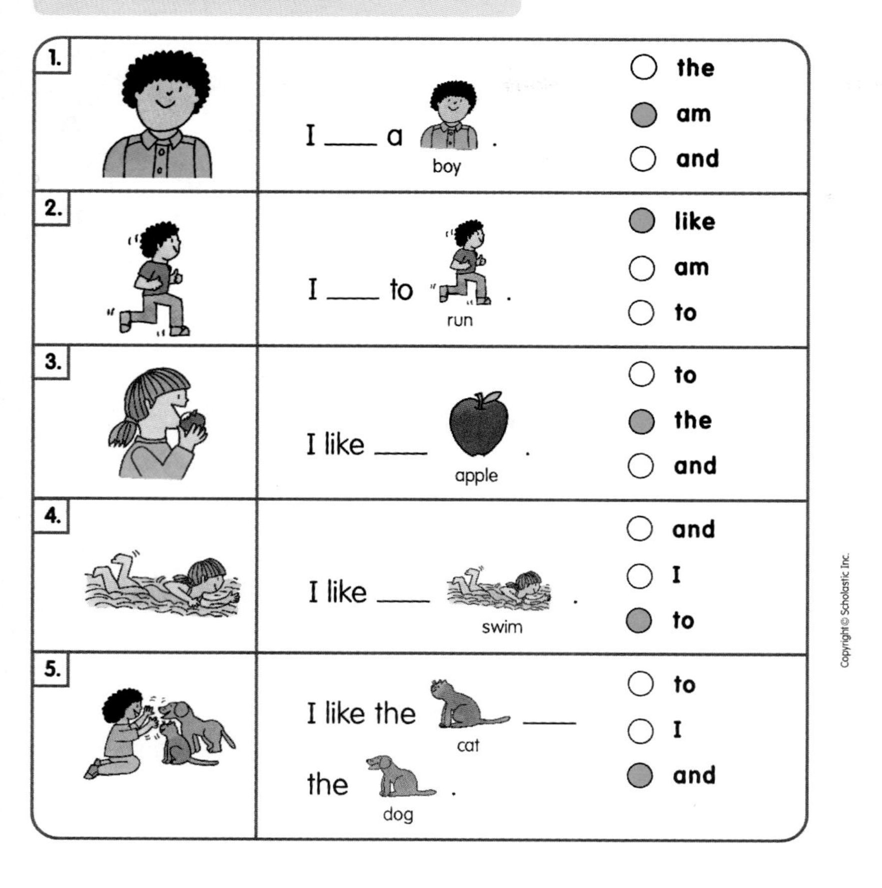
TEST YOURSELF

Look at each picture. Then fill in the bubble next to the word that best finishes each sentence.

1.	I ___ a boy.	○ the ● am ○ and
2.	I ___ to run.	● like ○ am ○ to
3.	I like ___ apple.	○ to ● the ○ and
4.	I like ___ swim.	○ and ○ I ● to
5.	I like the cat ___ the dog.	○ to ○ I ● and

Copyright © Scholastic Inc.

22 Assess: High-Frequency Words and /m/m, /a/a, /l/l

Read and Review Words With /m/m, /a/a, /l/l

QUICKCHECK ✔

Can children:

✔ **read words with /m/m, /a/a, /l/l?**

✔ **recognize high-frequency words?**

If YES go to Read and Write.

TEACH

Review Sound-Spellings Review with children the following sound-spelling relationships: **/m/*m***, **/a/*a***, and **/l/*l***. State aloud one of these sounds. Have a volunteer write on the chalkboard the spelling that stands for the sound. For example, the letter ***a*** stands for /a/. Continue with all the sounds. Then display pictures of objects whose names begin with one of these sounds. Have children write the letter that each picture name begins with as the picture is displayed.

Review High-Frequency Words Review the high-frequency words ***I, like, to, and, the***. Write sentences on the chalkboard that contain each high-frequency word. State aloud one word, and have volunteers circle the word in the sentences. Continue by having children generate sentences for each word.

READ AND WRITE

Build Sentences Write the following words on note cards: ***I, like, to, and, the, Al***. Display the cards. Have children make sentences using these words and, if necessary, pictures. Explain to children that they can draw pictures for any words they do not know. To get them started, offer this example: ***Al and I like the*** [picture of cat]. You may wish to have children work in small groups to complete the activity.

Complete Activity Page Read aloud the directions on page 22. Have children complete the page independently.

Supporting All Learners

VISUAL LEARNERS

CLASS DICTIONARY Begin a class dictionary with children by listing words with **/m/*m*, /a/*a*,** and **/l/*l***. Continue adding to the dictionary in subsequent lessons.

EXTRA HELP

Children may need help understanding the standardized test format. Use one or two items to model how to: study the picture; read the sentence that goes with the picture; read the answer choices and try each one in the sentence; fill in the bubble beside the answer that fits the sentence.

Have children write the letters ***m, a,*** and ***l*** on large self-sticking notes. Ask children to find classroom objects or pictures whose names begin with **/m/, /a/,** or **/l/** and then to attach their notes to them. When children are finished, say the name of each object, emphasizing its initial sound.

Lesson 14

pages 23–24

Read Words in Context

TEACH

Assemble the Story Ask children to remove pages 23–24. Have children fold the pages in half to form the Take-Home Book.

Preview the Story Preview *Al,* a story about two boys with the same name. Have a volunteer read aloud the title. Invite children to browse through the first two pages of the story and to comment on it. Suggest that they point out unfamiliar words. Read them aloud as you model how to blend them. Then have children predict what the story might be about.

READ AND WRITE

Read the Story Read the story aloud, or have volunteers take turns reading aloud a page at a time. Discuss anything of interest on each page. The following prompts may help children who need extra support while reading:

- **What letter sounds do you know in the words?**
- **Do the pictures give you any clues to what words are?**

Reflect and Respond Have children share their reactions to the story. How else might the two Als be alike or different?

Develop Fluency Reread the story as a choral reading, or have partners reread the story. Children can reread the story independently to develop fluency and increase reading rate. The stories are designed to be read again and again so that children can blend words and have multiple exposures to high-frequency words.

Ask children how they figure out unfamiliar words. Children can look for words with **/m/*m*, /a/*a*,** and **/l/*l*** in other stories and apply what they know about these sound-spelling relationships. Continue to review these relationships for children needing additional support.

Supporting All Learners

KINESTHETIC LEARNERS

ACT IT OUT! Invite pairs of children to act out the story. Encourage them to use movements to pantomime actions such as pointing to themselves, patting a cat or dog, and hugging each other. Girls may wish to substitute a female name for *Al*, perhaps choosing a name that begins with **/a/**.

TACTILE LEARNERS

FUN WITH CLAY Children can use clay to illustrate the story *Al* by molding figures or scenes. Then have children mold the letter forms ***A, a, L, l, M,*** and ***m***. Guide children to put the letters together to make words from the story.

AUDITORY LEARNERS

ALL ABOUT AL Have groups of three children take turns playing the parts of the two Als and an interviewer. The interviewer should ask each Al at least one question. Some questions might include how the boys like having a story told about them, how they like having the same name, or what each boy likes about his pet.

Reflect On Reading

ASSESS COMPREHENSION

To assess their understanding of the story, ask children questions such as:

- **How many boys are named Al?** ***(two)***
- **What animal does each boy like?** ***(One likes the cat, the other likes the dog.)***
- **How are the two boys alike?** ***(same name; they are boys; they have pets)***

HOME-SCHOOL CONNECTION

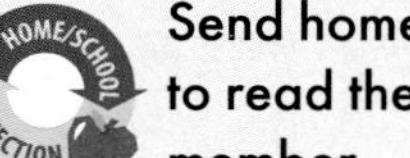

Send home *Al*. Encourage children to read the story to a family member.

WRITING CONNECTION

Have children write or dictate additional sentences to add to the story *Al*. You may wish to get children started by asking them what else the Als might like.

Phonics Connection

Phonics Readers:
#7 Mmm, I Like!
#8 Dad
#9 He Likes, She Likes

EXTRA HELP

Pairs of children can take turns retelling the story to each other, then rereading it together to check the accuracy of their retellings.

CHALLENGE

Have children list the words with ***m, a,*** and ***l*** in the story *Al*. Children can add other words with ***m, a,*** and ***l*** to the list. Then children can write their own sentences or story using these words.

Lesson 15

pages 25-26

Recognize and Write /t/t

QUICKCHECK ✔

Can children:

- ✔ **write capital and small *Tt*?**
- ✔ **recognize /t/?**
- ✔ **identify the letter that stands for /t/?**

If YES go to Read and Write.

TEACH

Develop Phonemic Awareness

Oddity Task Explain to children that you will read a list of three words. Two of the words begin with /**t**/; the other does not. Children are to choose the word that does not belong—the word that does not begin with /**t**/. Use the following word lists:

- ten top sun
- tap sock tin
- tub tack cup

Repeat this procedure for words that end with /**t**/. Start with: ***get, bat,*** and ***pan***. Continue with lists of words such as ***wig, not, cat; foot, log, sit;*** and ***pet, coat, leaf***.

Write the Letter Explain to children that the letter ***t*** stands for /**t**/, the sound they were listening for. Write ***Tt*** on the chalkboard. Point out the capital and small forms of the letter. Then model how to write the letter. Have children write both forms of the letter in the air. Have volunteers write the letter on the chalkboard.

Connect Sound-Symbol Write the word ***top*** on the chalkboard, and have a volunteer circle the letter ***t***. Remind children that the letter ***t*** stands for /**t**/. Ask children to suggest other words that begin or end with /**t**/. List these words on the chalkboard, and have volunteers circle the letter ***t*** in each.

READ AND WRITE

Complete Activity Pages Read aloud the directions on pages 25–26. Review each picture name with children.

Name

Ten begins with the t sound.

T T T

t t t

Take ten hats.

Supporting All Learners

VISUAL/TACTILE LEARNERS

T* IS FOR *TEDDY Read aloud "Teddy Bear" on page 9 in the *Big Book of Rhymes and Rhythms, 1A*. Have children find all the words with /**t**/ in the rhyme. Write these words on the chalkboard. Invite volunteers to add other /**t**/ words to the list.

KINESTHETIC/TACTILE LEARNERS

HUNTING FOR WORDS Ask children to work individually or in pairs to hunt for words that begin or end with /**t**/ in books, on bulletin boards, and in other places in the classroom. Children can write the words they find on the ABC cards in the Phonics and Word Building Kit or on paper.

AUDITORY LEARNERS

T* TIME!** Draw a large tea pot on chart paper. Write the following words on cards or in outlines of tea bags: ***tap, sat, tip, tin, cat, tack, top, fat, foot, pet, let, log, fit, kit, pig, get, flip, can, tan, fan, pool, back, ham, flop, and ***wig***. Turn the cards or tea bags facedown. Have children take turns picking cards/tea bags and reading aloud the words on them. If a word contains /**t**/, children should tape the card/tea bag inside the tea pot.

Look at each picture. If the picture name begins with the t sound as in **ten**, write t on the first line. If it ends in t as in **cat**, write t on the second line.

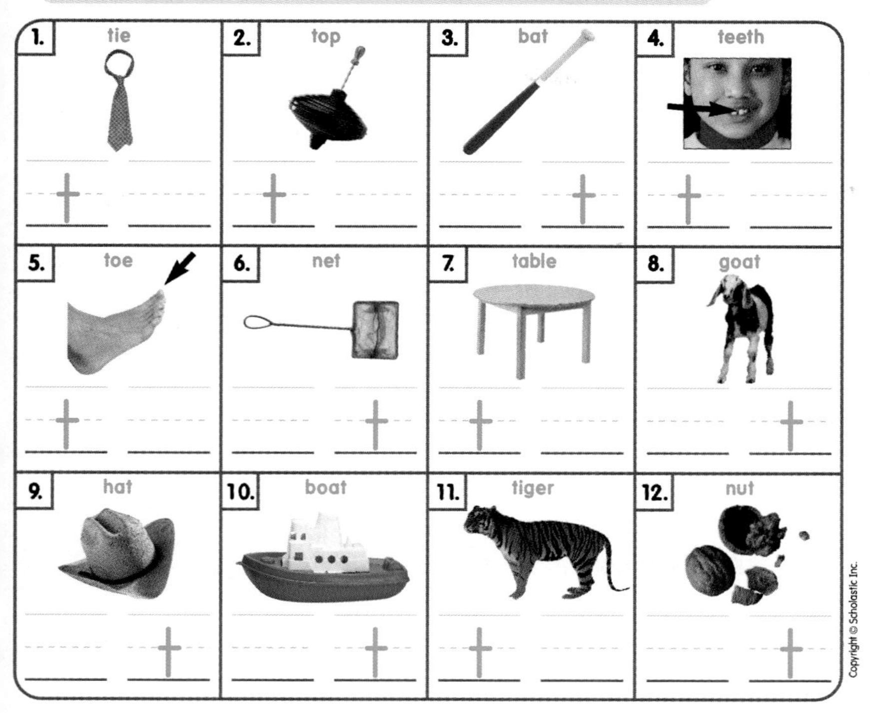

Write the letter t to finish each word.
Circle the word that names the picture.

13. a t

14. ma t

Integrated Curriculum

WRITING CONNECTION

WRITE AND BLEND Write the following word parts on the chalkboard: ***pe__, __en, ma__, __ap***. Have volunteers add the letter ***t*** to each one. Model for children how to blend each word. Have children write each word on a sheet of paper. Then have children use the words in sentences. Write the dictated sentences on chart paper, or have children write the sentences.

SOCIAL STUDIES CONNECTION

NAME GAME Help children list the names of everyone in the class. Write the list on chart paper. Have volunteers identify the name or names with ***/t/t***. Ask children to suggest other names, real or made-up, with ***/t/t***. Have children take turns pointing to a name and using it in a sentence.

Phonics Connection

Literacy Place: *Hello!*
Teacher's SourceBook, pp. T166–167;
Literacy-at-Work Book, pp. 21–22

My Book: *Teddy Bear Day*

Big Book of Rhymes and Rhythms, 1A: "Teddy Bear," p. 9

CHALLENGE

Have children write a story using words that begin or end with ***t***. Begin with a ***t*** title such as *Ten Tiny Pets*. Children can illustrate their stories and share them with each other.

Invite children acquiring English to "be a T" by standing tall, keeping their feet together and stretching out their arms at each side. For each word they hear you say that begins with ***/t/***, they can stand like ***T***'s; otherwise, they can sit. Say these and other words: ***ten, tip, top, man, tub, toy, like, tag, cup, tail***.

EXTRA HELP

For children needing additional phonemic awareness training, see the Scholastic Phonemic Awareness Kit. The oral blending exercises will help children orally string together sounds to form words. This is necessary for children to be able to decode, or sound out, words while reading.

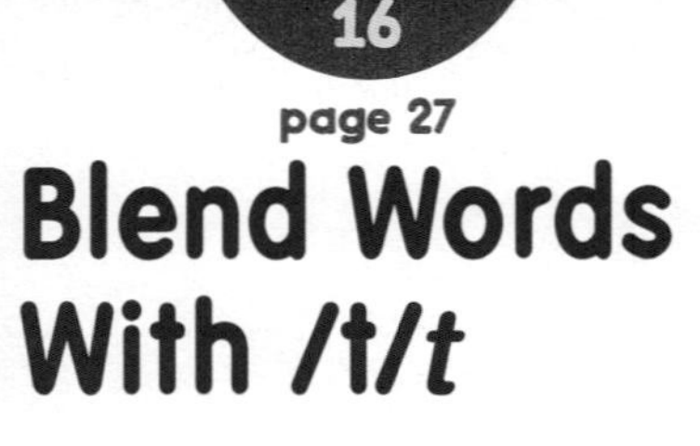

page 27

Blend Words With /t/t

QUICKCHECK ✔

Can children:

✓orally blend word parts?

✓identify /t/?

✓blend words with /t/t?

If YES go to Read and Write.

TEACH

Develop Phonemic Awareness

Oral Blending Say the following word parts. Ask children to blend them. Provide corrective feedback and modeling, when necessary.

/t/...ap	/t/...op	/t/...en
/t/...ip	/t/...an	/t/...ug

Connect Sound-Symbol Review that the letter ***t*** stands for /t/ as in ***tap***. Write ***tap*** on the chalkboard. Have a volunteer circle the ***t***. Then model how to blend the word.

THINK ALOUD

I can put the letters *t*, *a*, and *p* together to make the word *tap*. Let's say the word slowly as I move my finger under the letters. Listen to how I string together the sound that each letter stands for to make the word: *taaaap, taap, tap*.

Continue by helping children to blend the following words: ***tan*** and ***pat***.

READ AND WRITE

Blend Words To practice using the sound taught and to review previously taught sounds, list the following words and sentence on a chart. Have volunteers read each aloud.

- **at** **mat**
- **I like the mat.**

Complete Activity Page Read aloud the directions on page 27. Review the art with children.

Name

Color all the pictures whose names begin with t.

Write t words from the picture.

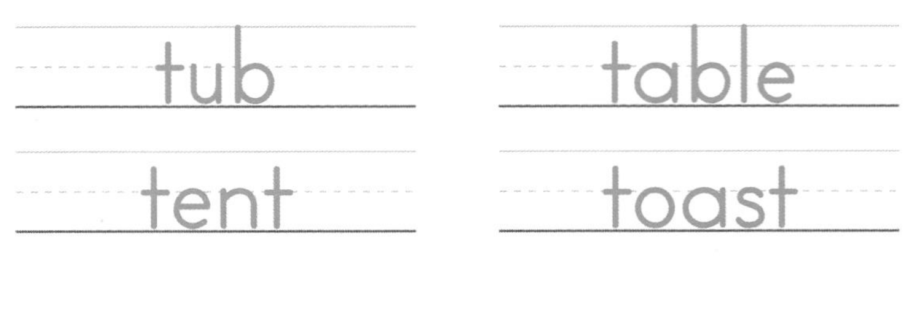

Supporting All Learners

KINESTHETIC/VISUAL LEARNERS

SCAVENGER HUNT Invite children to go on a scavenger hunt to collect things or pictures of things whose names begin or end with /t/. Display the collected items or pictures, and have children write or dictate the name of each on a card. Children can sort the labeled items into two groups—those that begin with /t/ and those that end with /t/.

- **Who Took the Farmer's Hat?**
  by Joan L. Nosdet
  Children will laugh at the farmer's wind-blown hat.
  (/t/t)

INTRODUCE HIGH-FREQUENCY WORDS

Read the words in the box.
Write one word on each blank.

said	see
he	up
she	

1. said
2. he
3. she
4. see
5. up

Use the words in the box to finish each sentence.

6. "I like to [mop]," he said.
7. I see the [steps].
8. I [mop] up the [steps].

Copyright © Scholastic Inc.

28 Recognize and Write High-Frequency Words

Lesson 17

page 28

Recognize High-Frequency Words

QUICKCHECK ✔

Can children:

- ✔ **recognize and write the high-frequency words *said, he, she, see, up*?**
- ✔ **complete sentences using the high-frequency words?**

If YES go to Read and Write.

TEACH

Introduce the High-Frequency Words

Write the high-frequency words ***said, he, she, see,*** and ***up*** in sentences on the chalkboard. Read the sentences aloud. Underline the high-frequency words, and ask children if they recognize them. Use these sentences:

1. **He said, "I am Al."**
2. **She said, "Hello, Al."**
3. **"Look up," said Al.**
4. **"See the ★★ ."**

Then ask volunteers to dictate sentences using the high-frequency words. You may wish to begin with the following sentence starters: ***He said*** ___ or ***She and I see the*** ___. Write children's sentences on the chalkboard.

READ AND WRITE

Practice Write each high-frequency word on a note card. Read each word aloud as you display the cards. Then do the following:

- **Mix the cards.**
- **Display one card at a time, and ask children to state each word aloud.**
- **Have children spell each word aloud, clapping on each letter.**
- **Ask children to write each word in the air as they state aloud each letter. Then have them write each word on a sheet of paper.**

Complete Activity Page Read aloud the directions on page 28. Have children complete the page independently.

Supporting All Learners

KINESTHETIC LEARNERS

WORD-CARD GRAB BAG Write the high-frequency words for this lesson and the previous lesson on index or note cards, one word per card. Place the cards in a bag. Have small groups of children take turns choosing a card, saying the word, and making up a sentence that uses the word.

Relate new words to known words and known concepts. For ***he*** and ***she,*** you might make a two-column chart using ***he*** and ***she*** as the headings. Work with children acquiring English to identify and list the names of classmates under the appropriate heading. For ***up,*** you may pantomime the action of going up stairs, or, if children know the word ***down,*** you may express the meaning of ***up*** through contrast.

Lesson 18

pages 29–30

Recognize and Write /s/s

QUICKCHECK ✔

Can children:

✔ **write capital and small *Ss*?**

✔ **recognize /s/?**

✔ **identify the letter that stands for /s/?**

If YES go to Read and Write.

TEACH

Develop Phonemic Awareness

Oddity Task Explain to children that you will show them three pictures. Two of the picture names begin with /s/; the other does not. Children are to say the picture names that begin with /s/. Use the following picture cards:

• seven	saw	mop
• six	sock	tiger
• sun	dress	seal
• book	soap	sand

Write the Letter Explain to children that the letter *s* stands for /s/, the sound they hear at the beginning of ***sun***. Write *Ss* on the chalkboard. Point out the capital and small forms of the letter. Then model for children how to write the letter.

Have children write both forms of the letter in the air with their fingers. You may also wish to have volunteers practice writing the letter on the chalkboard.

READ AND WRITE

Connect Sound-Symbol Write the word ***sad*** on the chalkboard. Have a volunteer circle the letter *s*. Remind children that the letter *s* stands for /s/. Ask children to suggest other words that begin with /s/. List these words on the chalkboard, and have volunteers circle the letter *s* in each one.

Complete Activity Pages Read aloud the directions on pages 29–30. Review each picture name with children.

HANDWRITING

Name

Sock begins with the s sound.

S s

S S S

s s s

See the socks.

Supporting All Learners

AUDITORY LEARNERS

RHYME TIME Read aloud "Sing a Song of Sixpence" from the *Big Book of Rhymes and Rhythms, 1A*. Have children find all the words with **/s/** in the rhyme. Write these words on the chalkboard. Invite volunteers to add other **/s/** words to the list.

TACTILE LEARNERS

LETTER FUN Review letter formation by providing children with one or more of the following materials: trays of sand; modeling clay; or string, glue, and cardboard. Children can form letters in the sand, shape them out of the clay, or glue pieces of string onto the cardboard in the shape of letters. If children form letters out of clay or cardboard, have them trace the letter with their fingers. Provide corrective feedback, as needed.

VISUAL LEARNERS

S* IS FOR *SOCK Have children draw and cut out sock shapes. Then have children write words with **/s/s** on the socks, one word per sock. Hang the socks on a line with clothes pins, or display them on a bulletin board. Challenge children to identify the word on each sock.

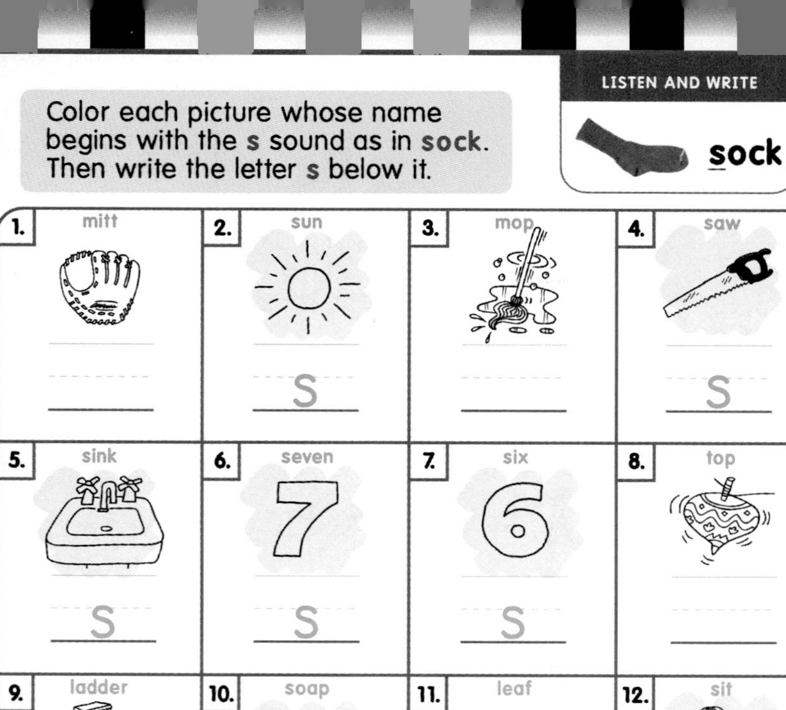

Copyright © Scholastic Inc.

Write the letter s to finish each word. Read the words to a friend.

13. s at 14. s ee 15. s un

30 Recognize /s/s

Integrated Curriculum

SPELLING CONNECTION

ADD AN Ss Write the following word parts on the chalkboard: ***_at, _un, _it, _et.*** Have each child write one of the word parts. Then have children add the letter ***s*** to each word part. Children can take turns showing and saying each word that they made. Have each child choose another word part and repeat the procedure.

MATH CONNECTION

WORD COUNT Have children count the number of words with ***/s/s*** in the classroom or in the school. Children can write the words in a list or journal. Then have children count the words to see who collected the most.

Phonics Connection

Literacy Place: *Hello!*
Teacher's SourceBook, pp. T222–223; Literacy-at-Work Book, pp. 29–30

My Book: *Suzy's Swim*

Big Book of Rhymes and Rhythms, 1A: "Sing a Song of Sixpence," p. 10

Chapter Book: *I Am Sam*, Chapter 10

CHALLENGE

Draw on chart paper a large ladder with four rungs. Invite children to "climb" the ladder by naming words that begin with ***s***. Write or have children write the words on the ladder, one word per rung, beginning with the bottom rung, and working toward the top. Then draw a second word ladder. Have children look for words that begin with ***s*** and add them to the ladder.

EXTRA HELP

Invite children to illustrate words that begin with ***s*** such as ***soap, sink,*** and ***sun***. When they have completed their drawings, help children label each picture. Then have children underline the letter ***s*** in each picture name.

Lesson 19

pages 31–32

Recognize and Write Words With */s/s* and *-ad, -at*

QUICKCHECK ✔

Can children:

- ✓ **orally blend word parts?**
- ✓ **identify /s/?**
- ✓ **blend and build words with /s/*s* and phonograms *-ad, -at*?**
- ✓ **build words with phonograms *-ad, -at*?**

If YES go to Read and Write.

TEACH

Develop Phonemic Awareness

Oral Blending Say the following word parts. Ask children to blend them. Provide corrective feedback and modeling when necessary.

/s/...at	/s/...am	/s/...ad
/s/...oap	/s/...even	/s/...un

Connect Sound-Symbol Review that the letter *s* stands for /s/ as in *sat*. Write the word *sat* on the chalkboard, and have a volunteer circle the letter *s*. Then model how to blend the word.

Introduce the Phonograms Write *-ad* and *-at* on the chalkboard. Point out the sounds these phonograms stand for. Add the letter *s* to the beginning of *-ad*, and model for children how to blend the word. Have children repeat the word *sad* aloud as you blend the word again. Then add *m* to the beginning of *-at*, and model again how to blend the word *mat*.

READ AND WRITE

Blend Words List the following words and sentence. Have volunteers read each aloud.

- Sam sat
- mad sad
- I am sad.

Complete Activity Pages Read aloud the directions on page 31–32. Have children complete the pages independently.

LISTEN AND WRITE

Name

Color all the pictures whose names begin with s.

Supporting All Learners

TACTILE/VISUAL LEARNERS

CENTER TIME! For additional practice, set up Center 1, "What's in a Name?" from *Quick-and-Easy Learning Centers: Phonics*. Working with names that begin with ***s, m, a,*** and ***l***, you may tailor activities 1 and/or 2 to the children's needs. Activity 3 may also be done using the name ***Sam*** as an example.

KINESTHETIC/VISUAL LEARNERS

WORD SORT Write the words ***sat, sad, mat,*** and ***mad*** on index cards. Challenge children to sort them two ways. First, they can sort them by words that begin with **/s/** and words that begin with **/m/**. Then they can sort them by words that contain ***-at*** and words that contain ***-ad***. Whenever children successfully sort the words, give them a Smiley Face Response Card from the Phonemic Awareness Kit.

AUDITORY LEARNERS

PUPPET TIME To help children blend words, use the blending puppet and the exercises in the Phonemic Awareness Kit. These oral blending activities help children to hear how the sounds in words are strung together to form the words. This awareness of how words work is critical to early reading success.

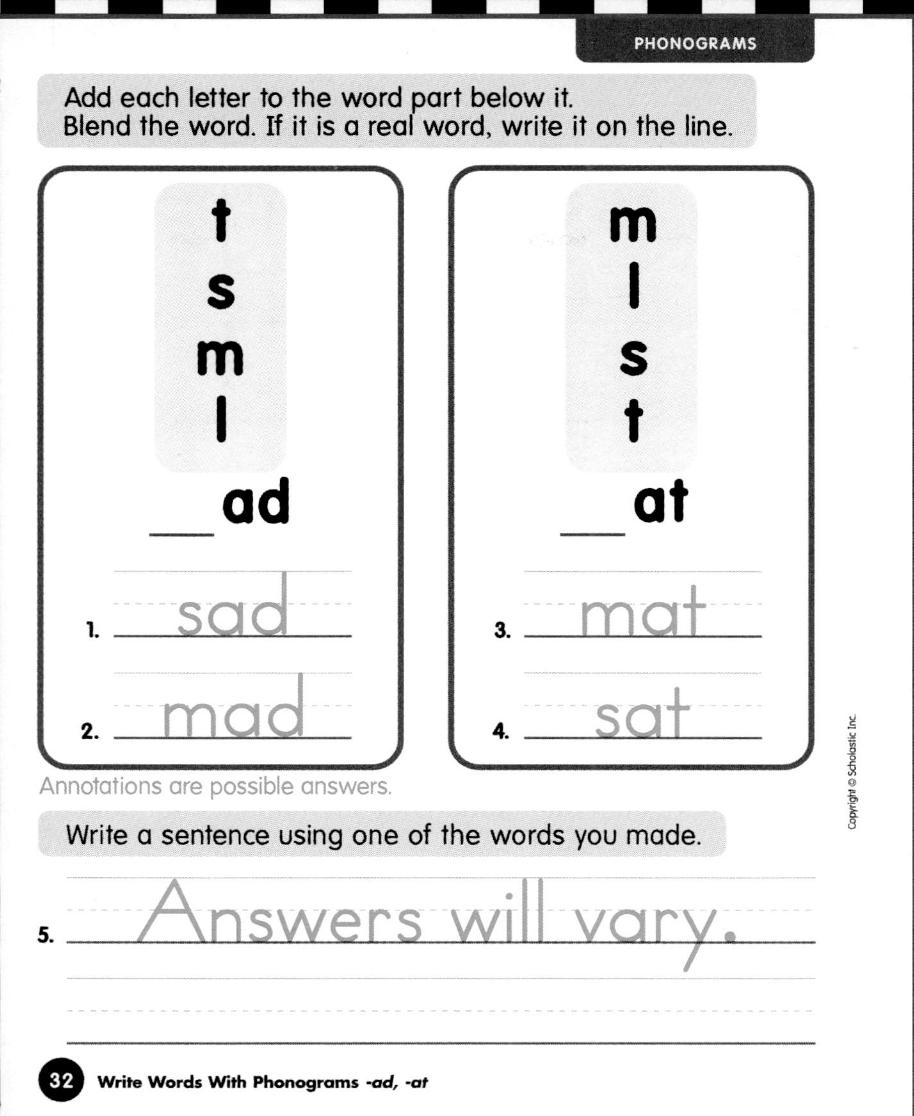

Integrated Curriculum

SPELLING CONNECTION

RHYME TIME Ask children to suggest words that rhyme with ***sad*** and ***mat***. Children can list or dictate a list of these words on the chalkboard, one column for ***-at*** and one column for ***-ad***. Have children underline the phonogram ***-at*** or ***-ad*** in each word. You may wish to add to the lists words such as ***cat, hat, rat, dad, had,*** and ***mad***.

TECHNOLOGY CONNECTION

MY BOOK FUN Have children read the following My Book: *Suzy's Swim*. For information on using the My Books on the computer, see the WiggleWorks Plus My Books Teaching Plan.

CHALLENGE

Have children use the ABC cards and pocket chart from the Phonics and Word Building Kit to make words with ***s*** and words with the phonograms ***-ad*** and ***-at***. Have children list each word they make, then use the words in sentences.

EXTRA HELP

For children having difficulties building words, use the magnetic letters from the Phonics and Word Building Kit. Form a word such as ***sat***. Model how to blend the word. Then mix the letters, and have the child form the word. Continue with other decodable words.

- **Silly Sally**
  by Audrey Wood
  Sally and her companions parade into town.
  (/*s*/*s*)
- **Sitting in My Box**
  by Dee Lillegard
  What's inside the box? Read and discover! **(/*s*/*s*)**

Lesson 20

pages 33-34

Recognize and Write /o/o

QUICKCHECK ✔

Can children:

✓ orally segment words?

✓ write capital and small *Oo*?

✓ recognize /o/?

✓ identify the letter that stands for /o/?

If YES go to Read and Write.

TEACH

Develop Phonemic Awareness

Oral Segmentation Have children listen to a list of words. For each word, they will say the first sound and then the rest of the word. For example, say ***sat***. Explain that the first sound in ***sat*** is /s/ and the rest of the word is ***at***. Continue by having children say the first sound and the rest of the word for the following:

- mop
- sad
- man
- lot
- sock
- lock

Write the Letter Explain to children that the letter ***o*** stands for /o/ as in ***hop***. Write ***Oo*** on the chalkboard. Point out the capital and small forms of the letter. Then model how to write the letter.

Have children write both forms of the letter in the air with their fingers. Have volunteers practice writing the letter on the chalkboard.

READ AND WRITE

Connect Sound-Symbol Write the word ***hop*** on the chalkboard, and have a volunteer circle the letter ***o***. Remind children that ***o*** stands for /o/. Ask children to suggest other words that contain /o/. List them on the chalkboard.

Then write the following word parts on the chalkboard: ***l_t, d_t, d_g, f_g***. Have volunteers add the letter ***o*** to each one. Model for children how to blend each word.

Complete Activity Pages Read aloud the directions on pages 33–34. Review each picture name with children.

Name

O O O

o o o

I see the octopus.

Supporting All Learners

VISUAL/AUDITORY LEARNERS

RHYME TIME Read aloud "Higglety, Pigglety, Pop." Have children find all the words with **/o/** in the rhyme. Write these words on the chalkboard. Invite volunteers to add other **/o/** words to the list.

KINESTHETIC LEARNERS

PICTURE SORT Give children the following picture cards: **bat, box, can, cat, dog, fan, fox, hat, log, man, mop,** and **top**. Ask children to sort the cards into two piles: one containing words with **/o/** as in ***top*** and one containing words with **/a/** as in ***hat***.

TACTILE LEARNERS

BUILD WORDS Write the word parts ***-at*** and ***-ad*** on the chalkboard. Have children add a letter to the beginning of each phonogram to make a new word. Continue by having children replace the initial consonant in the first word to build a second word. For example, children might build ***sad,*** then ***mad***.

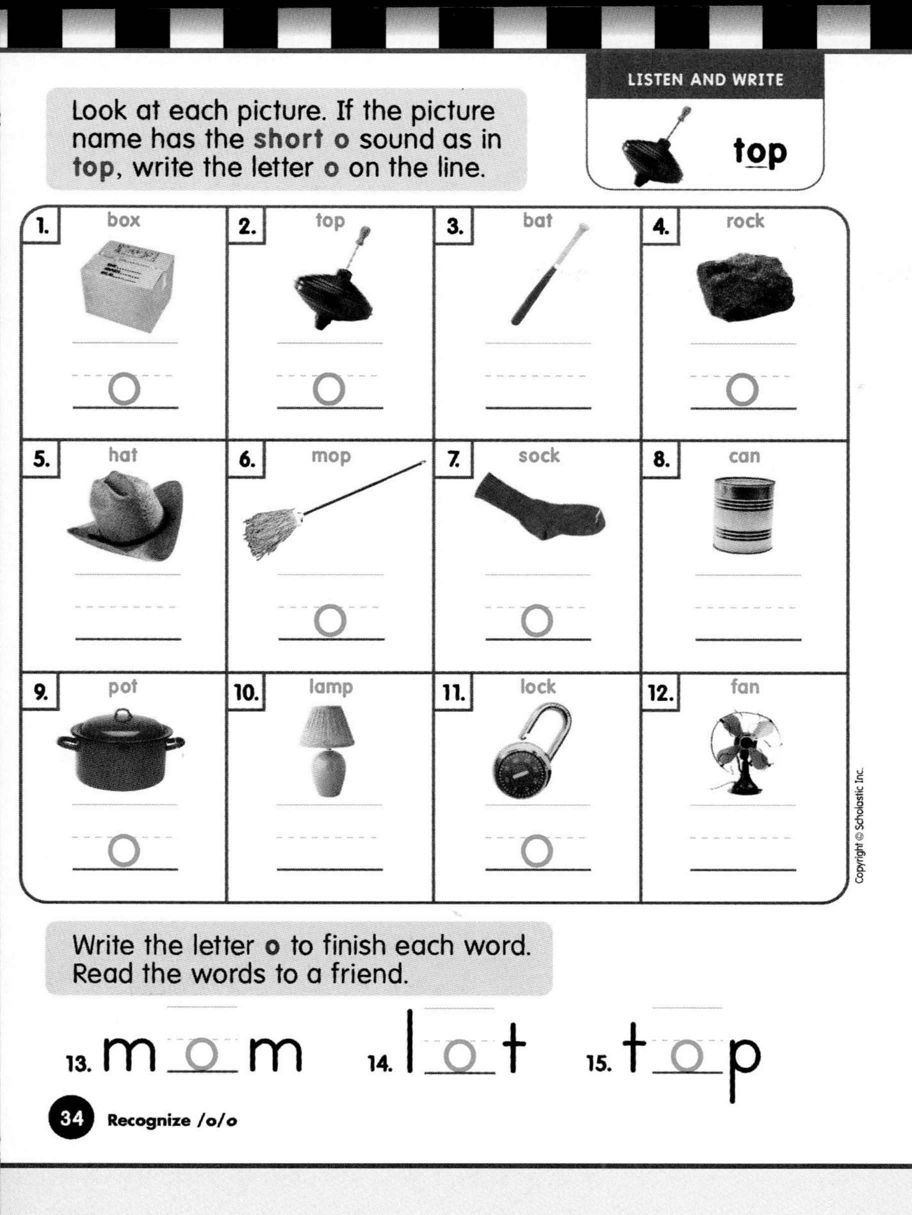

Integrated Curriculum

WRITING CONNECTION

***O* NOTES** Have children generate a list of words with **/o/**. Record these words on the chalkboard or on chart paper. Then have children write notes or letters to each other using words with **/o/**.

TECHNOLOGY CONNECTION

MAGNET BOARD FUN Have children build ***o*** words on the Magnet Board. See the WiggleWorks Plus My Books Teaching Plan for information on this and other Magnet Board activities.

Phonics Connection

Literacy Place: *Hello!*
Teacher's SourceBook, pp. T224–225;
Literacy-at-Work Book, pp. 31–32

My Book: *Fox and Rabbit*

Big Book of Rhymes and Rhythms, 1A: "Higglety, Pigglety, Pop," p. 11

Chapter Book: *I Am Sam*, Chapter 11

ESL

Display pictures of objects whose names contain **/o/**. These pictures might include the following: **frog, dog, pot, log, box, fox, mop,** and **dot**. Prompt children acquiring English to take turns identifying each picture name and dictating a sentence with that word. Write the sentences on the chalkboard, and help children to read them chorally.

EXTRA HELP

Review blending the words in the practice activity by providing children with magnetic letters. Begin by showing children how to place the letters ***ot*** in order on their desks. Say ***ot***, and have children repeat it. Then, as you model blending ***w*** and ***ot***, have children follow along by listening for **/w/** and simultaneously moving the letter ***w*** into position in front of ***ot***. Continue with other **/o/** words.

OBSERVATION

As you note whether children sort the cards correctly, you may also note how quickly they do it, as well as when they hesitate. Add this information to your anecdotal records.

Lesson 21

pages 35-36

Blend and Build Words With /o/o and -ot, -op

QUICKCHECK ✔

Can children:

- ✔ **orally blend word parts?**
- ✔ **identify /o/?**
- ✔ **blend and build words with /o/o and phonograms -ot, -op?**

If YES go to Read and Write.

TEACH

Develop Phonemic Awareness

Oral Blending Say the following word parts. Ask children to blend them. Provide corrective feedback and modeling when necessary.

/l/...ot	**/m/...op**	**/f/...ox**
/s/...ock	**/n/...ot**	**/l/...ock**

Connect Sound-Symbol Review with children that the letter *o* stands for /o/ as in *lot*. Write *lot* on the chalkboard, and have a volunteer circle the *o*. Then model how to blend the word.

Introduce the Phonograms Write the phonograms ***-op*** and ***-ot*** on the chalkboard. Point out the sounds they stand for. Add ***m*** to the beginning of ***-op***, and model how to blend the word formed. Then add ***l*** to the beginning of ***-ot***. Have a volunteer model how to blend ***lot***.

READ AND WRITE

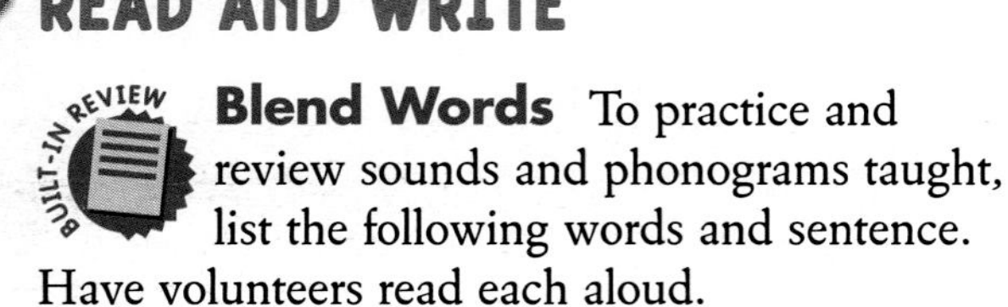

Blend Words To practice and review sounds and phonograms taught, list the following words and sentence. Have volunteers read each aloud.

- **mop** **top** **lot**
- **I mop a lot.**

Complete Activity Pages Read aloud the directions on pages 35–36. Have children complete the pages independently.

Name

Say the names of the pictures in each puzzle. Write the letter **o** or **a** to complete the picture names.

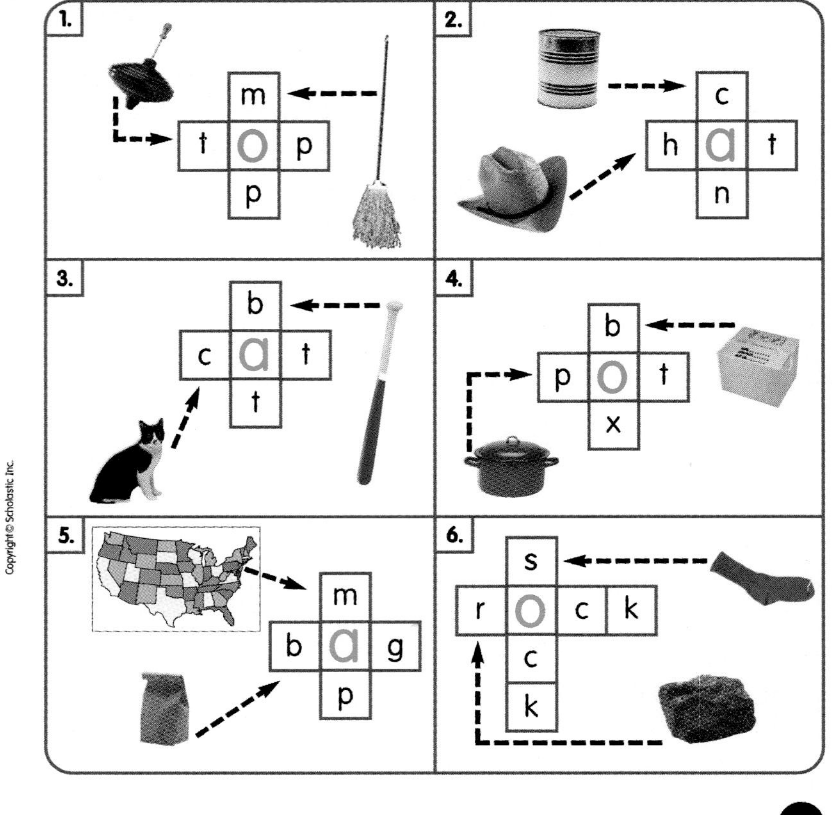

Supporting All Learners

TACTILE LEARNERS

OUTRAGEOUS O's Invite children to create or decorate their names and featured letters, such as ***o***'s, by using a variety of art materials that might range from pasta and dough to glitter and sprinkles. Create a display of letters and name art.

VISUAL LEARNERS

RHYME TIME Use the pocket ABC cards to spell ***hop*** in the pocket chart. Have partners take turns making new words. One child can replace the first letter in a word while the other child writes that word. When children have completed their lists of words, have them compare the words they made.

KINESTHETIC LEARNERS

BUILD WORDS Write ***-ot*** on the chalkboard. (If available, use a pocket chart and letter cards.) Have children add a letter to the beginning of the phonogram to make a new word. Continue by having children replace the initial consonant in the first word to build a second word. For example, children might build ***hot*** from ***lot*** and ***dot*** from ***hot***.

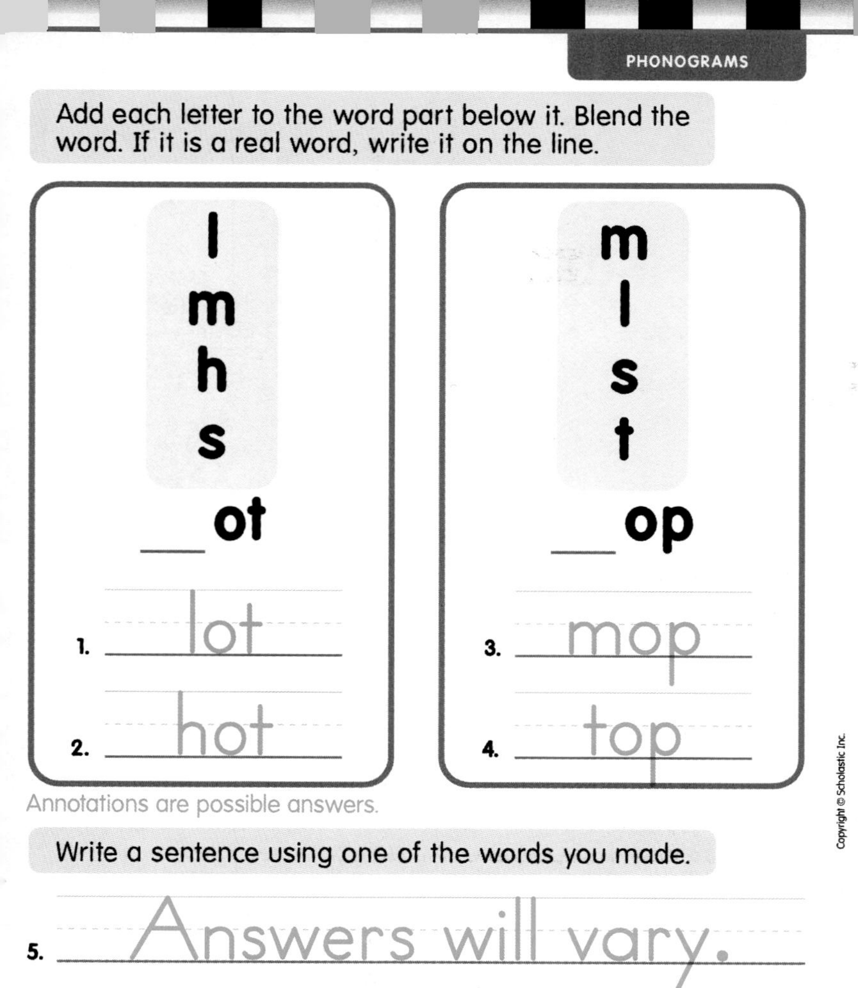
PHONOGRAMS

Add each letter to the word part below it. Blend the word. If it is a real word, write it on the line.

l m h s __ot	m l s t __op
1. lot	3. mop
2. hot	4. top

Annotations are possible answers.

Write a sentence using one of the words you made.

5. Answers will vary.

36 Write Words With Phonograms *-ot, -op*

Copyright © Scholastic Inc.

Integrated Curriculum

WRITING CONNECTION

STORY TIME Ask children to suggest words that rhyme with ***mop*** and ***lot***. List or have children list these words on the chalkboard in separate columns. Then have volunteers underline the phonogram ***-ot*** or ***-op*** in each word. Children can create a story using the words listed. Write the dictated story on chart paper.

SCIENCE CONNECTION

SAND WORDS Children can work with partners to take turns writing and reading words with ***-ot*** and ***-op*** in sand. You may wish to have children keep a list of the words they make.

CHALLENGE

Have partners or small groups of children work together to make word puzzles with the phonograms ***-ot*** and ***-op***. Children can use the following puzzle as a model:

mop – m + t = top

The short ***o*** sound in English is similar to the Spanish ***a*** sound, as in ***casa*** (house). Have children work with partners to make picture cards to help them remember that ***casa*** and ***rock,*** for example, have the same vowel sound.

The Book Shop

- **Hop on Pop**
  by Dr. Seuss
  This silly book is full of playful rhymes. ***(-op)***
- **Mop Top**
  by Don Freeman
  A boy doesn't want a haircut in this humorous story. ***(-op)***

Lesson 22

page 37

Write and Read Words With /t/t, /s/s, /o/o

QUICKCHECK ✔

Can children:

✔ **spell words with /t/t, /s/s, /o/o,?**

✔ **read a poem?**

If YES go to Read and Write.

TEACH

Link to Spelling Review with children the following sound-spelling relationships: **/t/*t*, /s/*s*, /o/*o***. State aloud one of these sounds. Have a volunteer write on the chalkboard the spelling that stands for the sound. Continue with all the sounds. You may also wish to review **/m/*m*, /a/*a***, and **/l/*l***.

Phonemic Awareness Oral Segmentation
State aloud the word ***sat***. Have children orally segment the word. (/s/ /a/ /t/) Ask them how many sounds the word contains. *(3)* Draw three connected boxes on the chalkboard, and have a volunteer write the spelling that stands for each sound in the word ***sat*** in the appropriate box. Continue with the words ***mat*** and ***lot***.

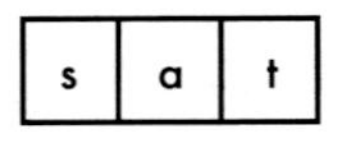

READ AND WRITE

Dictate Dictate the following words and sentence. The first set of words are decodable based on the sounds previously taught. The second are high-frequency words. Have children write the words and sentence on a sheet of paper. When children are done, write the words and sentence on chart paper, and have children make any necessary corrections on their papers.

- **sat** **lot** **at**
- **said** **up** **and**
- **She sat up.**

Complete Activity Page Read aloud the directions on page 37. Have children circle the words with **/m/**.

POETRY

Name

Read the poem. Write a sentence telling who likes to mop.

MOP, MOP, MOP!

He likes to mop,

Mop, mop, mop.

She likes to mop,

Mop, mop, mop.

I like to mop,

A LOT!

Mop, Mop, Mop!

Answers will vary.

Supporting All Learners

KINESTHETIC/AUDITORY LEARNERS

LET'S TELL A STORY Have children display the Phonemic Awareness Kit picture cards with names that contain **/t/*t*, /s/*s***, and **/o/*o***. Have a volunteer start a story about one of the pictures by saying a sentence about it. Have another child add a sentence to the story. Children can continue in this way to make up a story. Write their sentences on chart paper.

EXTRA HELP

For children needing additional support segmenting words, use Elkonin boxes and counters. Have children drag one counter into each box as you say each sound in the word. Information on this procedure can be found on page T31.

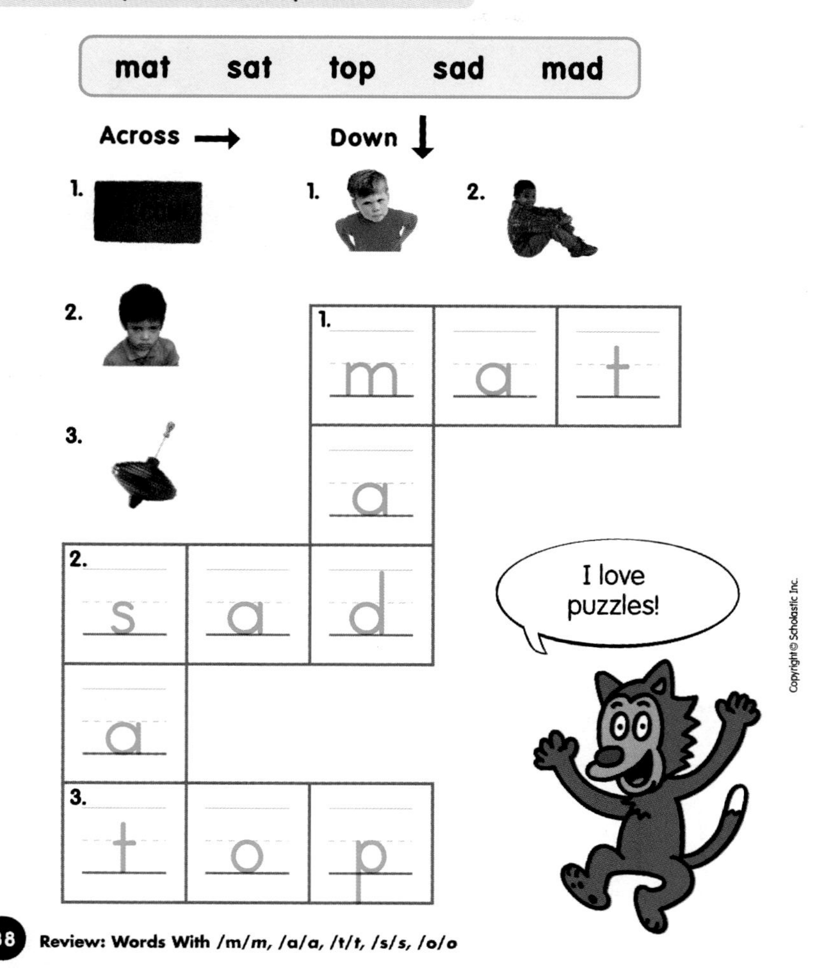
REVIEW

Look at the pictures. Write the word for each picture in the puzzle.

mat sat top sad mad

Across →

1.

2.

3.

Down ↓

1.

2.

1. m a t

2. s a d

3. t o p

38 Review: Words With /m/m, /a/a, /t/t, /s/s, /o/o

page 38

Review Words With /m/m, /a/a, /t/t, /s/s, /o/o

QUICKCHECK ✔

Can children:

- ✔ **read words with sounds previously taught?**
- ✔ **write words with sounds previously taught?**

If YES go to Read and Write.

TEACH

Phonemic Awareness Oral Segmentation

Say the word ***mat***. Have children orally segment the word. (**/m/ /a/ /t/**) Ask them to say the word slowly and to tap lightly on their desk for each sound the word contains. Draw three connected boxes on the chalkboard, and have a volunteer write the spelling that stands for each sound in the word ***mat*** in the appropriate box.

m	a	t

Then give each child three counters and one copy of the Segmentation Reproducible Master on page T29. Explain to children that you are going to read aloud a word. Children are going to count how many sounds they hear in the word, placing one counter on each box on the reproducible master. For example, if you say ***sat,*** children should place three counters on the page, one on each box. Use these groups of words in order: ***am, at, Al; Sam, sat, sad; mat, mad;*** and ***mop, top.*** Ask children which sound changes in each new word in the group.

READ AND WRITE

Complete Activity Page Read aloud the directions on page 38. Review the picture names with children.

Supporting All Learners

TACTILE LEARNERS

WORD ART Have children use sand, string, pasta shapes, collage materials, and glue to make words with ***/m/m, /a/a, /t/t, /s/s, /o/o.*** Children can write the words first, then glue materials to the letters. When children have finished their words, display them on a wall or bulletin board.

EXTRA HELP

Help children to extend all the continuous sounds in each word in the oral segmentation activities. For example, guide children to say ***ssssaaaat*** for each sound in the word ***sat.*** Help children to understand that there are either two or three distinct sounds in each word you name.

Lesson 24

pages 39-40

Read and Review Words with /t/t, /s/s, /o/o

QUICKCHECK ✔

Can children:

✔ **read words with /t/t, /s/s, /o/o?**

✔ **recognize high-frequency words?**

If YES go to Read and Write.

TEACH

Review Sound-Spellings Review with children the sound-spelling relationships from the past few lessons. These include /t/t, /s/s, and /o/o. Say one of these sounds. Have a volunteer write on the chalkboard the spelling that stands for the sound. Continue with all the sounds. Then display pictures of objects whose names begin with one of these sounds. Have children write the letter that each picture's name begins with.

Then review *-ot* and *-op*. Have children suggest words containing each phonogram. Volunteers can write the words on the chalkboard.

Review High-Frequency Words Review the high-frequency words ***said, he, she, see,*** and ***up.*** Write on the chalkboard sentences containing each high-frequency word. Have volunteers circle the high-frequency words in the sentences.

READ AND WRITE

Build Words Distribute the following letter cards to children: ***a, d, m, o, s, t.*** Provide time for children to build as many words as possible using the letter cards.

Build Sentences Write the following words on note cards: ***said, he, she, see, sat, the, like, I, mop, up.*** Have children make sentences using the words. To get them started, suggest the sentence ***She sat up.***

Complete Activity Pages Read aloud the directions on pages 39–40. Have children complete the pages independently.

REVIEW

Name

Check each word as you read it to a partner.
Circle any words you need to practice.

I can read!

☐ a ☐ at ☐ mat
☐ am ☐ sat ☐ sad
☐ mad ☐ lot ☐ Sam
☐ mop ☐ top ☐ Mom

Lookout Words!

☐ I ☐ like ☐ to ☐ and ☐ the
☐ said ☐ he ☐ she ☐ see ☐ up

Supporting All Learners

AUDITORY LEARNERS

RHYME AND RIDDLE GAME To review words, play a rhyme and riddle game with decodable words and high-frequency words. For example, say, ***I rhyme with*** hat. ***You sit on me. I start with /m/. What am I?*** or ***I rhyme with*** me. ***You do this with your eyes. What am I? What sound do I start with?***

VISUAL LEARNERS

PORTFOLIO

WORD HUNT Have children look for words with **/t/t, /s/s,** and **/o/o** in books, magazines, and newspapers in the classroom. Children can collect as many of these words as they can, and then list the words. Children can also underline ***t, s,*** or ***o*** in each word.

Fill in the bubble next to the sentence that tells about each picture.

1.	● The [cat] sat. ○ She likes the [cat].
2.	● He sees a top. ○ He sees a mop.
3.	○ She said, "Mom!" ● She said, "Up!"
4.	● The [dog] sat on a mat. ○ The mad [dog] sat.
5.	○ He sat up. ● The sad [girl] sat.
6.	● He likes to mop. ○ He likes to hop.

Integrated Curriculum

SPELLING CONNECTION

MORE WORDS Write the phonograms ***-ot*** and ***-op*** on the chalkboard or on chart paper. Give children letter cards. Have children hold up the letter cards to the phonograms to make words. Each time a real word is made, have a volunteer write it on a list.

MATH CONNECTION

GRAPH TIME Children can look in their school for words with ***/t/t***, ***/s/s***, and ***/o/o*** and words with the phonograms ***-ot*** and ***-op***. Have children count these words, then make a graph such as the following.

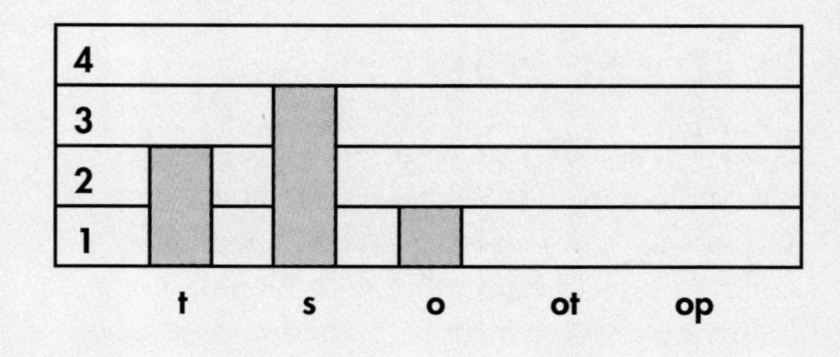

CHALLENGE

Ask children to compose sentences about one or more of the picture cards in the Phonemic Awareness Kit whose name begins with ***t***, ***s***, or ***o***. Challenge children to use their sentences to develop a story.

To help children acquiring English review decodable and high-frequency words, prepare a set of word cards and matching picture cards. Use words such as: ***mop, top, mat, sat, sad, mad, he, she, see, up***. For each word card, draw a simple related picture or paste a cutout picture on a separate card. Children can take turns reading each word aloud and matching it to its picture.

I CAN READ! OPTIONS

The I Can Read! page can be used for one or all of the following:

- paired reading
- individual assessment
- choral reading
- homework practice
- program placement

ASSESSMENT

To gain additional insight into children's progress, talk with them individually about each item they missed on page 40. Find out which words and sounds present difficulty. Provide additional practice with these sounds.

Lesson 25

pages 41-42

Read Words in Context

TEACH

Assemble the Story Ask children to remove pages 41–42. Have them fold the pages in half to form the Take-Home Book.

Preview the Story Preview *At the Top*, a story about two boys who climb to the top of a hill. Have a volunteer read aloud the title. Invite children to browse through the first two pages of the story and to comment on anything they notice. Ask children to point out any unfamiliar words. Read these words aloud as children repeat them. Then have children predict what they think the story might be about.

READ AND WRITE

Read the Story Read the story aloud, or have volunteers read aloud a page at a time. Discuss anything of interest, and encourage children to help each other with blending. The following prompts may help children who need extra support:

- **What letter sounds do you know in the word?**
- **What do the pictures tell you about the story?**

Reflect and Respond Have children share their reactions to the story. What did they like about the story? Did they predict the ending?

Develop Fluency Reread the story as a choral reading, or have partners reread the story independently. Provide time for children to reread the story on subsequent days to develop fluency and increase reading rate.

Encourage children to look for words with /**t**/*t*, /**s**/*s*, and /**o**/*o* in other stories and to apply what they have learned about these sound-spelling relationships to decode words. Continue to review these relationships for children needing additional support.

Supporting All Learners

KINESTHETIC LEARNERS

PUPPET SHOW Have pairs of children make finger puppets to act out the story.

AUDITORY LEARNERS

STORY TIME Pairs or small groups of children can take turns reading *At the Top* to each other. Have children tape-record their reading. Place the recording in the classroom listening center for all to enjoy.

VISUAL LEARNERS

SAM AND MATT AGAIN Children can draw pictures of Sam and Matt doing something else together. Then children can write or dictate captions to go with their pictures. Give children an opportunity to share their pictures and captions, then display them.

2

"Up, up, up!" said Sam.
"Up to the top!" said Matt.

"I am at the top!" said Sam.
"I see a lot!" said Matt.

3

Reflect On Reading

ASSESS COMPREHENSION

To assess their understanding of the story, ask children questions such as:

- **When the story begins, where are the boys?** ***(near the bottom of a hill)***
- **Where are the boys when the story ends?** ***(at the top of the hill)***
- **What do the boys do after they climb the hill?** ***(They sit; they enjoy the view; they eat a snack.)***

HOME-SCHOOL CONNECTION

Send home *At the Top*. Encourage children to read the story to a family member.

WRITING CONNECTION

STORY EXTENSION Have children write or dictate sentences to add to the story *At the Top*. Write the sentences on chart paper. You may also wish to have children illustrate their sentences.

Phonics Connection

Phonics Readers:
#10, *To Tad*
#11, *Dad and Sam*
#12, *A Lot on Top*

CHALLENGE

Have children work in groups of three to count words with ***t, s,*** and ***o*** in the story *At the Top*. One child in each group can count the words with **/t/*t*,** while the second child counts the words with **/s/*s*,** and the third child counts the words with **/o/*o*.**

EXTRA HELP

Pairs of children can practice reading the story aloud by taking the part of Sam or Matt. Children can read the last sentence together.

Lesson 26

pages 43-44

Recognize and Write /h/*h*

QUICKCHECK ✔

Can children:

- ✔ **recognize rhyme?**
- ✔ **write capital and small *Hh*?**
- ✔ **recognize /h/?**
- ✔ **identify the letter that stands for /h/?**

If YES go to Read and Write.

TEACH

Develop Phonemic Awareness

Rhyme Write the rhyme "Once I Saw" on chart paper. Reread the rhyme several times, encouraging children to join in. During later readings, replace the word *goats* with ***bats***, and the word *coats* with ***hats***. Read the rhyme that uses the new words.

Once I Saw

Once I saw three goats,
And they had three coats,
Tra-la-la-la-la-la-la-la,
Funny little goats.

Write the Letter Explain to children that the letter ***h*** stands for **/h/,** the sound they hear at the beginning of ***hot***. Write ***Hh*** on the chalkboard. Point out the capital and small forms of the letter. Then model how to write the letter. Have children write both forms of the letter in the air with their fingers.

READ AND WRITE

Connect Sound-Symbol Write the word ***hot*** on the chalkboard, and have a volunteer circle the letter ***h***. Remind children that the letter ***h*** stands for /**h**/. Ask children to suggest other words that begin with /**h**/. List these words on the chalkboard.

Complete Activity Pages Read aloud the directions on pages 43–44. Review each picture name with children

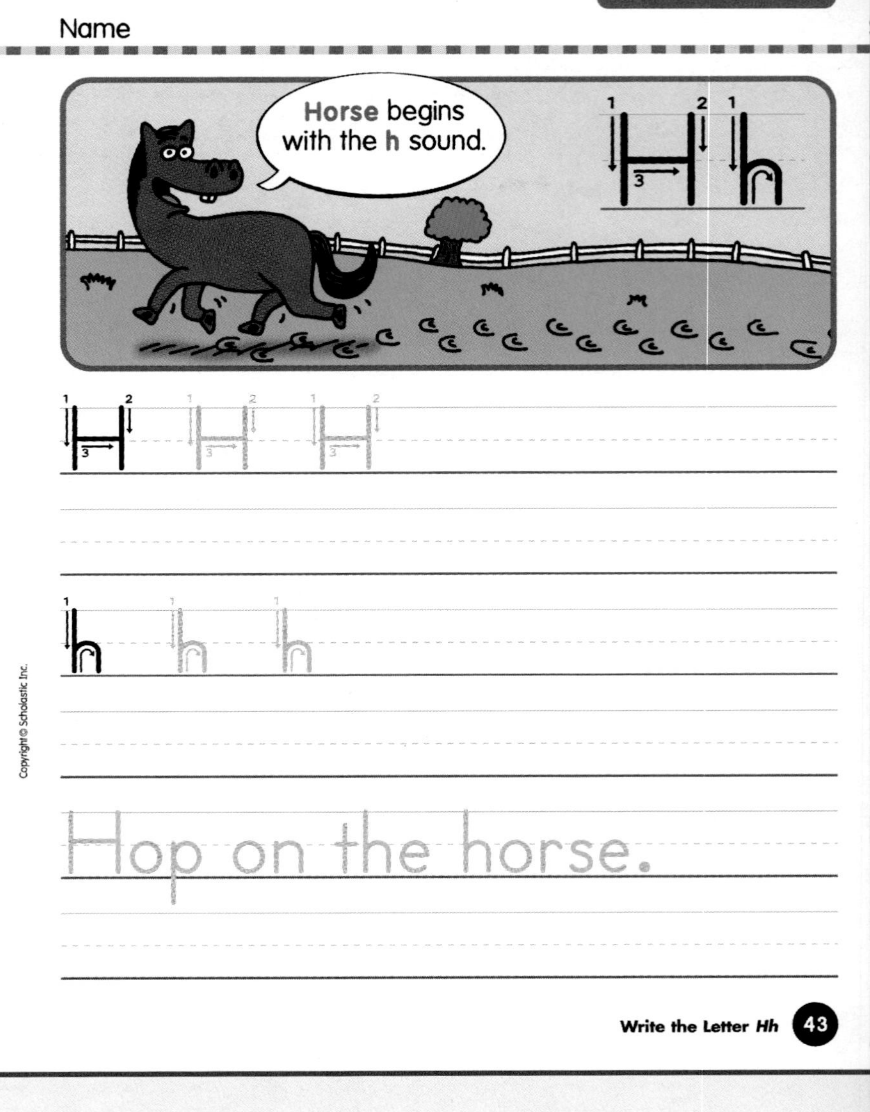

Supporting All Learners

TACTILE LEARNERS

HURRAY FOR *H!* Read aloud "Hillary Hume" from the *Big Book of Rhymes and Rhythms*. Have children find all the words with **/h/** in the rhyme. When they find a word, have them shout "Hurray!" Write these words on the chalkboard. Have children trace over the ***h*** in each word. Invite volunteers to add other **/h/** words to the list.

AUDITORY LEARNERS

PLAY A NAME GAME Say ***Hi, my name is Harry, and I like hippos and hats*** or a similar alliterative sentence. Use children's names that begin with **/h/** and names of characters that begin with **/h/** from familiar rhymes and books. Invite children to supply words that begin with the same sound as the names. Help children make silly or serious sentences.

AUDITORY LEARNERS

LISTEN FOR *H* Have children listen to "To Market" and "Hokey Pokey" on the Sounds of Phonics Audiocassette from the Phonics and Word Building Kit. Invite children to identify all the words they hear that begin with **/h/** in each song.

Integrated Curriculum

SPELLING CONNECTION

SPELLING GRAB BAG Write the following word parts on pieces of paper, each word part on at least two pieces of paper: ***_op, _at, _am***. Place the pieces of paper in a bag. Have each child choose one piece of paper and add the letter ***h*** to the word part. Have children blend each word. Children can take turns holding up their words and modeling how to blend them. Continue this activity with the following word parts: ***_ot, _ip, _im***.

ART CONNECTION

AN *Hh* POSTER Have children make a poster of things with names that begin with **/h/h**. Children can cut out pictures from magazines or newspapers, or draw pictures of things with names that begin with **/h/h**. Label or have children dictate a label for each picture.

Phonics Connection

Literacy Place: *Problem Patrol*
Teacher's SourceBook, pp. T38–39;
Literacy-at-Work Book, pp. 9–10

My Book: *Who Has It?*

Big Book of Rhymes and Rhythms, 1A: "Hillary Hume," pp. 12–13

Chapter Book: *A Lot of Hats,* Chapter 1

CHALLENGE

Give children a blank game board with nine squares. Dictate words that begin with the letter ***h*** or begin or end with the letter ***t***. Ask children to write each word in any space they wish. Then state aloud one word. Have children place a marker on the space that contains the word. The first child to get three markers in a row wins.

To help children acquiring English connect the sound **/h/** with ***Hh,*** have each child make a poster like this: Write ***Hh*** in the center. Cut from discarded magazines two or three pictures whose names begin with **/h/** (such as ***hand, hat, house,*** or ***heart***) and paste them on the poster around the ***Hh***. Have children say the picture names on their poster as they trace the ***Hh***.

Lesson 27

page 45

Blend Words With /h/*h*

QUICKCHECK ✔

Can children:

- ✔ **orally blend word parts?**
- ✔ **identify /h/?**
- ✔ **blend words with /h/*h*?**

If YES go to Read and Write.

TEACH

Develop Phonemic Awareness

Oral Blending Say the following words parts, and ask children to blend them. Provide corrective feedback and modeling when necessary.

/h/...it	**/h/...am**	**/h/...at**
/h/...appy	**/h/...and**	**/h/...ot**

Connect Sound-Symbol Review with children that *h* stands for /h/ as in ***hat***. Write the word ***hat*** on the chalkboard. Have a volunteer circle the *h*. Then model how to blend the word.

THINK ALOUD

I can put the letters *h, a,* and *t* together to make the word *hat.* Let's say the word slowly as I move my finger under the letters. Listen to how I string together the sound that each letter stands for to make the word: *haaaat, haat, hat.*

Continue by helping children blend the following words: ***hot*** and ***had***.

READ AND WRITE

BUILT-IN REVIEW

Blend Words To practice using the sound taught and to review previously taught sounds, list the following words and sentence on a chart. Have volunteers read each aloud. Model blending when necessary.

- **had** **ham** **hop**
- **hat** **sat** **mat**
- **"He had the hat," said Sam**

Complete Activity Page Read aloud the directions on page 45. Review each picture name with children.

Name

Circle the word that names each picture. Then write the word on the line.

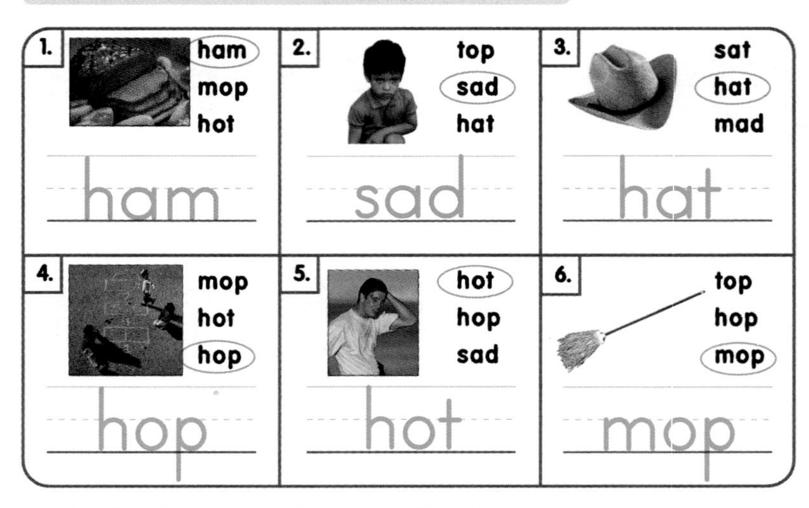

1. ham / mop / hot — ham	2. top / sad / hat — sad	3. sat / hat / mad — hat
4. mop / hot / hop — hop	5. hot / hop / sad — hot	6. top / hop / mop — mop

Use one of the words from above to finish each sentence.

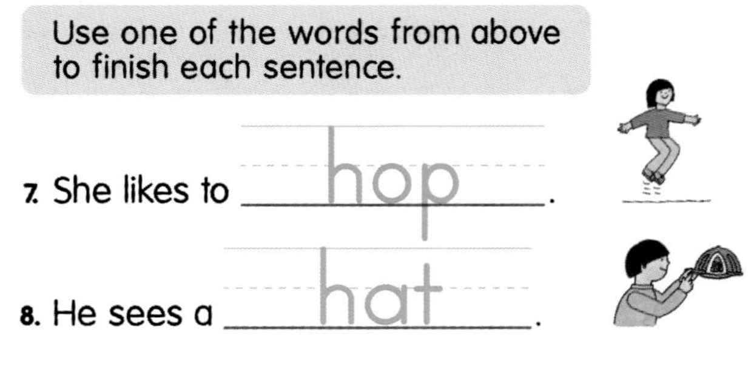

7. She likes to hop.
8. He sees a hat.

Supporting All Learners

TACTILE LEARNERS

WORD WALL On a large note card, write the letter ***h*** and the key word ***horse***. Display the card on the Word Wall. Have children suggest words with **/h/*h*,** write them on small note cards, and add them to the wall. Remind children to look for words that begin with ***h*** during their reading. Add these words to the wall. Periodically, have children use their fingers to trace the letter ***h*** in words on the Word Wall.

The Book Shop

- **Who Hid It?**
  by Taro Gomi
  Children will delight in this picture-puzzle story. **(/h/*h*)**
- **Hot Hippo**
  by Mwenge Hadithi
  In this African story, Hippo wishes he could live in water instead of on land. **(/h/*h*)**

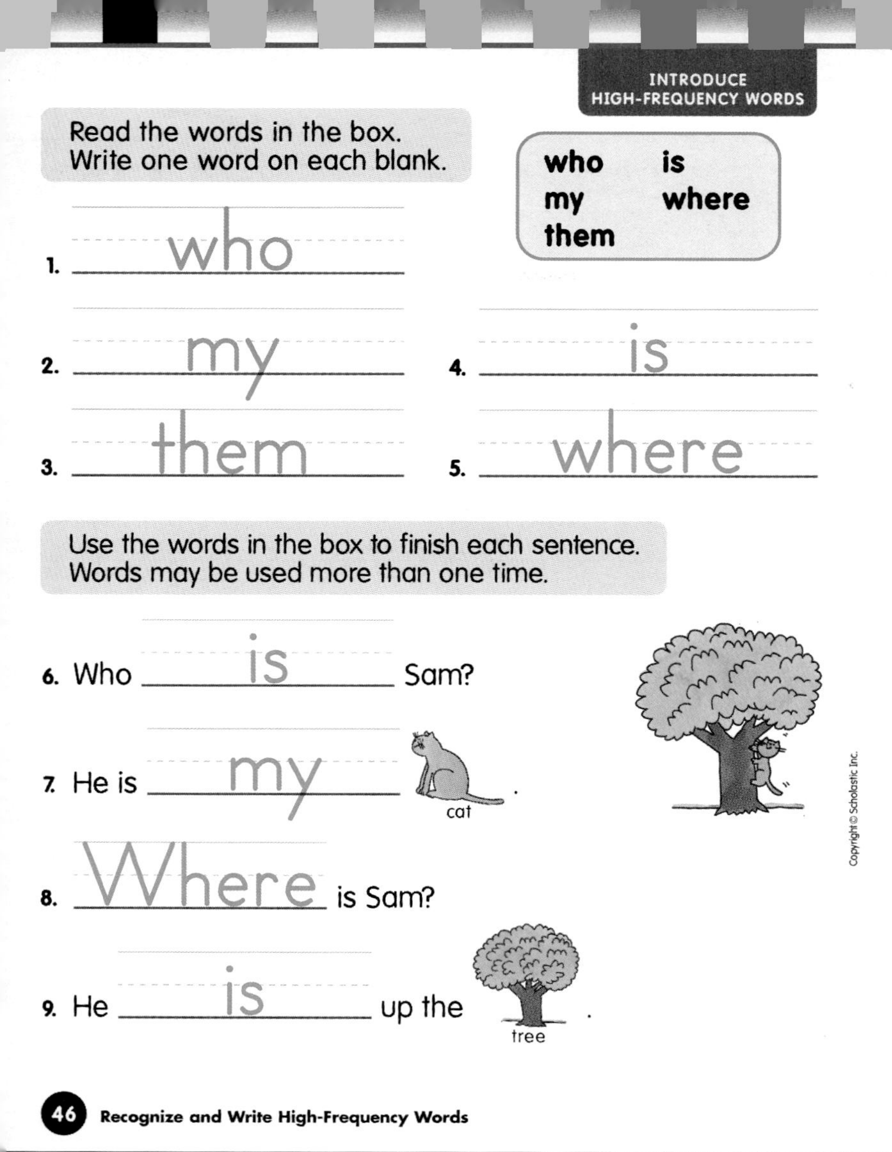
INTRODUCE HIGH-FREQUENCY WORDS

Read the words in the box.
Write one word on each blank.

who	is
my	where
them	

1. who
2. my
3. them
4. is
5. where

Use the words in the box to finish each sentence.
Words may be used more than one time.

6. Who is Sam?
7. He is my cat.
8. Where is Sam?
9. He is up the tree.

Copyright © Scholastic Inc.

46 Recognize and Write High-Frequency Words

Lesson 28

page 46

Recognize High-Frequency Words

QUICKCHECK ✔

Can children:

✔ **recognize and write the high-frequency words *who, my, them, is, where*?**

✔ **complete sentences using the high-frequency words?**

If YES go to Read and Write.

TEACH

Introduce the High-Frequency Words

Write the high-frequency words ***who, my, them, is,*** and ***where*** in sentences on the chalkboard. Read the sentences aloud. Underline the high-frequency words. Ask children if they recognize them. You may wish to use the following sentences:

1. **Where is the cat?**
2. **Tell them yes.**
3. **Who hid the hat?**
4. **My dog runs fast.**

Ask volunteers to dictate sentences using the high-frequency words. You may wish to begin with the following sentence starters: ***Where is*** ______ or ***My*** ______ ***is*** ______.

READ AND WRITE

Practice Write each high-frequency word on a note card. Read each word aloud. Then do the following:

- Mix the cards.
- Display one card at a time, and ask children to state each word aloud.
- Have children spell each word aloud, clapping on each letter.
- Ask children to write each word in the air as they state aloud each letter. Then have them write each word on a sheet of paper.

Complete Activity Page Read aloud the directions on page 46. Have children complete the page independently.

Supporting All Learners

AUDITORY/VISUAL LEARNERS

WORD WALL Add the note cards to the Word Wall for future reference. Encourage children to refer to the Word Wall when reading and writing. Periodically review the words on the wall. Select words at random as children read them aloud. You may also wish to have partners quiz each other by selecting words on the wall to read aloud.

EXTRA HELP

Review the following high-frequency words: ***who, my, them, is, where, he, she, said, and, like, the***. Write each word on the chalkboard. Point to one word at a time. Have volunteers read aloud the word and use it in a sentence. Then have children create flash cards for the words. Pairs of children can quiz each other using the flash cards.

Lesson 29

pages 47–48

Recognize and Write /i/*i*

QUICKCHECK ✓

Can children:

✓ **write capital and small *Ii*?**

✓ **recognize /i/?**

✓ **identify the letter that stands for /i/?**

If YES go to Read and Write.

TEACH

Develop Phonemic Awareness

Oddity Task Explain to children that you will read a list of three words. Two of the words contain /**i**/; the other does not. Children are to choose the word that does not belong—the word that does not contain /**i**/. Use the following word lists:

• **him**	**did**	**top**
• **sit**	**cat**	**lid**
• **big**	**fin**	**sun**
• **hot**	**pick**	**win**

Write the Letter Explain to children that the letter ***i*** stands for /**i**/ as in ***pig***. Write ***Ii*** on the chalkboard. Point out the capital and small forms of the letter. Then model for children how to write the letter.

Have children write both forms of the letter in the air with their fingers. You may also wish to have volunteers practice writing the letter on the chalkboard.

READ AND WRITE

Connect Sound-Symbol Write the word ***hit*** on the chalkboard. Have a volunteer circle the letter ***i***. Remind children that the letter ***i*** stands for /**i**/. Ask children to suggest other words that contain /**i**/. List these words on the chalkboard. Have volunteers circle the letter ***i*** in each.

Complete Activity Pages Read aloud the directions on pages 47–48. Review each picture name with children.

HANDWRITING

Name

Inch begins with the short i sound.

I i

Is an inch big?

Supporting All Learners

AUDITORY LEARNERS

RHYME TIME Read aloud "Higglety, Pigglety, Pop." Have children find all the words with /**i**/ in the rhyme. Write these words on the chalkboard. Invite volunteers to add other /**i**/ words to the list.

TACTILE LEARNERS

LEARNING CENTER Provide a wide assortment of alphabet letters. You may wish to use the magnetic letters in the Phonics and Word Building Kit as well as other letter sets. Invite children to spread out the letters and match the capital ***I*** with its small counterpart. Encourage children to refer to a classroom ABC chart if they need to. Then have children make words with /**i**/***i***.

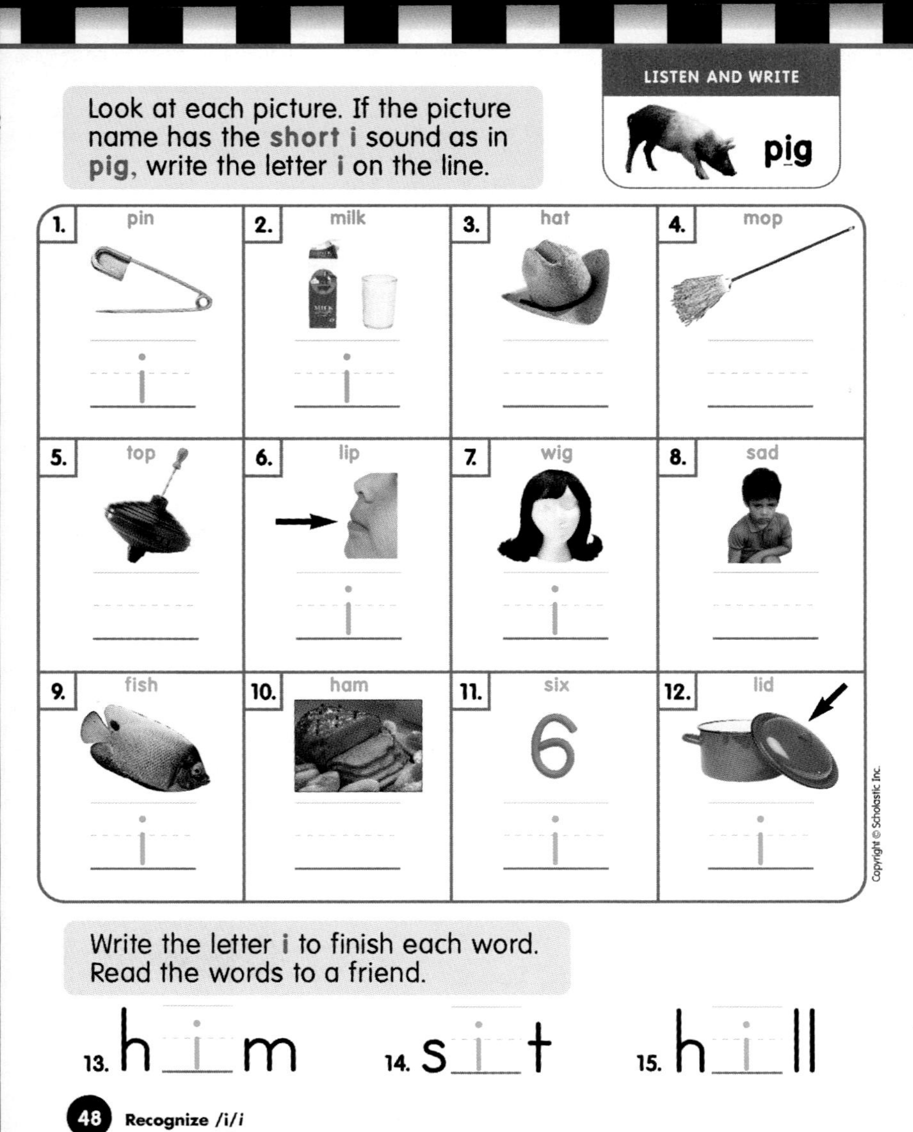

LISTEN AND WRITE

pig

Look at each picture. If the picture name has the **short i** sound as in **pig**, write the letter **i** on the line.

1. pin i
2. milk i
3. hat
4. mop
5. top
6. lip i
7. wig i
8. sad
9. fish i
10. ham
11. six i
12. lid i

Write the letter **i** to finish each word. Read the words to a friend.

13. h i m 14. s i t 15. h i ll

Copyright © Scholastic Inc.

48 Recognize /i/i

Integrated Curriculum

SPELLING CONNECTION

MAKE WORDS Write the following words parts on the chalkboard: ***f__t, s__t, h__m, h__p, l__p***. Ask children to write these word parts on paper. Have children add the letter ***i*** to each word part. Then have volunteers model how to blend each word.

TECHNOLOGY CONNECTION

MAGNET BOARD FUN Have children build ***-id*** and ***-ip*** words on the Magnet Board. See the WiggleWorks Plus My Books Teaching Plan for information on this and other Magnet Board activities.

Phonics Connection

Literacy Place: *Problem Patrol*
Teacher's SourceBook, pp. T40–41;
Literacy-at-Work Book, pp. 10–11

My Book: *What Is It?*

Big Book of Rhymes and Rhythms, 1A: "Higglety, Pigglety, Pop," p. 11

Chapter Book: *A Lot of Hats,* Chapter 2

EXTRA HELP

Display the following picture cards from the Phonemic Awareness Kit: **bat, cat, dig, fan, fish, hat, man, pig, six,** and **wig**. Have pairs or small groups of children sort the pictures into two piles: one pile with pictures whose names contain **/i/,** and one pile with pictures whose names contain **/a/.**

CHALLENGE

Display the ABC card for ***Ii*** from the Phonics and Word Building Kit. Have children whose names contain ***i*** write their names on the ABC card. Then have children write and suggest other words to add to the list. Display the card for future reference when reading and writing.

Lesson 30

pages 49–50

Blend and Build Words With /i/*i* and -*id*, -*ip*

QUICKCHECK ✔

Can children:

✔ **orally blend word parts?**

✔ **identify /i/?**

✔ **blend and build words with /i/*i* and phonograms -*id*, -*ip*?**

If YES go to Read and Write.

TEACH

Develop Phonemic Awareness

Oral Blending Say the following word parts. Ask children to blend them. Provide corrective feedback and modeling when necessary.

/s/...it	/s/...ick	/m/...ix
/l/...id	/m/...itt	/l/...ick

Connect Sound-Symbol Review with children that *i* stands for /i/ as in the word *sit*. Write *sit* on the chalkboard. Have a volunteer circle the *i*. Model how to blend the word. Continue by helping children to blend *hit*.

Introduce the Phonogram Write -*id* and -*ip* on the chalkboard. Point out the sounds these phonograms stand for. Add the letter *l* to the beginning of -*id*. Model how to blend the word formed. Have children repeat *lid* aloud as you blend it again. Then add *s* to the beginning of -*ip*. Model how to blend *sip*.

READ AND WRITE

BUILT-IN REVIEW

Blend Words To practice using the sounds and phonograms taught, list the following words and sentence on a chart. Have volunteers read each aloud. Model blending when necessary.

- hid lid did
- hip lip hop
- Who hid the ham?

Complete Activity Pages Read aloud the directions on pages 49–50. Review the pictures with children.

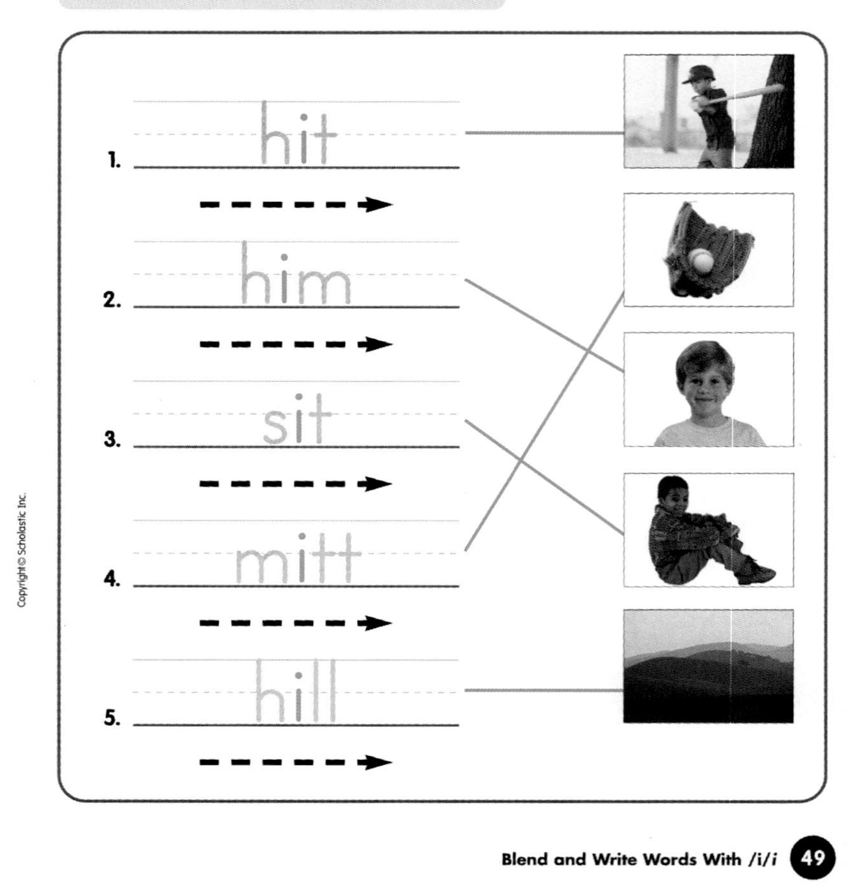

Supporting All Learners

KINESTHETIC/VISUAL LEARNERS

WORD INDEX Invite children to go on a scavenger hunt to collect things or pictures of things with names that have the short *i* sound, such as ***stick, pig, wig,*** and ***mitten***. Display the collected items or pictures, and have children write or dictate the name of each on an index card.

VISUAL LEARNERS

CHANGING LETTERS Have children replace the initial consonant in each of the following words to build another word. For example, children might build the following sequence of words:

hip
sip
tip
lip

Add each letter to the word part below it.
Blend the word. If it is a real word, write it on the line.

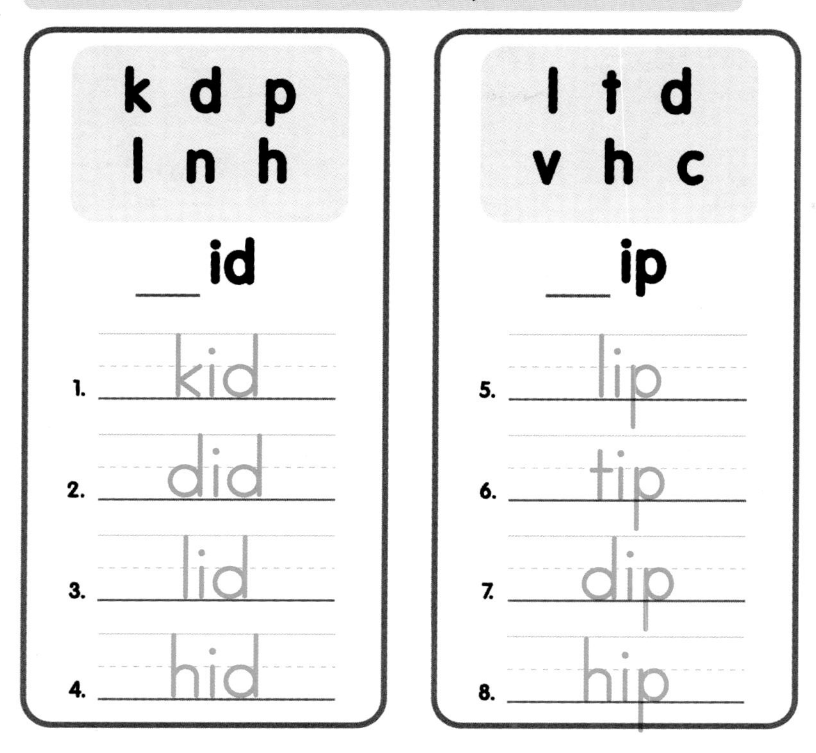

k d p
l n h

__id

1. kid
2. did
3. lid
4. hid

l t d
v h c

__ip

5. lip
6. tip
7. dip
8. hip

Write a sentence using one of the words you made.

9. Answers will vary.

Integrated Curriculum

SPELLING CONNECTION

RHYME TIME Ask children to suggest words that rhyme with ***lid*** and ***sip***. List these words on the chalkboard in separate columns. Have children underline the phonogram ***-id*** or ***-ip*** in each word. Be sure children include the words ***did, hid, kid, dip, hip, lip, rip,*** and ***tip***. Have volunteers blend these words as they repeat them aloud. Then have children look for the word parts ***-id*** and ***-ip*** as they read.

SOCIAL STUDIES CONNECTION

WORD SEARCH Have children look for words with ***-id*** and ***-ip*** on signs, labels, and books in the school or neighborhood. Children can write the words they find on sheets of paper to share with the class.

CHALLENGE

Have children create a story using short ***i*** words. Begin with a short ***i*** title such as **The Silly Pig**. Write the dictated story on chart paper, and return to it in subsequent lessons.

Invite children to illustrate words with **/i/**, such as ***pig, mitt,*** and ***ship***. When they have completed their drawings, help them label each picture. Then have children underline the letter in each picture name that stands for **/i/**.

EXTRA HELP

Write the word part ***-ip*** on the chalkboard. Have children add a letter to the beginning of the phonogram to make a new word.

The Book Shop

- **It Looked Like Spilt Milk**
  by Charles Shaw
  In this visual treat, a cloud looks like many things.
  (/i/i)
- **Bit by Bit**
  by Steve Sanfield
  A jacket changes, bit by bit. **(/i/i)**

Lesson 31

pages 51–52

Recognize and Write /p/p

QUICKCHECK ✔

Can children:

✔ write capital and small *Pp*?

✔ recognize /p/?

✔ identify the letter that stands for /p/?

If YES go to Read and Write.

TEACH

Develop Phonemic Awareness

Alliteration Explain to children that you are going to read some silly sentences with words that begin with the same sound. Children will say the sound they hear. For example, say: ***Lana the lamb likes to leap.*** Guide children to say the sound they hear at the beginning of ***Lana, lamb, likes,*** and ***leap.*** Continue with these sentences:

- **María mailed some mittens to Mike.**
- **Harry the hog hid in the house.**
- **Silly Sam sat on his socks.**
- **Pam the penguin painted a picture.**

Write the Letter Explain that ***p*** stands for /p/, the sound at the beginning of ***pin*** and at the end of ***mop.*** Write ***Pp*** on the chalkboard. Point out the capital and small forms of the letter. Model how to write the letter.

Have children write both forms of the letter in the air with their fingers. Have volunteers practice writing the letter on the chalkboard.

READ AND WRITE

Connect Sound-Symbol Write ***pat*** and ***top*** on the chalkboard. Have a volunteer circle the ***p*** in each word. Remind children that ***p*** stands for /p/. Ask children to suggest other words that begin or end with /p/. List these words on the chalkboard. Have volunteers circle the letter ***p*** in each word.

Complete Activity Pages Read aloud the directions on pages 51–52. Review each picture name with children.

HANDWRITING

Name

Pencil begins with the p sound.

P p

P P P

p p p

Pick up a pencil.

Write the Letter *Pp* 51

Supporting All Learners

VISUAL/AUDITORY LEARNERS

LETTER HUNT Read aloud "Pease Porridge Hot" on page 15 in the *Big Book of Rhymes and Rhythms, 1A.* Have children find all the words with **/p/** in the rhyme. Write these words on the chalkboard. Invite volunteers to add other **/p/** words to the list.

VISUAL/KINESTHETIC LEARNERS

LETTER CARD SHUFFLE Distribute the letter card **p** to each child. Then have children draw three connected boxes on a sheet of paper. Tell children that you are going to say a list of words. All of the words contain **/p/.** Some words begin with **/p/,** and some words end with **/p/.** If children hear **/p/** at the beginning of the word, they are to place the letter card in the first box. If they hear **/p/** at the end, they are to place the card in the last box. Use the following words: ***pig, pat, hop, hip, sip, top, pot, pad,*** and ***mop.***

Look at each picture. If the picture name begins with the p sound as in pig, write the letter p on the first line. If it ends in p as in top, write p on the second line.

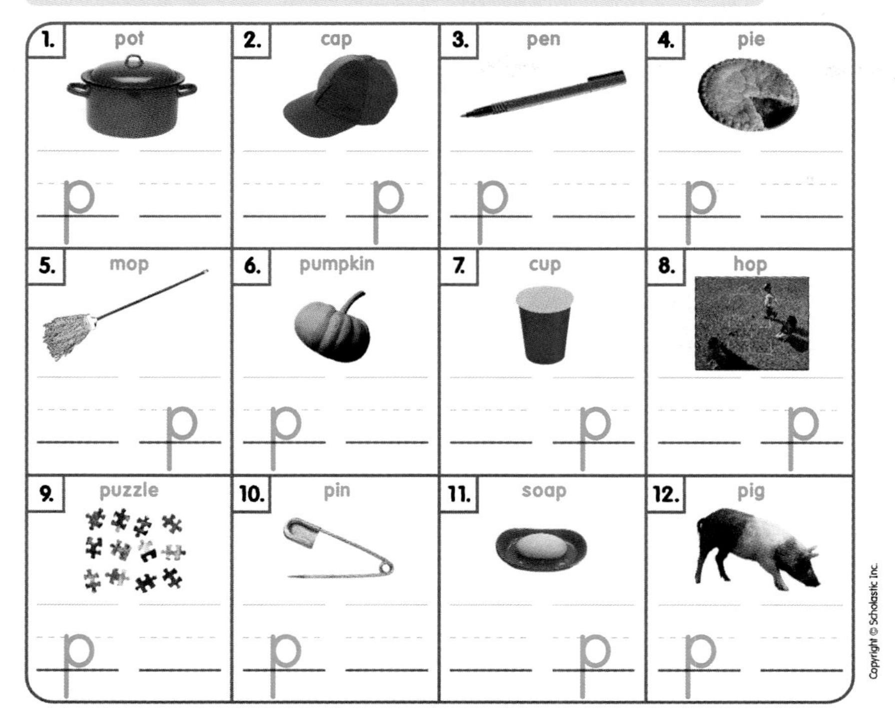

Write the letter p to finish each word.
Circle the word that names the picture.

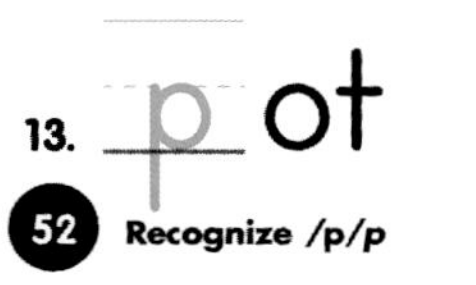

13. p ot

14. to p

Integrated Curriculum

SPELLING CONNECTION

Write the following word parts on the chalkboard: ***__ot, __ig, ho__, ma__, ti__***. Have children hold up letter cards next to each word part to make new words. Then have children write the letter ***p*** in each blank. Model for children how to blend each word. Children can write the words they made on a sheet of paper.

SCIENCE CONNECTION

PEBBLE WORDS Have children write large words with **/p/*p*** on construction paper, one word per page. Children can then glue pebbles to each ***p*** in each word. Have children trace the ***p*** as they say the word.

Phonics Connection

Literacy Place: *Problem Patrol*
Teacher's SourceBook, pp. T84–85;
Literacy-at-Work Book, p. 15

My Book: *Pizza, Please!*

Big Book of Rhymes and Rhythms, 1A: "Pease Porridge Hot," p. 15

Chapter Book: *A Lot of Hats*, Chapter 3

EXTRA HELP

Play "Pease Porridge" and "Paw Paw Patch" on the Sounds of Phonics Audiocassette (Creative Expression) from the Phonics and Word Building Kit. Have children listen for words that begin and end with **/p/**. Children should raise their hands each time they hear a word that begins or ends with **/p/**. Write these words on the chalkboard. After each selection, repeat the words for children, emphasizing **/p/** in each.

Encourage children acquiring English to write words that begin and end with **/p/** on the Magnet Board. Then children can click on the words and hear them read aloud.

Lesson 32

page 53

Blend Words With /p/p

QUICKCHECK ✔

Can children:

- ✔ **orally blend word parts?**
- ✔ **identify /p/?**
- ✔ **blend words with /p/p?**

If YES go to Read and Write.

TEACH

Develop Phonemic Awareness

Oral Blending State aloud the following word parts, and ask children to orally blend them. Provide corrective feedback and modeling when necessary.

/p/...ot	**/p/...at**	**/p/...enguin**
/p/...an	**/p/...ig**	**/p/...encil**

Connect Sound-Symbol Review that ***p*** stands for /**p**/ as in ***pat***. Write the word ***pat*** on the chalkboard. Have a volunteer circle the ***p***. Then model how to blend the word.

THINK ALOUD

I can put the letters *p, a,* and *t* together to make the word *pat*. Let's say the word slowly as I move my finger under the letters. Listen to how I string together the sound that each letter stands for to make the word: *paaaat, paat, pat.*

READ AND WRITE

Blend Words To practice using the sound taught and to review previously taught sounds, list the following words and sentence on a chart. Have volunteers read each aloud. Model blending when necessary.

- **hop** **pad** **lip**
- **hip** **top** **pat**
- **She said, "Hop to the top!"**

Complete Activity Page Read aloud the directions on page 53. Review each picture name with children.

Name

Circle the word that names each picture.
Then write the word on the line.

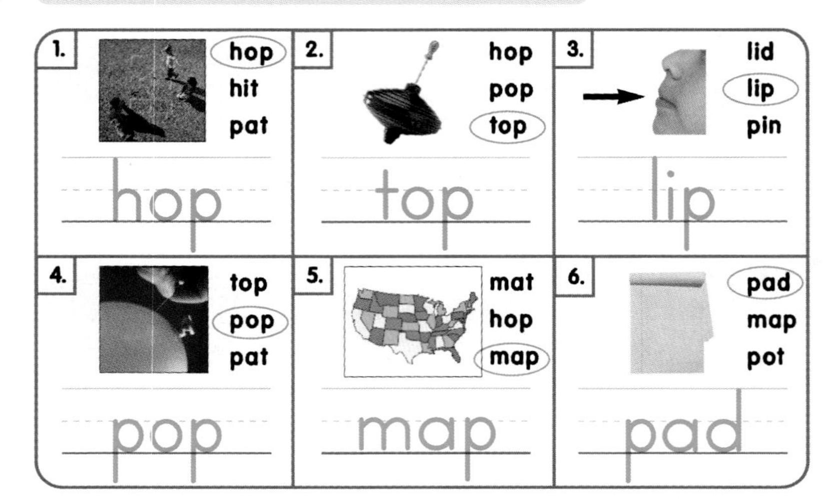

Write the word that names each picture.

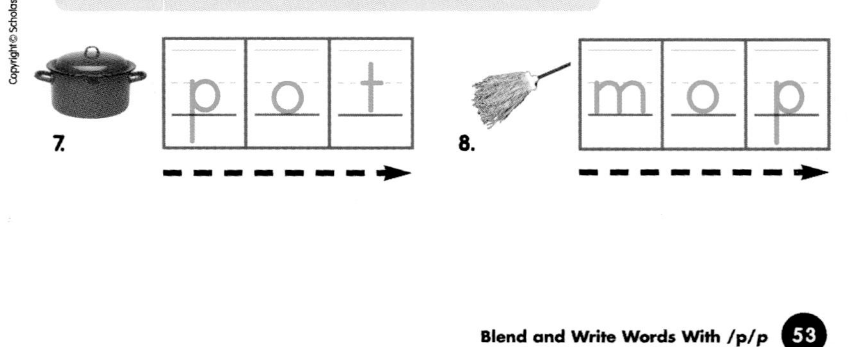

Supporting All Learners

KINESTHETIC LEARNERS

BE-A-WORD Give one alphabet card with a letter previously taught to each child in the class. (Many children may have the same letter.) Invite children to get together in small groups to form words. Have children come to the front of the room and form the word they made. List the words formed on the chalkboard.

The Book Shop

- ***Juan Tuza and the Magic Pouch***
  by Francisco X. Mora
  Pepe and Juan Tuza receive a reward. **(/p/p)**

- ***The Piggy in the Puddle***
  by Charlotte Pomerantz
  This poem is about a pig (and parents) and a puddle. **(/p/p)**

POETRY

Read the poem. Write a sentence telling who likes to hop.

Answers will vary.

Supporting All Learners

KINESTHETIC LEARNERS

PORTFOLIO

PUT THEM TOGETHER Have children work with the magnetic letters, pocket ABC cards, and pocket chart from the Phonics and Word Building Kit. Children can make words with **/h/*h*, /i/*i*,** and **/p/*p***. Have children list the words they make.

EXTRA HELP

Children needing additional support segmenting words can use Elkonin boxes and counters. A description of this procedure can be found on page T31.

CHALLENGE

Have children generate a list of ***h*** words and record them on the chalkboard. Then have children create a story using as many ***h*** words on the chalkboard as they can. Provide time for children to share their stories.

page 54

Write and Read Words With /h/*h*, /i/*i*, /p/*p*

QUICKCHECK ✔

Can children:

✔ **spell words with /h/*h*, /i/*i*, /p/*p*?**

✔ **read a poem?**

If YES go to Read and Write.

TEACH

Link to Spelling Review the following sound-spelling relationships: **/h/*h*, /i/*i*, /p/*p***. Say one of these sounds. Have a volunteer write on the chalkboard the spelling that stands for the sound. For example, the letter ***h*** stands for **/h/**. Continue with all the sounds. You may also review **/s/*s*, /t/*t*, /a/*a***, and **/o/*o***.

Phonemic Awareness Oral Segmentation
Say ***hip***. Have children orally segment the word. (**/h/ /i/ /p/**) Ask them how many sounds the word contains. **(3)** Draw three connected boxes on the chalkboard. Have a volunteer write the spelling that stands for each sound in ***hip*** in the appropriate box. Continue with ***ham*** and ***hop***.

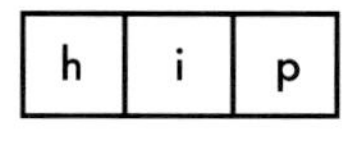

h	i	p

READ AND WRITE

Dictate Dictate the following words and sentence. The first set of words are decodable based on the sounds previously taught. The second set are high-frequency words. Have children write the words and sentence. Then write them on the chalkboard. Have children make any necessary corrections.

- **hit** **hid** **pan**
- **and** **like** **who**
- **I hid the hat.**

Complete Activity Page Read aloud the directions on page 54. Children can read the poem independently or with a partner. Have children circle the words that contain **/i/**.

Lesson 34

pages 55–56

Read and Review Words With /h/h, /i/i, /p/p

QUICKCHECK ✔

Can children:

✔ **read words with /h/h, /i/i, /p/p?**

✔ **recognize high-frequency words?**

If YES go to Read and Write.

TEACH

Review Sound-Spellings Review with children the sound-spelling relationships from the past few lessons. These include **/h/h, /i/i,** and **/p/p**. Say one of these sounds. Have a volunteer write the spelling that stands for the sound on the chalkboard. For example, the letter ***i*** stands for /i/. Continue with all the sounds.

You can also review ***-id*** and ***-ip***. Write each phonogram on the chalkboard. Have children suggest words containing each. Ask volunteers to write the words on the chalkboard.

Review High-Frequency Words Review the high-frequency words ***who, my, them, is,*** and ***where***. Write sentences containing each word. Say one word. Have volunteers circle it in the sentences. Then have children generate sentences for each word.

READ AND WRITE

Build Words Distribute the following letter cards to children: ***a, d, h, i, l,*** and ***p***. If children have their own set of cards, have them locate the letter set. Have children build as many words as possible. Suggest that they record their words. Then have partners compare their lists. Continue building words with the following letter card set: ***m, n, o, a, d,*** and ***t***.

Build Sentences Write these words on note cards: ***who, my, is, where, hat, hid, has, top,*** and ***the***. Have children make sentences using the words. Suggest the sentence ***Who hid my hat?***

Complete Activity Pages Read aloud the directions on pages 55–56. Review the art with children.

REVIEW

Name

Check each word as you read it to a partner. Circle any words you need to practice.

I can read!

☐ ham	☐ pot	☐ hat
☐ lit	☐ hit	☐ hip
☐ pop	☐ lip	☐ pat
☐ top	☐ hot	☐ hid
☐ sit	☐ him	☐ hop

Lookout Words!

☐ who	☐ them	☐ where	☐ he	☐ and
☐ my	☐ is	☐ like	☐ she	☐ the

Supporting All Learners

VISUAL LEARNERS

CLASS DICTIONARY Have children make a list of words with **/h/h, /i/i,** and **/p/p**. Gather the lists and compile them. Add the lists to your class dictionary. In addition, you may distribute copies of each list, and have children add them to their individual dictionaries.

VISUAL LEARNERS

WORD SEARCH Have children look for the high-frequency words and words with **/h/h, /i/i,** and **/p/p** in books, newspapers, and magazines. Children can make a list of the words they find, then take turns using the words in sentences.

AUDITORY/VISUAL LEARNERS

CLIMB THE LADDER Draw a large ladder on chart paper. Invite children to "climb" the ladder by naming words that begin with ***h, i,*** or ***p***. Write the words on the ladder, one word per rung, beginning with the bottom rung, and working toward the top. Remind children to look for words that begin with ***h, i,*** or ***p*** when they read. Make a second word ladder on which to add the words children find.

TEST YOURSELF

Fill in the bubble next to the sentence that tells about each picture.

1.	● The pot is hot. ○ He hid the pot.
2.	○ Tim is hot. ● Tim is sad.
3.	○ Where is my hat? ● Who is at the top?
4.	○ Sam hit the mop. ● Sam hid the cap.
5.	● Where is the ham? ○ Where is the lid to the pot?
6.	● I like to hop. ○ I like to sit.

Integrated Curriculum

WRITING CONNECTION

ANSWER THIS Have children use the high-frequency words to write or dictate questions such as ***Where is my hat?*** Then children can challenge others to answer the questions by writing or dictating sentences using high-frequency words.

TECHNOLOGY CONNECTION

For additional practice with **/h/h, /i/i,** and **/p/p,** have children read the following My Books: ***Who Has It?, What Is It?,*** and ***Pizza, Please!*** For information on using the My Books on the computer, see the WiggleWorks Plus My Books Teaching Plan.

I CAN READ! OPTIONS

The I Can Read! page can be used for one or all of the following:

- paired reading
- individual assessment
- choral reading
- homework practice
- program placement

EXTRA HELP

For children who are struggling with the sound-spelling relationships or high-frequency words reviewed, use the activities and materials in the Phonics and Word Building Kit for additional support. For example, use the red magnetic letters to build words. Once a word is built, mix the letters and have children rebuild it.

One way to help children acquiring English participate with greater confidence in the Review activities is to go over the key letter sounds, phonograms, and words beforehand. For the Build Words and Build Sentences activities, you may wish to pair these children with those whose primary language is English, then let the same partners work together to complete the pages.

Lesson 35

pages 57–58

Read Words in Context

TEACH

Assemble the Story Ask children to remove pages 57–58. Have children fold the pages in half to form the Take-Home Book.

Preview the Story Preview this story. Have a volunteer read aloud the title *Who Hid the Sock?* Invite children to browse through the first two pages of the story and to comment on it. Have them point out unfamiliar words. Read these words aloud, and model how to blend them. Then have children predict what the story might be about.

READ AND WRITE

Read the Story Read the story aloud, or have volunteers take turns reading aloud a page at a time. Discuss anything of interest. Encourage children to help each other with blending. The following prompts may help children who need extra support:

- **What letter sounds do you know in the word?**
- **Are there any word parts you know?**

Reflect and Respond Have children share their reactions to the story. What did they like about it? What surprised them?

Develop Fluency Reread the story as a choral reading or have partners reread the story independently. Provide time for children to reread the story in subsequent days to develop fluency and increase reading rate.

When reading other stories, ask children to share how they figure out unfamiliar words. Encourage children to look for words with **/h/*h*, /i/*i*,** and **/p/*p*** and to apply what they have learned about these sound-spelling relationships to decode words. Continue to review these sound-spelling relationships for children needing additional support.

Supporting All Learners

KINESTHETIC LEARNERS

PUPPET SHOW Have children create their own sock puppets. Have them write sentences telling about their puppets. Then have small groups of children create a brief puppet show using their puppets.

VISUAL/KINESTHETIC LEARNERS

ANIMAL POSTER Children can make a poster of farm animals, including hens, pigs, and horses. Have children draw or cut out magazine pictures of these and other farm animals. Then children can label each animal. Ask children to take turns identifying animals with names that contain **/h/*h*, /i/*i*,** and **/p/*p***.

Reflect On Reading

ASSESS COMPREHENSION

To assess their understanding of the story, ask children questions such as:

- **What did the little boy lose? (*his sock*)**
- **Where did the little boy look for his missing sock? (*He looked under his bed, under the mop, under the hat, and under the pot.*)**
- **Who had the missing sock? (*Pam*)**
- **What did Pam do with the sock? (*She made a puppet out of it.*)**

HOME-SCHOOL CONNECTION

Send home a note to families suggesting that they play I Spy with their children to help them listen for sounds and write their corresponding spellings. For example, one family member might say ***I spy with my little eye something that you can write with. Its name begins with /p/.*** The child is to guess the name of the object. (***pen***) The child must then write the object's name or the letter that begins the object's name. You may also wish to send home *Who Hid the Sock?* Children can read the story to family members.

CHALLENGE

Children can write their own stories about another missing item such as a book. Have children write about where they might look for the lost item, who might have it, and what might have happened to it.

EXTRA HELP

Children can review words with **/h/h, /i/i,** and **/p/p** by reading aloud the Phonics Readers *Who Has the Hat?, The Big Hit,* and *Pop! Pop!*

Phonics Connection

Phonics Readers:
#13, Who Has the Hat?
#14, The Big Hit
#15, Pop! Pop!

Lesson 36

pages 59-60

Recognize and Write /f/f

QUICKCHECK ✔

Can children:

- ✔ **write capital and small *Ff*?**
- ✔ **recognize /f/?**
- ✔ **identify the letter that stands for /f/?**

If YES go to Read and Write.

TEACH

Develop Phonemic Awareness

Rhyme Write the rhyme "Looby Loo" on chart paper. (The rhyme appears on page 8 of the *Big Book of Rhymes of Rhythms, 1A.*) Track the print as you read. Reread it several times, encouraging children to join in. Then replace the words ***looby loo*** and ***looby light*** with ***fooby foo*** and ***fooby fight***. Read the rhyme with the new words. Have children reread it with you.

Looby Loo

Here we go looby loo,
Here we go looby light,
Here we go looby loo,
All on a Saturday night.

Write the Letter Explain to children that the letter *f* stands for /f/, the sound at the beginning of ***fan***. Write ***Ff*** on the chalkboard. Point out the capital and small forms of the letter. Then model how to write them.

Have children write both forms of the letter in the air. You may also wish to have volunteers practice writing the letter on the chalkboard.

READ AND WRITE

Connect Sound-Symbol Write *fog* on the chalkboard. Have a volunteer circle the *f*. Remind children that *f* stands for /f/. Ask children to suggest other words that contain /f/. List them on the chalkboard. Have volunteers circle each *f*.

Complete Activity Pages Read aloud the directions on pages 59–60. Review each picture name with children.

HANDWRITING

Name

Fan begins with the f sound.

F f

F F F

f f f

Fix the fan.

Supporting All Learners

TACTILE LEARNERS

STARTS WITH "F" Read aloud "Five Little Fishes" on page 16 in the *Big Book of Rhymes and Rhythms, 1A*. Have children find all the words with /f/ in the rhyme. Write these words on the chalkboard. Children can trace over the words with their fingers or the chalk. Then have children add other /f/ words to the list.

KINESTHETIC LEARNERS

LETTER-SOUND HOP On a large sheet of paper, draw ten circles. Select five letters children have been learning. Write each letter in two of the circles on the sheet. Tape the sheet to the floor. On separate note cards, write words that begin with each letter. Select a child, call out a word, and have the child jump to the circle that contains the letter that begins the word. Continue until all the words have been called.

LISTEN AND WRITE

fan

Look at each picture. If the picture name begins with the **f** sound as in **fan**, write the letter **f** on the line.

1. foot — f
2. five — f
3. leaf
4. house
5. fish — f
6. fence — f
7. finger — f
8. four — f
9. fork — f
10. boat
11. feather — f
12. fox — f

Copyright © Scholastic Inc.

Write the letter **f** to finish each word. Read the words to a friend.

13. f it 14. f at 15. f ill

60 **Recognize /f/f**

Integrated Curriculum

SPELLING CONNECTION

Use the pocket ABC cards and pocket chart from the Phonics and Word Building Kit to make the following word parts: ***__at, __ill, __it.*** Have children hold up the ABC card for ***f*** to each word part. Model for children how to blend each word. Then have children say each word and write it on a sheet of paper. Children can also try holding up other ABC cards to the word parts to make more words.

ART CONNECTION

FOOT ART Children can draw or trace then cut outlines of feet. Have children write a word with ***/f/f*** in each foot cut-out, then illustrate that word. Help children display the feet in a trail around the room and, if possible, out into the hall. Children can revisit the feet to see how many ***f*** words they recognize.

Phonics Connection

Literacy Place: *Problem Patrol*
Teacher's SourceBook, pp. T86–87;
Literacy-at-Work Book, pp. 21–22

My Book: *My Feet*

Big Book of Rhymes and Rhythms, 1A: "Five Little Fishes," p. 16

Chapter Book: *A Lot of Hats,* Chapter 4

ESL Use a combination of picture clues and gestures to help children acquiring English understand the words in "Five Little Fishes." For example, as you read the word ***five,*** you may hold up five fingers, or write the number five; as you read ***little,*** you may make a gesture with your thumb and forefinger indicating small or tiny; and as you read ***fishes,*** you may draw outline shapes of two fishes on the chalkboard.

EXTRA HELP

To give children additional practice, set up Center 4, Wheels of Phonics in *Quick-and-Easy Learning Centers: Phonics*. This center features circular games that call children's attention to a host of phonics features. Activities such as the Alphabet Wheel, Guess My Rule, and Snail's Pace may be adapted by limiting them to the sounds, phonograms, and high-frequency words that children have learned so far.

pages 61-62

Blend Words With /f/f

QUICKCHECK ✔

Can children:

- ✔ **orally blend word parts?**
- ✔ **identify /f/**
- ✔ **blend and build words with /f/f?**

If YES go to Read and Write.

TEACH

Develop Phonemic Awareness

Oral Blending Say the following words parts. Ask children to blend them. Provide corrective feedback and modeling when necessary.

/f/...it	/f/...an	/f/...at
/f/...ox	/f/...un	/f/...unny

Connect Sound-Symbol Review that *f* stands for /f/ as in *fit*. Write *fit* on the chalkboard. Model how to blend the word.

I can put *f, i,* and *t* together to make the word *fit*. Let's say the word slowly as I move my finger under the letters. Listen to how I string together the sound that each letter stands for: *ffffiiiit, ffiit, fit.*

READ AND WRITE

Blend Words To practice using the sounds taught, list the following words and sentence on a chart. Have volunteers read each aloud. Model blending.

- fan　　man　　tan
- fat　　fit　　if
- Where is the fan?

Build Words Distribute these letter cards to children: ***a, f, m, o, p,*** and ***t***. Have children build as many words as possible. Children can continue building words with these letter cards: ***h, i, a, t, l, p***.

Complete Activity Pages Read aloud the directions on pages 61–62. Review the picture names with children.

Name

Read each word. Write a word from the box that rhymes with the word. You may write each word more than one time.

fit
fill
fat

1. sit — fit	2. hat — fat	3. hill — fill
4. pat — fat	5. till — fill	6. pit — fit
7. hit — fit	8. sat — fat	9. will — fill

Write a sentence using two words from above.

Answers will vary.

Supporting All Learners

KINESTHETIC LEARNERS

WORD WALL On a large note card, write the letter ***f*** and the key word ***fan***. Display the card on the Word Wall. Have children suggest words with **/f/f**, write them on small note cards, and add them to the wall. Remind children to look for words that begin with ***f*** during their reading. Add these words to the wall.

AUDITORY LEARNERS

STAMP YOUR FEET Read aloud or have children read aloud *Fish Is Fish* by Leo Lionni or *Fun/No Fun* by James Stevenson. Have children stamp their feet whenever they hear a word with **/f/**.

VISUAL LEARNERS

STAMP Have children write ***fan*** on the chalkboard. Tell children that you want them to replace **/f/**, the first sound in ***fan,*** with **/m/** to make a new word. Ask children what letter stands for **/m/**. Then replace the letter ***f*** in ***fan*** with the letter ***m***. Have a volunteer blend the new word. Continue by having children change letters in this sequence of words: ***man, tan, tap, top***.

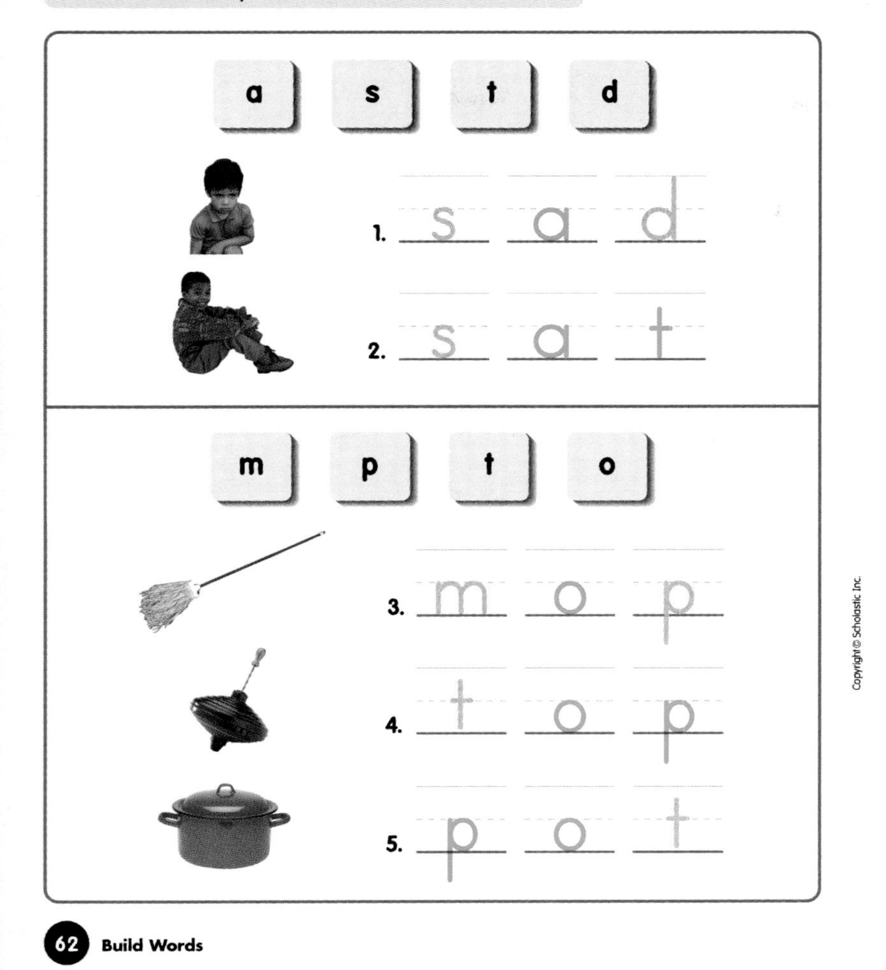
BUILD WORDS

Look at each picture. Use the letter tiles to write each picture name.

a s t d

1. s a d
2. s a t

m p t o

3. m o p
4. t o p
5. p o t

62 Build Words

Integrated Curriculum

SPELLING CONNECTION

CLIMB THE LADDER Explain to children that you will give them the beginning and ending word of a three-word ladder. They are to change one letter in each word to complete the word ladder. Use the following words:

- fat (mat) map
- fit (hit) hot
- lip (tip) top

Note: The words in parentheses are possible answers.

SCIENCE CONNECTION

ALL ABOUT FEET Have children read the My Book *My Feet* by Cass Hollander to each other or all together. Have children talk about why a duck, a cat, a bear, or other animal in the story might have trouble with "people feet" instead of their own feet. Then have children draw pictures of other animals that might look funny with "people feet."

EXTRA HELP

During the Blend Words section, it might be necessary to model for children how to blend the words the first time and then have them blend the words independently.

Display pictures of objects whose names begin with **/f/**. These pictures may include the following: **fan, fish, four, face,** and **finger**. Prompt children acquiring English to take turns identifying each picture name. Then have children take turns dictating sentences with these words. Write the sentences on the chalkboard, and help children to read them chorally.

- **Fish Is Fish**
  by Leo Lionni
  A fish learns to be happy. **(/f/f)**
- **Fun/No Fun**
  by James Stevenson
  An author/artist lists what was not fun in his childhood. **(/f/f)**

Lesson 38

pages 63-64

Recognize and Write /n/n

QUICKCHECK ✔

Can children:

- ✔ **recognize rhyme?**
- ✔ **write capital and small *Nn*?**
- ✔ **recognize /n/?**
- ✔ **identify the letter that stands for /n/?**

If YES go to Read and Write.

TEACH

Develop Phonemic Awareness

Rhyme Explain to children that you are going to show three pictures and say their names. Children are to choose the two picture names that rhyme. Use these picture cards:

• man	can	dig
• sun	bun	red
• hen	queen	green

Write the Letter Explain that ***n*** stands for **/n/**, the sound at the beginning of ***nut*** and at the end of ***man***. Write ***Nn*** on the chalkboard. Point out the capital and small forms of the letter. Then model how to write the letter.

Have children write both forms of the letter in the air with their fingers. Volunteers can write the letter on the chalkboard.

READ AND WRITE

Connect Sound-Symbol Write *not* and ***pan*** on the chalkboard. Remind children that the letter ***n*** stands for **/n/**. Ask children to suggest other words that begin or end with **/n/**. List them on the chalkboard. Have volunteers circle the ***n*** in each.

Then write the following word parts on the chalkboard: ***pi_, ma _, _et, _ap, ta_***. Have volunteers add the letter ***n*** to each. Model for children how to blend each word. Then have children write each word.

Complete Activity Pages Read aloud the directions on pages 63–64. Review each picture name with children.

HANDWRITING

Name

N N N

n n n

We eat nuts.

Supporting All Learners

VISUAL/AUDITORY LEARNERS

LETTER FUN Read aloud "Engine, Engine, Number Nine" on page 17 in the *Big Book of Rhymes and Rhythms, 1A*. Have children find all the words with **/n/** in the rhyme. Write these words on the chalkboard. Invite volunteers to add other **/n/** words to the list. Ask children to count how many words in their list are from the rhyme.

KINESTHETIC/TACTILE LEARNERS

BUILD AND TRACE Display the pocket ABC cards for ***o, h, i, p, s, f*** and the pocket chart from the Phonics and Word Building Kit. Invite children to make words and then trace the letters with their fingers. Children can then take turns building and tracing words with two or three other sets of letter cards such as ***t, l, n, a, o, p,*** and ***s, d, i, m, a, h***.

LISTEN AND WRITE

Look at each picture. If the picture name begins with the **n** sound as in **nut**, write the letter **n** on the line. If it ends with **n** as in **tin**, write **n** on the second line.

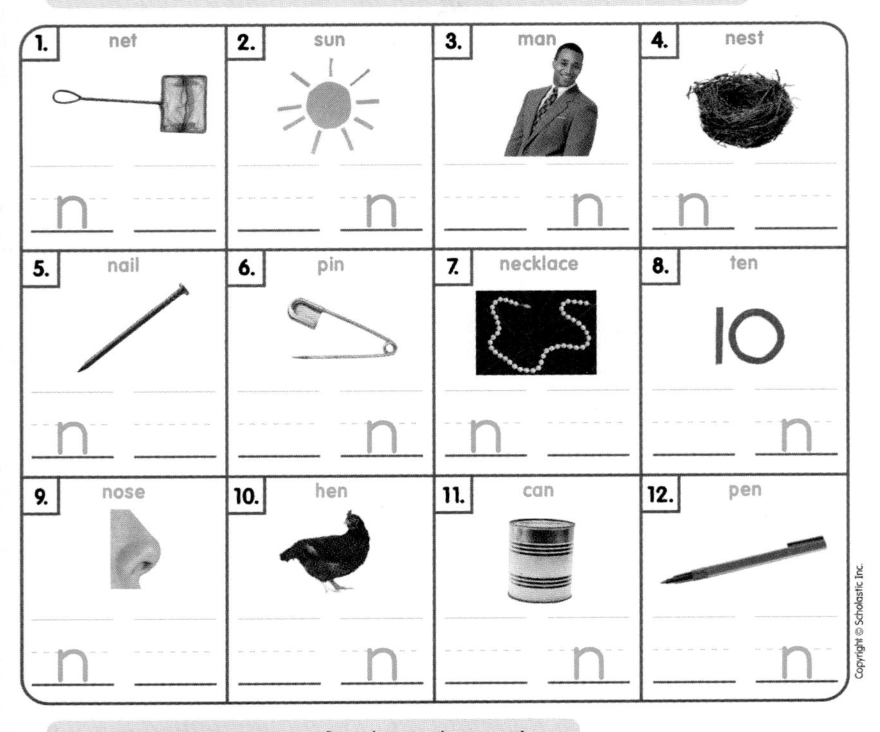

Write the letter **n** to finish each word. Circle the word that names the picture.

13. pi n

14. n ot

Integrated Curriculum

WRITING CONNECTION

Have children generate a list of words that begin with **/n/n**. Record these words on the chalkboard. Then have children create a story using as many **/n/** words as possible. Write the dictated story on chart paper for group and individual reading. Return to the story, rereading it in subsequent days.

SOCIAL STUDIES CONNECTION

WORD HUNT Have children look on signs in their neighborhood for words with **/n/n**. Children can create a list of neighborhood ***Nn*** words, then use the words to write sentences about their neighborhood or town.

Phonics Connection

Literacy Place: *Problem Patrol*
Teacher's SourceBook, pp. T142–143;
Literacy-at-Work Book, pp. 32–33

My Book: *Oh! No!*

Big Book of Rhymes and Rhythms, 1A: "Engine, Engine, Number Nine," p. 17

Chapter Book: *A Lot of Hats*, Chapter 5

CHALLENGE

Invite children to write a story using words that begin or end with **/n/**. Suggest that they begin by generating a list of **/n/** words and possible story titles such as "The Noisy Neighbor" or "Nan Has Fun."

EXTRA HELP

Write words that begin and end with ***n***, as well as words that begin and end with ***p*** on index cards, one word per card. Review each word with children. Then give one word card to each child. Call out a word. The child who has the card should stand and spell aloud the word.

Lesson 39

pages 65-66

Blend and Build Words With /n/*n* and *-og*, *-ig*

QUICKCHECK ✔

Can children:

✔ **orally blend word parts?**

✔ **identify /n/?**

✔ **blend and build words with /n/*n* and phonograms *-og, -ig*?**

If YES go to Read and Write.

TEACH

Develop Phonemic Awareness

Oral Blending Say the following word parts and ask children to blend them. Provide corrective feedback and modeling when necessary.

/n/...et	/n/...ot	/n/...urse
/n/...ut	/n/...est	/n/...umber

Connect Sound-Symbol Remind children that *n* stands for /n/ as in ***not***. Write ***not*** on the chalkboard. Volunteers can circle the *n*. Model how to blend the word. Then help children blend the words ***pan*** and ***fin***.

Introduce the Phonogram Write *-og* and *-ig* on the chalkboard and point out the sounds they stand for. Add *f* to the beginning of *-og*. Model how to blend *fog*. Have children repeat *fog*. Then add *p* to the beginning of *-ig*. Model how to blend *pig*.

READ AND WRITE

Blend Words List the following words on a chart, and have volunteers read them aloud.

• in	pin	log
• fog	man	fan

Complete Activity Pages Read aloud the directions on pages 65–66. Review each picture name with children.

Supporting All Learners

AUDITORY LEARNERS

TIME TO LISTEN Have children listen to "To Market" and "There Was a Pig Who Went Out to Dig" on the Sounds of Phonics Audiocassette (Community Involvement) in the Phonics and Word Building Kit. Both of these songs feature the phonogram ***-ig***. Ask children to raise their hands every time they hear a word with ***-ig***.

KINESTHETIC/TACTILE LEARNERS

WORD WALL Ask children to write words with ***-og*** and ***-ig*** on large note cards. Have children write the beginning consonant in black and the phonogram ***-og*** or ***-ig*** in red. Add the cards to the Word Wall.

AUDITORY/VISUAL LEARNERS

BUILD WORDS Write ***-og*** and ***-ig*** on the chalkboard. (If available, use a pocket chart and letter cards.) Have children add a letter to the beginning of each to make a new word. Then children can replace the initial consonant to make a new word. For example; ***fog*** becomes ***log***, and ***log*** becomes ***hog***.

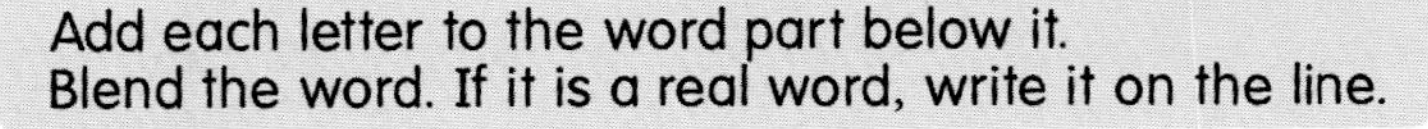

Add each letter to the word part below it.
Blend the word. If it is a real word, write it on the line.

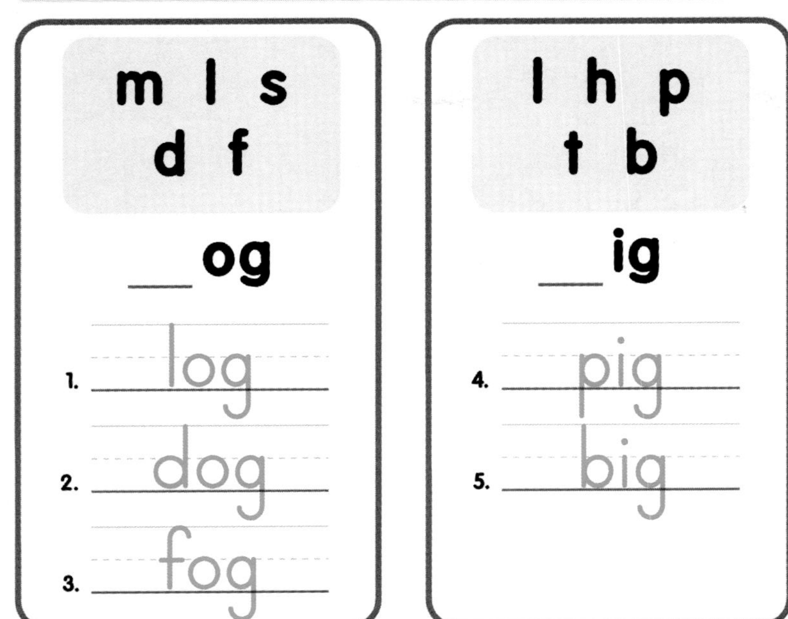
m l s
d f
__og
1. log
2. dog
3. fog

l h p
t b
__ig
4. pig
5. big

Write a sentence using one of the words you made.

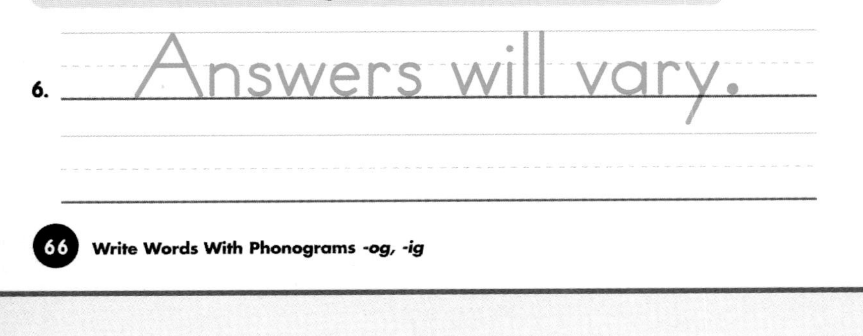
6. Answers will vary.

Integrated Curriculum

SPELLING CONNECTION

WORD BUILDING FUN Write the phonograms ***-og*** and ***-ig*** on the chalkboard. Write ***fog*** beside ***-og*** and ***pig*** beside ***-ig***. Ask children to suggest words that rhyme with ***fog*** and ***pig***. List these words on the chalkboard in the appropriate row. Then have volunteers underline the phonogram ***-og*** or ***-ig*** in each word. You may wish to add the words ***dog, hog, log,*** and ***dig*** to the lists if children did not suggest them.

MATH CONNECTION

CHART LETTERS Have children collect and count words with *-og* and ***-ig***. Children can record the results of their counts in a chart such as the one shown.

6		
5		
4		
3		
2		
1		
	Words With ***-og***	Words With ***-ig***

EXTRA HELP

For children needing additional practice, see the Scholastic Phonemic Awareness Kit. The oral segmentation exercises will help children to break apart words sound by sound. This is necessary for children to be able to encode, or spell, words while writing.

CHALLENGE

Have children write rhyming sentences using words with ***-og*** or ***-ig***. You may wish to provide an example such as ***The big pig wears a wig to dig***. Children can illustrate their sentences, share them, then display them.

- **Noisy Nora**
  by Rosemary Wells
  Children will enjoy this lilting, whimsical story of noisy little Nora, a mouse who can't seem to get any attention from her busy family. **(/n/n)**

Lesson 40

pages 67-68

Recognize and Write /k/c

QUICKCHECK ✔

Can children:

✔ **write capital and small *Cc*?**

✔ **recognize /k/?**

✔ **recognize that the letter *c* can stand for /k/?**

If YES go to Read and Write.

TEACH

Develop Phonemic Awareness

Oddity Task Explain to children that you will show them three pictures and say their names. Children are to choose the picture name that begins with a different sound. Use the following picture cards:

- cup table can
- hat cake coat
- cube cat nine

Write the Letter Explain to children that *c* can stand for /k/, the sound at the beginning of *cat*. Write *Cc* on the chalkboard. Point out the capital and small forms of the letter. Then model for children how to write the letter.

Have children write both forms of the letter in the air with their fingers. Invite volunteers to practice writing the letter on the chalkboard.

READ AND WRITE

Connect Sound-Symbol Write *cup* on the chalkboard. Have a volunteer circle the letter *c*. Remind children that *c* can stand for /k/. Ask children to suggest other words that contain /k/. List these words on the chalkboard. Have volunteers circle the letter *c* in each word that contains this spelling for /k/.

Complete Activity Pages Read aloud the directions on pages 67–68. Review each picture name with children.

Name

C C C

c c c

Can a cat hop?

Supporting All Learners

AUDITORY/VISUAL LEARNERS

LETTER SEARCH Read aloud "The Cat's in the Cupboard" on pages 18–19 in the *Big Book of Rhymes and Rhythms, 1A*. Have children find all the words with **/k/** in the rhyme. Write these words on the chalkboard. Invite volunteers to add other **/k/c** words to the list.

VISUAL/AUDITORY LEARNERS

PICTURE CLUES Give children the following picture cards: **cup, cat, can,** and **coat**. Have children say each picture name. Then challenge children to come up with three clues, including one sound or spelling clue, that might help someone guess what the word is without seeing the picture. Once children have come up with their clues, they can try them out on classmates.

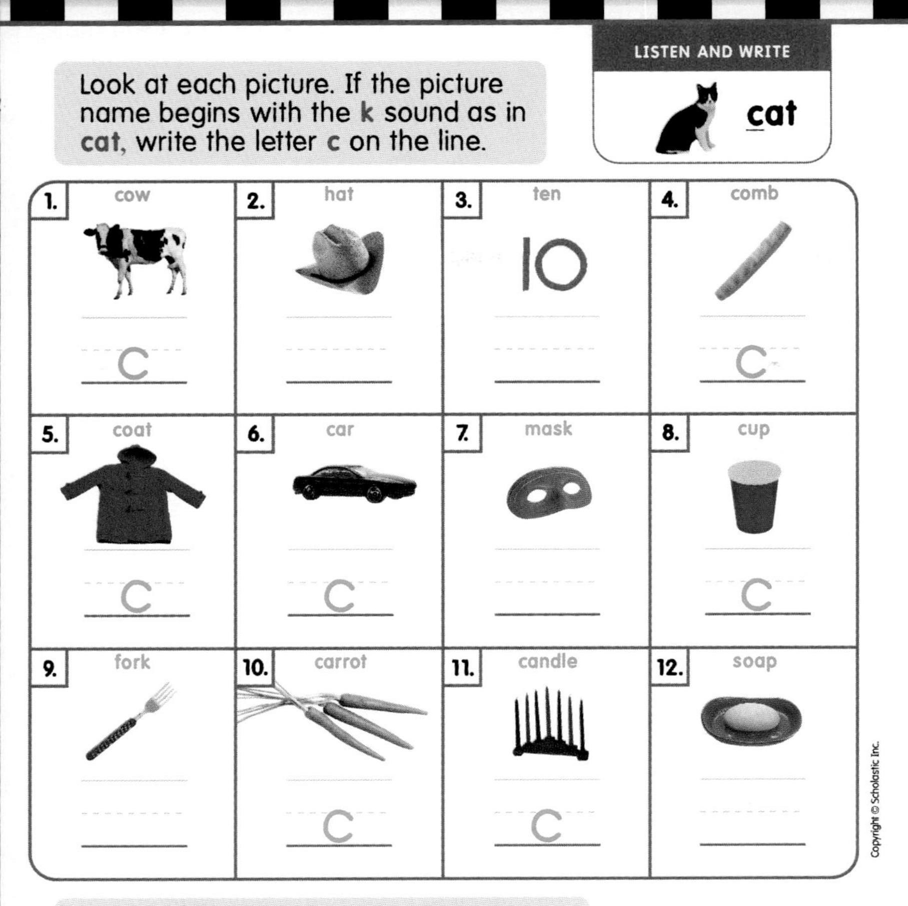

LISTEN AND WRITE

cat

Look at each picture. If the picture name begins with the k sound as in cat, write the letter c on the line.

1. cow	2. hat	3. ten	4. comb
c			c
5. coat	6. car	7. mask	8. cup
c	c		c
9. fork	10. carrot	11. candle	12. soap
	c	c	

Write the letter c to finish each word. Circle the word that names the picture.

13. c ap 14. c an

Integrated Curriculum

SPELLING CONNECTION

ADD A *Cc* Write the following sentences and word parts on the chalkboard:

My __at __an run fast.
Tim __an get a new __ap.
The __up is by the __ake.
__an you see the __at?

Have children add the letter *c* to each word part. Model for children how to blend each word. Then read each sentence aloud with children. Have children write each word they made on a sheet of paper.

Phonics Connection

Literacy Place: *Problem Patrol*
Teacher's SourceBook, pp. T144–145;
Literacy-at-Work Book, pp. 34–35

My Book: *Who Is Coming?*

Big Book of Rhymes and Rhythms, 1A: "The Cat's in the Cupboard," pp. 18–19

Chapter Book: *A Lot of Hats*, Chapter 6

ESL Show children the following picture cards: **cup, coat, can, five, fish,** and **feet**. Help children identify the picture names. Then have children repeat the picture names and sort the cards into two groups: one with words that begin with **/k/** as in ***cat,*** and one with words that begin with **/f/** as in ***fan***.

EXTRA HELP

Write the words ***pat, can, pan, fat, cat,*** and ***fan*** on index cards, one word per card. Children can work in pairs to sort the cards in two ways. First, they can sort them by words that begin with **/f/, /k/,** and **/p/**. Then, they can sort them by words that rhyme.

Lesson 41

page 69

Blend Words With /k/c

QUICKCHECK ✔

Can children:

✔ **orally blend word parts?**

✔ **identify /k/?**

✔ **blend words with /k/c?**

If YES go to Read and Write.

TEACH

Develop Phonemic Awareness

Oral Blending Say the following word parts, and ask children to blend them. Provide corrective feedback and modeling when necessary.

/k/...at	**/k/...ot**	**/k/...an**
/k/...one	**/k/...orn**	**/k/...up**

Connect Sound-Symbol Remind children that the letter *c* can stand for /**k**/ as in ***cat***. Write the word *cat* on the chalkboard, and have a volunteer circle the letter ***c***. Model how to blend the word.

I can put the letters *c, a,* and *t* together to make the word *cat*. Let's say the word slowly as I move my finger under the letters. Listen to how I string together the sound that each letter stands for to make the word.

Then help children blend the words ***can*** and ***cap***.

READ AND WRITE

Blend Words For more practice with /**k**/*c*, list the following words and sentence on a chart. Have volunteers read each aloud. Model blending when necessary.

- **cat cot cap can**
- **I can see the cat.**

Complete Activity Page Read aloud the directions on page 69. Have children complete the page independently.

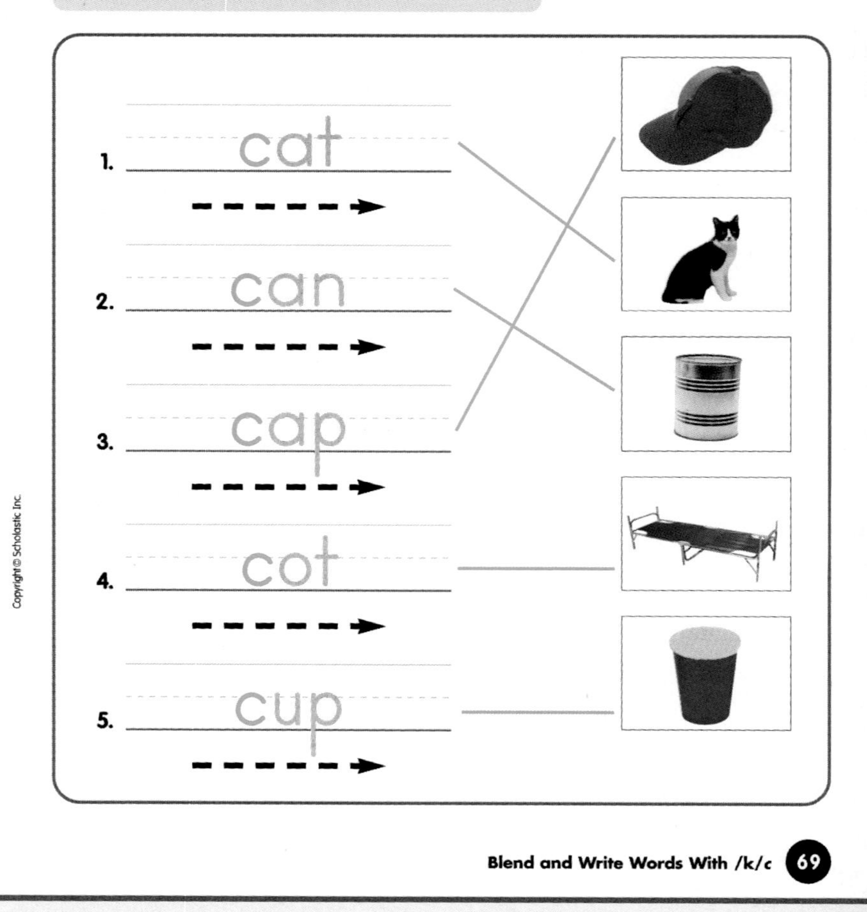

Supporting All Learners

ESL Have children write the letter **c** on three large self-sticking notes. Then challenge them to find three classroom objects, or pictures whose names begin with **/k/c**. Ask children to place the notes on those objects or pictures. When they are finished, provide corrective feedback. Point out any words that begin with **/k/** but are spelled with **k**. Tell children that **/k/** can be spelled in more than one way.

The Book Shop

- **Black Crow, Black Crow**
  by Ginger Fogleson Guy
  A young girl talks to a crow outside her window. **(/k/c)**

- **Cookies Week**
  by Cindy Ward
  A cat named Cookie creates mischief for every day of the week. **(/k/c)**

POETRY

Read the poem. Finish each sentence with a word from the poem.

THE FOG

I can see the [frog].
It sits on a log.
I see the fog.
Fog.
Fog.
Fog.
Who hid the [frog]?
Who hid the log?
Is it the fog?
Fog.
Fog.
Fog.

The [frog] sits on a log.

The fog hid it.

page 70

Write and Read Words With /f/f, /n/n, /k/c

QUICKCHECK ✔

Can children:

✔ **spell words with /f/f, /n/n, and /k/c?**

✔ **read a poem?**

If YES go to Read and Write.

TEACH

Link to Spelling Review with children the following sound-spelling relationships: **/f/*f*, /n/*n*, /k/*c***. Say one of these sounds. Have a volunteer write on the chalkboard the letter that stands for the sound. Continue with all the sounds. You may also wish to review **/h/*h*, /i/*i***, and **/p/*p***.

Phonemic Awareness Oral Segmentation
Say the word ***fin***. Have children orally segment it. Ask them how many sounds ***fin*** contains. Draw three connected boxes on the chalkboard, and have a volunteer write the spelling that stands for each sound in ***fin***. Continue with the words ***can*** and ***not***.

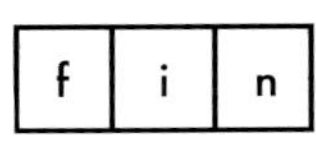

f	i	n

READ AND WRITE

Dictate Dictate the following words and sentence. The first set of words are decodable based on the sounds previously taught. The second are high-frequency words. Have children write the words and sentence on a sheet of paper. When children are done, write the words and sentence on the chalkboard, and have children correct their papers.

- **cat** **fan** **pin**
- **he** **my** **is**
- **Where is the pan?**

Complete Activity Page Read aloud the directions on page 70. Have children circle the words in the poem that contain **/f/** as in ***fan***.

Supporting All Learners

VISUAL LEARNERS

PORTFOLIO

Have children draw pictures of things whose names contain **/f/f, /n/n,** and **/k/c,** one picture per page. Children should label their drawings on the back of the picture. Then volunteers can take turns holding up their pictures while children ask questions about the drawings and try to guess the names of the things drawn.

CHALLENGE

Have children generate a list of words that begin and end with **/n/**. Record these words on the chalkboard. Then have children create a story using as many **/n/** words on the chalkboard as they can. Write the dictated story on chart paper for group and individual reading. Return to the story, rereading it in subsequent lessons.

Lesson 43

pages 71-72

Read and Review Words With /f/f, /n/n, /k/c

QUICKCHECK ✔

Can children:

✔ **read words with /f/f, /n/n, and /k/c?**

✔ **recognize high-frequency words?**

If YES go to Read and Write.

TEACH

Review Sound-Spellings Review with children the sound-spelling relationships from the past few lessons including **/f/*f*, /n/*n*,** and **/k/*c***. Say one of these sounds, and have a volunteer write on the chalkboard the spelling that stands for the sound. For example, the letter ***c*** can stand for **/k/**. Then display pictures of objects whose names begin with one of these sounds. Have children write the letter that each picture name begins with as the picture is displayed.

Write ***-og*** and ***-ig*** on the chalkboard, and have children suggest words containing each phonogram. Have volunteers write the words on the chalkboard.

Review High-Frequency Words On the chalkboard, write sentences containing the high-frequency words ***who, my, them, is,*** and ***where***. Say one word, and have volunteers circle the word in the sentences. Then have children generate sentences for each word.

READ AND WRITE

Build Words Distribute the following letter cards to children: ***c, f, n, o, a,*** and ***t***. Invite children to build as many words as possible using the letter cards. Children can record their words on a separate sheet of paper. Continue building words with the following letter cards: ***f, n, g, o, t,*** and ***l***.

Complete Activity Pages Read aloud the directions on pages 71–72. Review the art with children.

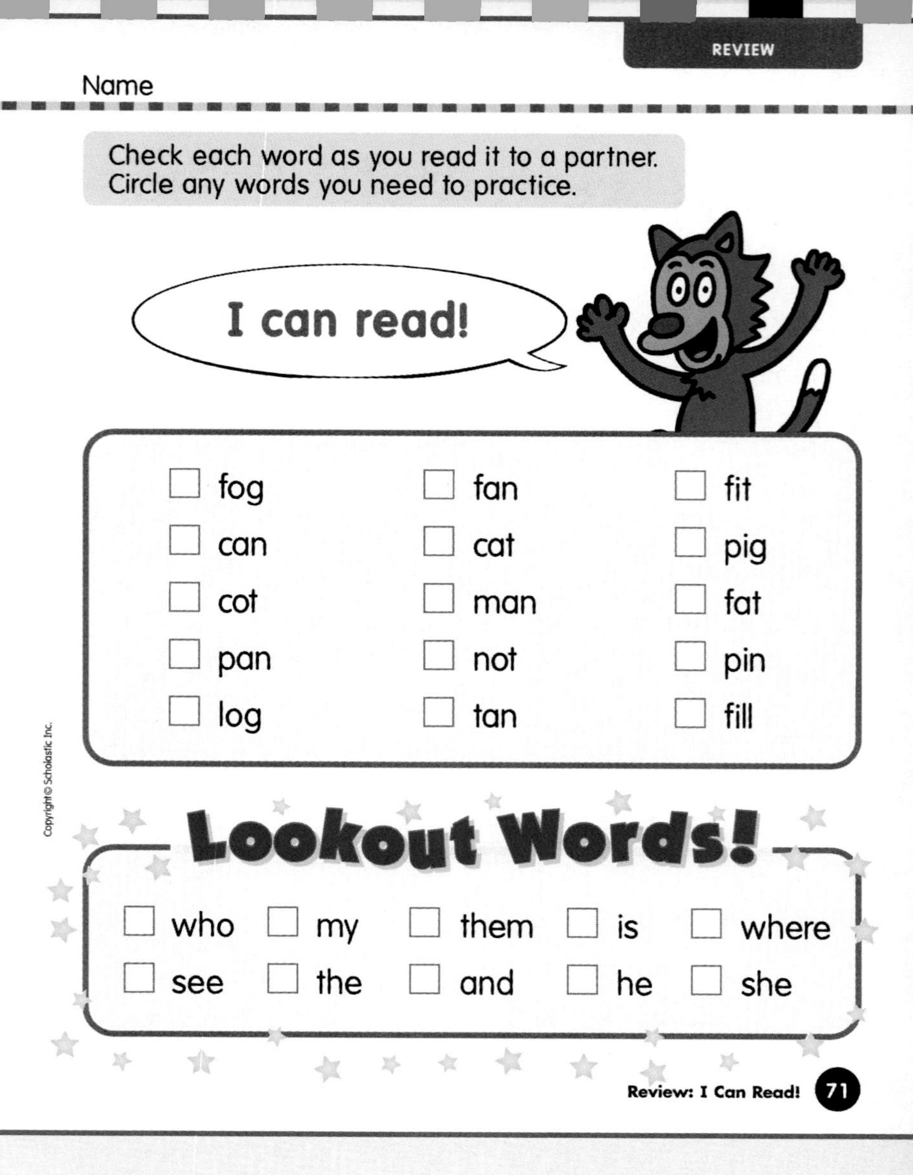

Supporting All Learners

VISUAL LEARNERS

PORTFOLIO

I CAN READ! BANNERS Children can create I Can Read! banners to display their reading progress. You may supply banner shapes, or have children create them. Have children write their name and the title I Can Read! on the banners. Children can list any words they are proud of being able to read, write, and spell on the banners.

AUDITORY LEARNERS

READ ALOUDS Have partners or small groups of children take turns reading aloud books to each other. Children listening can raise their hands whenever they hear words with **/f/f, /n/n,** and **/k/c** or high-frequency words.

Fill in the bubble next to the sentence that tells about each picture.

1.	● The man can nap. ○ The man can mop.
2.	● The cap fits him. ○ Fit the cap on the cat.
3.	● The fan is on. ○ It is a fin.
4.	○ We can fit. ● We can hop.
5.	○ She hid the fan. ● She hid the pin.
6.	● The pin is in the pan. ○ The pin is not in the pan.
7.	○ The cat can't sit. ● He can pat the cat.

EXTRA HELP

Reteach words frequently missed in the assessment, and reinforce blending by using pocket ABC cards and the pocket chart from the Phonics and Word Building Kit. Place three cards at a time, such as ***c, a,*** and ***n,*** in the wrong order, such as ***n, a, c,*** in the chart. Then say the target word ***can,*** and ask a volunteer to arrange the letters in the chart to make the word.

Write the following words on note cards: ***I, like, can, cat, pat, the, cap, my,*** and ***sit***. Display the cards, and have children make sentences using the words. To get them started, suggest the sentence ***I can sit***. Have children do a choral reading of the sentences.

Integrated Curriculum

WRITING CONNECTION

WRITE TIME Have children use words with **/f/f, /n/n,** and **/k/c** and the high-frequency words to write or dictate sentences about a friend or a pet. Children can illustrate their sentences. You may wish to have volunteers share their sentences.

TECHNOLOGY CONNECTION

MY BOOKS For additional practice with **/f/f, /n/n,** and **/k/c,** have children read the following My Books: *My Feet, Oh, No!,* and *Who Is Coming?* For information on using the My Books on the computer, see the WiggleWorks Plus My Books Teaching Plan.

I CAN READ! OPTIONS

The I Can Read! page can be used for one or all of the following:

- **paired reading**
- **individual assessment**
- **choral reading**
- **homework practice**
- **program placement**

ASSESSMENT

Besides circling words they missed, children may also self-assess by talking with you or one another about the words they weren't sure about. Children may also decide whether the words they missed have anything in common, such as beginning with the same consonant sound or ending with the same phonogram.

pages 73-74

Read Words in Context

TEACH

Assemble the Story Ask children to remove pages 73–74. Have them fold the pages in half to form the Take-Home Book.

Preview the Story Preview *Ann and Nick,* a nonfiction selection about a girl and her seeing-eye dog. Invite children to browse through the first two pages of the story and to comment on anything they notice. Ask them to point out unfamiliar words. Read these words aloud as you model how to blend them. Then have children predict what they think the story is about.

READ AND WRITE

Read the Story Read the story aloud, or have volunteers take turns reading aloud a page at a time. Discuss items of interest on each page, and encourage children to help each other with any blending difficulties. The following prompts may help children while reading:

- **What do the photographs tell you about the story?**
- **What letter sounds do you know in the word?**

Reflect and Respond Have children share their reactions to *Ann and Nick*. What did they learn from the story? Encourage children to think of more information to add to the story.

Develop Fluency You may wish to reread the story as a choral reading or have partners reread the story independently. Provide time for children to reread the story on subsequent days to develop fluency and increase reading rate.

When reading other stories, have children share how they figure out unfamiliar words and model how they blend words. Encourage children to look for words with **/f/*f*, /n/*n*,** and **/k/*c*** and apply what they have learned about these sound-spelling relationships to decode words. Continue to review these sound-spelling relationships.

Supporting All Learners

KINESTHETIC LEARNERS

QUESTIONS AND ANSWERS Have pairs of children think of two or three questions to ask Ann about Nick. Then children can act out a brief interview with her.

VISUAL/AUDITORY LEARNERS

ACT IT OUT Children can act out the story *Ann and Nick*. Have children take turns reading and acting out the part of Ann and acting out the part of Nick.

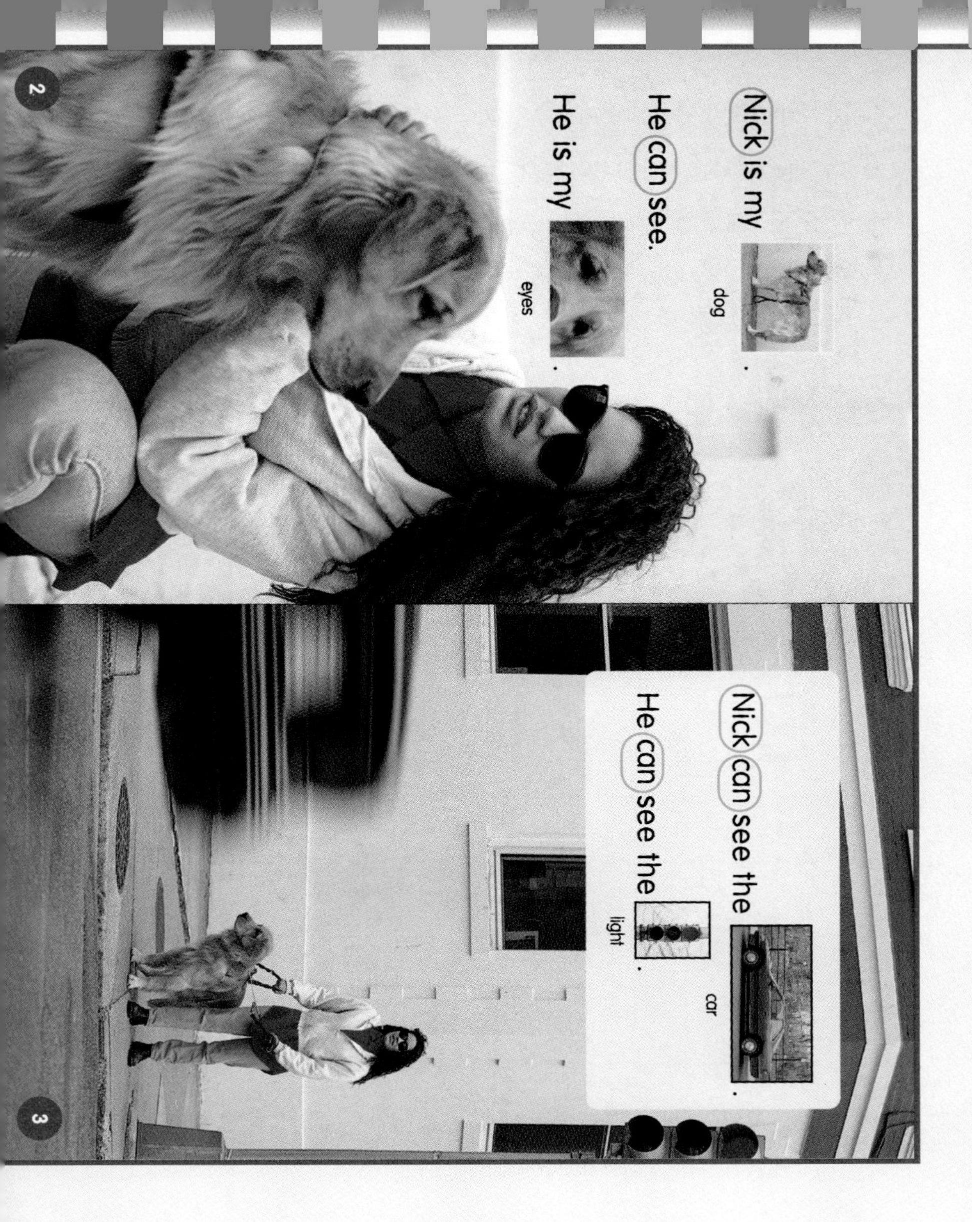

Reflect On Reading

ASSESS COMPREHENSION

To assess their understanding of the story, ask children questions such as:

- **Who is Nick?** ***(a seeing-eye dog, or a dog that is trained to help a person who cannot see)***
- **How does Nick help Ann?** ***(He is her eyes; he sees the lights and the cars.)***
- **How do Ann and Nick feel about each other?** ***(They love each other.)***

HOME-SCHOOL CONNECTION

Send home ***Ann and Nick***. Encourage children to read the story to a family member.

Phonics Connection

Phonics Readers:
#16, *Will It Fit?*
#17, *Where Is Nat?*
#18, *My Cat Can*

CHALLENGE

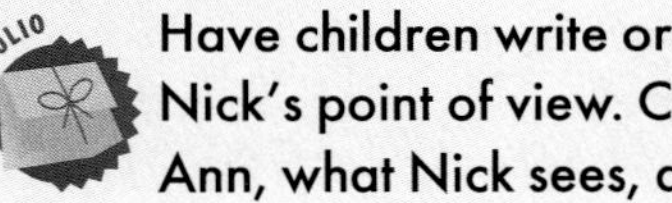

Have children write or dictate the story of Ann and Nick from Nick's point of view. Children can write about how Nick helps Ann, what Nick sees, and what Nick likes.

EXTRA HELP

Children can read the story together aloud or take turns reading the sentences in the story aloud. Then children can talk about the story and read it to others.

Lesson 45

pages 75-76

Recognize and Write /b/*b*

QUICKCHECK ✔

Can children:

✔ **orally segment words?**

✔ **write capital and small *Bb*?**

✔ **recognize /b/?**

✔ **identify the letter that stands for /b/?**

If YES go to Read and Write.

TEACH

Develop Phonemic Awareness

Oral Segmentation Explain that you are going to say a word. Ask children to say the first sound and then the rest of the word. For example, if you say ***mat,*** children will say **/m/*...at***. Continue with these words:

- sat
- fish
- man
- lake
- bat
- big
- box
- bake

Write the Letter Remind children that the letter ***b*** stands for /**b**/, the sound they hear at the beginning of ***bat***. Write ***Bb*** on the chalkboard. Point out the capital and small forms of the letter. Then model how to write the letter.

Have children write both forms of the letter on the chalkboard or in the air with their fingers.

READ AND WRITE

Connect Sound-Symbol Write the words ***big*** and ***rub*** on the chalkboard, and have a volunteer circle ***b*** in each word. Ask children to suggest other words that begin or end with /**b**/, and list them on the chalkboard. Have volunteers circle the letter ***b*** in each word.

Then write the following on the chalkboard: ***_at, ca _, _it, _ig, bi_***. Have volunteers add the letter ***b*** to each word part. Model how to blend each word.

Complete Activity Pages Read aloud the directions on pages 75–76. Review each picture name with children.

Supporting All Learners

AUDITORY LEARNERS

LISTEN FOR *B*'s! Have children write both forms of the letter ***Bb*** on a sheet of paper. Then read aloud "Betty Botter" on page 20 in the *Big Book of Rhymes and Rhythms, 1A*. Have children hold up their ***Bb's*** whenever they hear a word with **/b/** in the rhyme. Write these words on the chalkboard. Invite volunteers to add other **/b/** words to the list.

VISUAL/KINESTHETIC LEARNERS

GRAB BAG GAME Make word cards for words that begin or end with ***b***. Place them in a big grab bag. Have small groups of children take turns pulling cards out of the bag. Each child should say the word on the card, name each letter in the word, and then use the word in a sentence.

Look at each picture. If the picture name begins with the **b** sound as in **box**, write the letter **b** on the first line. If it ends in **b** as in **cab**, write **b** on the second line.

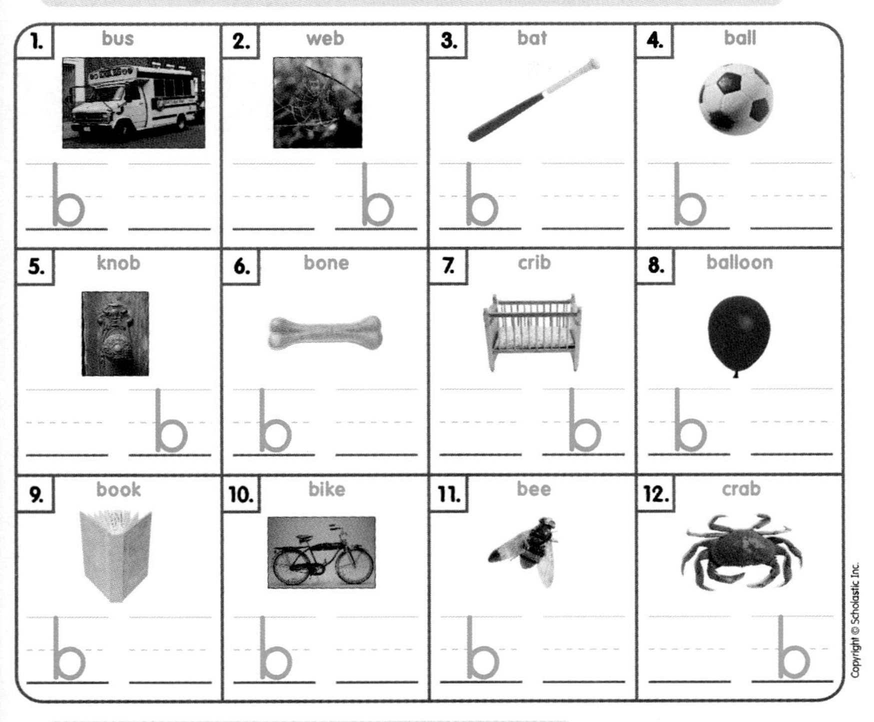

Write the letter **b** to finish each word. Circle the word that names the picture.

14. b ib

Integrated Curriculum

WRITING CONNECTION

STORY TIME Start a story or have a volunteer start a story by writing or dictating a sentence using words with **/b/b**. Then have children take turns adding sentences with **/b/** words to the story. You may wish to start the story with a sentence such as the following: ***Bob and Betty saw a big bat.***

MATH CONNECTION

HOW MANY? Have children count how many words with ***b*** are in "Betty Botter" (page 20 in the *Big Book of Rhymes and Rhythms,* 1A). Children can write the number on chart paper, under the title "Betty Botter." Then children can count words with ***b*** in other rhymes and books in the classroom. Have children write the number of ***b*** words on the chart paper, under the title of each rhyme or book. Have children compare how many ***b*** words each rhyme or book has.

ASSESSMENT

Observe children as they write each new letter. Provide corrective feedback before children get into a habit of forming a specific letter incorrectly.

EXTRA HELP

Have children work together to write or dictate as many words with **/b/** as they can think of. Then children can use the list of **/b/** words to make up silly tongue-twister sentences such as ***The big bee bit the blue bug***. Write the tongue-twister sentences on chart paper to share with the class.

CHALLENGE

Have partners play a What Is It? game. To begin, have one child think of something with **/b/b** in its name. Have the child write or dictate the name on a sheet of paper. Then the child's partner should ask questions to figure out the word. Children may ask questions such as ***Is the word big? Is the thing big? Is the /b/ at the beginning, middle, or end of the word?*** After children guess the name of the word, have each pair of children switch roles and play again.

Phonics Connection

Literacy Place: *Problem Patrol*
Teacher's SourceBook, pp. T196–197;
Literacy-at-Work Book, pp. 42–43

My Book: *Big and Bigger*

Big Book of Rhymes and Rhythms, 1A: "Betty Botter," p. 20

Chapter Book: *A Lot of Hats,* Chapter 7

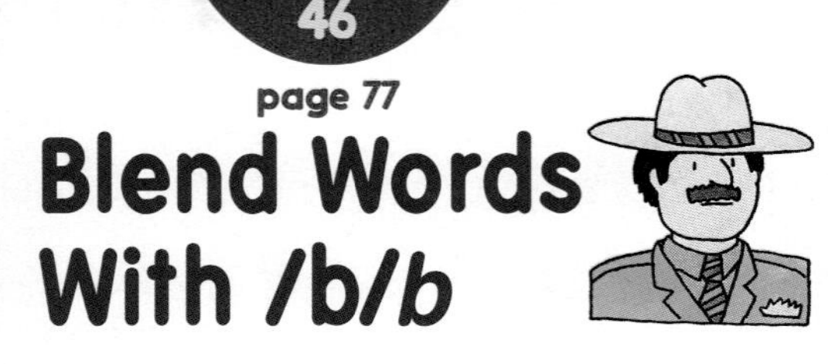

Lesson 46

page 77

Blend Words With /b/*b*

QUICKCHECK ✔

Can children:

✓ orally blend word parts?

✓ identify /b/?

✓ blend words with /b/*b*?

If YES go to Read and Write.

TEACH

Develop Phonemic Awareness

Oral Blending Say the following word parts, and ask children to blend them. Provide corrective feedback and modeling when necessary.

/b/...at	**/b/...ig**	**/b/...ox**
/b/...one	**/b/...ean**	**/b/...it**

Connect Sound-Symbol Remind children that the letter ***b*** stands for /**b**/ as in ***bat***. Write ***bat*** on the chalkboard, and ask a volunteer to circle the letter ***b***. Model how to blend the word.

THINK ALOUD

I can put the letters *b, a,* and *t* together to make the word *bat*. Let's say the word slowly as I move my finger under the letters. Listen to how I string together the sound that each letter stands for to make the word: *baaaat, baat, bat.*

Then help children blend the words ***big*** and ***cab***.

READ AND WRITE

Blend Words Ask volunteers to read the following words and sentence aloud. Model blending when necessary.

- **big** **bit** **bat**
- **cab** **lab** **Bob**
- **The bib is big.**

Complete Activity Page Read aloud the directions on page 77. Have children complete the page independently.

Name

Read each word. Find the word in the box that rhymes with it. Then write the word on the line.

bit big bat Bob bag bill

1. hat — bat	2. tag — bag	3. hit — bit
4. rob — Bob	5. hill — bill	6. pig — big

Use one of the words from above to finish each sentence.

7. I see a man in a big hat.

8. Sam is at bat.

Supporting All Learners

VISUAL/AUDITORY LEARNERS

PICTURE NAME GAME Take out the following picture cards from the Phonemic Awareness Kit: **ball, bat, boat, bone, book, box, boy, bun, bus.** Select one card, but do not show it to children. Tell children that you will say the first sound in the picture name and then the rest of the word. Challenge children to tell you the picture name before you show them the picture. Then have volunteers select picture cards and say the picture name in parts for the class to blend.

The Book Shop

- **Chicka Chicka Boom Boom**
  ABC book *by Bill Martin, Jr., and John Archambault*
  (/b/*b*)
- **Bear's Bargain**
  by Frank Asch
  Bear wants to be like Little Bird.
  (/b/*b*)

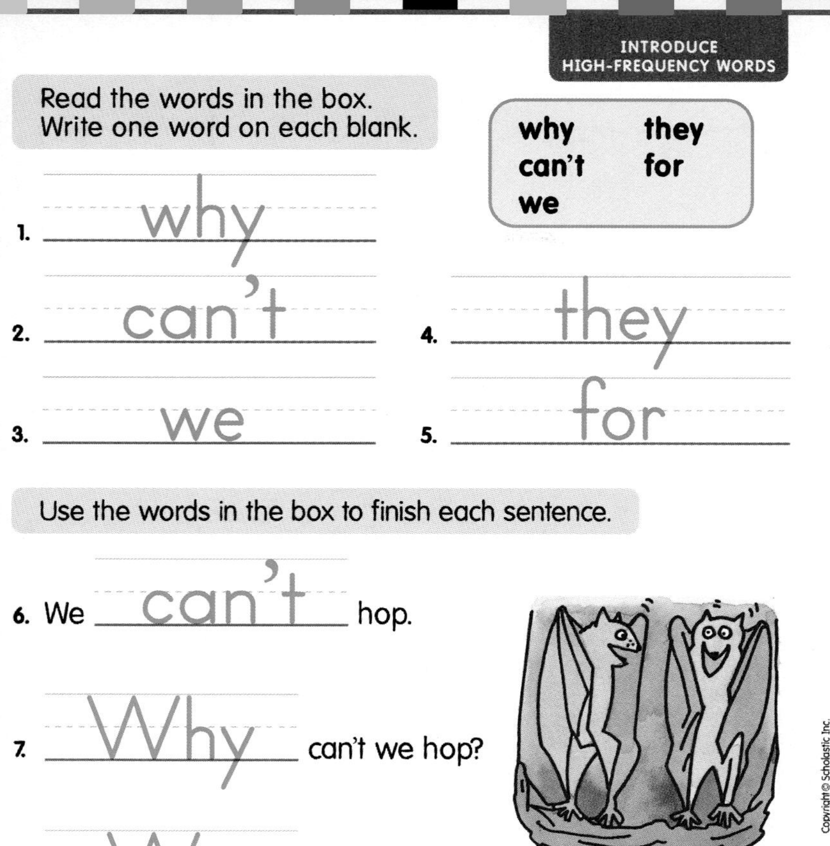
INTRODUCE HIGH-FREQUENCY WORDS

Read the words in the box.
Write one word on each blank.

why	they
can't	for
we	

1. why
2. can't
3. we
4. they
5. for

Use the words in the box to finish each sentence.

6. We can't hop.
7. Why can't we hop?
8. We are bats.

Supporting All Learners

VISUAL/TACTILE LEARNERS

WORD WALL Have children write the high-frequency words on note cards, one word per card. Then children can add the cards to the Word Wall. Remind children to refer to the Word Wall when reading or writing.

EXTRA HELP

For children who need additional practice reading the high-frequency words in context, use Phonics Reader #10, *To Tad,* to teach the word ***for*** and to review words such as ***is*** and ***to***.

Write the sentences with the high-frequency words on sentence strips, and underline each high-frequency word. Have children use the sentence strips for oral reading practice. Cut each strip in two parts, mix the parts, and then have children match them to make complete sentences.

Recognize High-Frequency Words

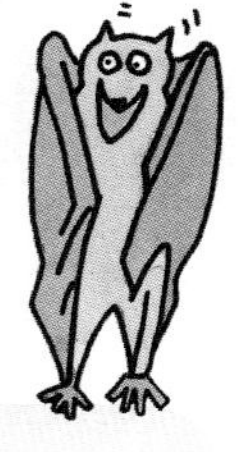

QUICKCHECK ✔

Can children:

✔ **recognize and write the high-frequency words *why, can't, we, they,* and *or*?**

✔ **complete sentences using the high-frequency words?**

If YES go to Read and Write.

TEACH

Introduce the High-Frequency Words

Use the high-frequency words ***why, can't, we, they,*** and ***for*** in sentences on the chalkboard. Read the sentences aloud. Underline the words, and ask children if they recognize them. If necessary, read the sentences again. You may wish to use the following sentences:

1. **The cat is for Pam.**
2. **They said, "We can pat the cat!"**
3. **"But we can't sit on it."**
4. **Why not?**

Then ask volunteers to dictate sentences using the words. Begin with the following sentence starters: ***We like*** ___ or ***Why can't they*** ___?

READ AND WRITE

Practice Write each high-frequency word on a note card. Read each word aloud as you display the cards. Then do the following:

- **Mix the cards.**
- **Display one card at a time, and ask children to state each word aloud.**
- **Have children spell each word aloud, clapping on each letter.**
- **Ask children to write each word in the air as they state aloud each letter. Then have them write each word on a sheet of paper.**

Complete Activity Page Read aloud the directions on page 78. Have children complete the page independently.

Lesson 48

pages 79-80

Recognize and Write /w/w

QUICKCHECK ✔

Can children:

✔ **orally segment words?**
✔ **write capital and small *Ww*?**
✔ **recognize /w/?**
✔ **identify the letter that stands for /w/?**

If YES go to Read and Write.

TEACH

Develop Phonemic Awareness

Oral Segmentation Explain to children that you are going to say a word. Ask them to say the first sound in the word and then the rest of the word. For example, if you say ***sad,*** children should respond by saying **/s/...*ad***. Continue with the following words:

- sick
- fan
- mop
- like
- will
- win
- wig
- wet

Write the Letter Explain to children that *w* stands for **/w/,** the sound at the beginning of ***wig***. Write ***Ww*** on the chalkboard. Point out the capital and small forms of the letter. Then model how to write the letter.

Have children write both forms of the letter on the chalkboard or in the air with their fingers.

READ AND WRITE

Connect Sound-Symbol Write the word ***web*** on the chalkboard. Have a volunteer circle the letter *w*. Remind children that the letter *w* stands for **/w/**. Ask children to suggest other words that begin with **/w/**. List these words on the chalkboard, and ask volunteers to circle the letter *w* in each one.

Then write the following on the chalkboard: ***_in, _ig, _ill***. Have volunteers add *w* to each. Model how to blend each word.

Complete Activity Pages Read aloud the directions on pages 79–80. Review each picture name with children.

Name

Wig begins with the w sound.

Ww

W W W

w w w

We want that wig.

Supporting All Learners

VISUAL/AUDITORY LEARNERS

FIND *Ww* WORDS Read aloud "How Much Wood?" on page 21 in the *Big Book of Rhymes and Rhythms, 1A*. Have children find all the words with **/w/** in the rhyme. Write these words on the chalkboard. Invite volunteers to add other **/w/** words to the list.

AUDITORY LEARNERS

LISTENING TIME Have children listen to "Wee Willie Winkie" and "Go In and Out the Window" on the Sounds of Phonics Audiocassette (Managing Information). Invite children to raise their hands each time they hear a word that begins with **/w/**.

LISTEN AND WRITE

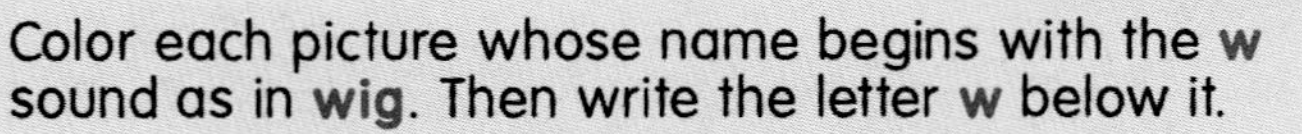
Color each picture whose name begins with the w sound as in wig. Then write the letter w below it.

Write the letter w to finish each word. Circle the word that names the picture.

13. w in

14. w ig

80 Recognize /w/w

Integrated Curriculum

SPELLING CONNECTION

ADD *Ww* Have children use the pocket ABC cards and pocket chart to make the word parts ***__in, __ig,*** and ***__ill.*** Then have children hold up the pocket ABC card for ***w*** in front of each word part. Children can write each word that is made.

SCIENCE CONNECTION

Have children select an animal whose name begins with ***w***, such as a woodchuck, walrus, or worm. Help children to find books about the animal. Ask children to draw or cut out a picture of the animal on a large sheet of colored construction paper. They can then write one fact learned about the animal at the bottom of the paper. Collect and bind the sheets of paper to form a booklet for the classroom library.

Phonics Connection

Literacy Place: *Problem Patrol*
Teacher's SourceBook, pp. T198–199;
Literacy-at-Work Book, pp. 44–45

My Book: *Who Walks?*

Big Book of Rhymes and Rhythms, 1A: "How Much Wood?" p. 21

Chapter Book: *A Lot of Hats,* Chapter 8

CHALLENGE

Cut a large sheet of paper into the shape of a wagon. Invite children to write, draw, or cut out words and pictures of objects and animals whose names begin with /w/. When the wagon is "full," review the words with children. Then have them write a sentence about the wagon.

EXTRA HELP

Have children look through newspapers for words with ***w***. If possible, have children circle each word and then underline each ***w***. You may also wish to write the words children find on the chalkboard or on chart paper.

Lesson 49

pages 81-82

Blend and Build Words With /w/w

QUICKCHECK

Can children:

- **orally blend word parts?**
- **identify /w/?**
- **blend and build words with /w/w?**

If YES go to Read and Write.

TEACH

Develop Phonemic Awareness

Oral Blending Say the following word parts, and ask children to blend them. Provide corrective feedback and modeling when necessary.

/w/. . . in	**/w/. . . ig**	**/w/. . . et**
/w/. . . orm	**/w/. . . allet**	**/w/. . . ater**

Connect Sound-Symbol Remind children that the letter *w* stands for /w/ as in ***win***. Write ***win*** on the chalkboard. Have a volunteer circle the letter *w* and model how to blend the word.

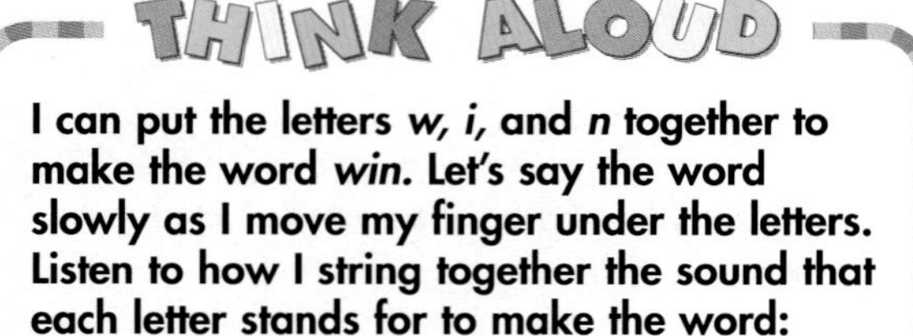

THINK ALOUD

I can put the letters *w, i,* and *n* together to make the word *win*. Let's say the word slowly as I move my finger under the letters. Listen to how I string together the sound that each letter stands for to make the word: *wiiiinnnn, wiinn, win.*

READ AND WRITE

Blend Words To practice using the sound taught and to review previously taught sounds, list the following on a chart. Have volunteers read each aloud. Model blending when necessary.

- **win** **wig** **will**
- **Why can't we win?**

Complete Activity Pages Read aloud the directions on pages 81–82. Review each picture name with children.

Name

Read each word. Find the word in the box that rhymes with it. Then write the word on the line.

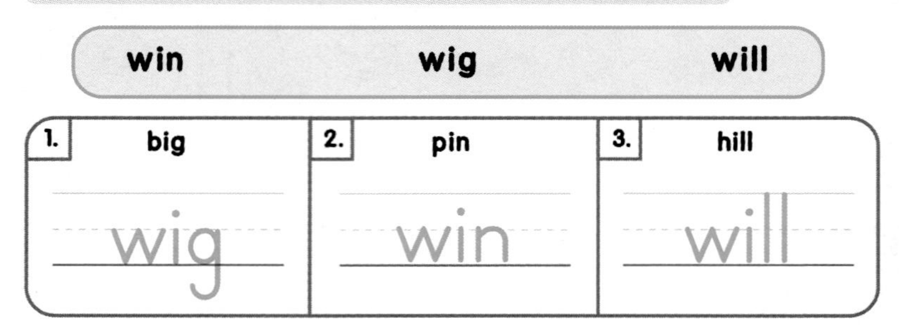

win	wig	will

1. big	2. pin	3. hill
wig	win	will

Use one of the words from above to finish each sentence.

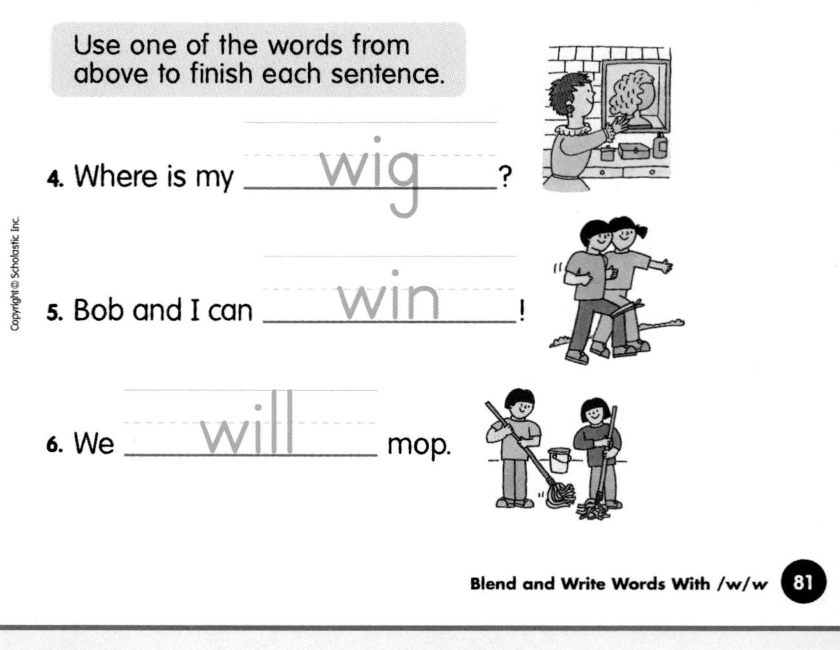

4. Where is my wig?

5. Bob and I can win!

6. We will mop.

Supporting All Learners

VISUAL/AUDITORY LEARNERS

LEARNING CENTER For additional practice, use Center 6, Read-Along Songs, Activity 1, from *Quick-and-Easy Learning Centers: Phonics*. This activity focuses on using familiar lyrics to foster sequencing skills using initial letter cues. This activity also builds sight word awareness.

TACTILE LEARNERS

DOUGH LETTERS Children can form words with **/w/** out of clay or dough. You can make dough by mixing four cups of flour, one cup of salt, and one and three-quarter cups of warm water.

VISUAL/TACTILE LEARNERS

BUILD WORDS Distribute the following letter cards to children: ***a, m, t, b, c, i, n, w***. If children have their own set of cards, have them locate the set. Ask children to build as many words as possible using the letter cards.

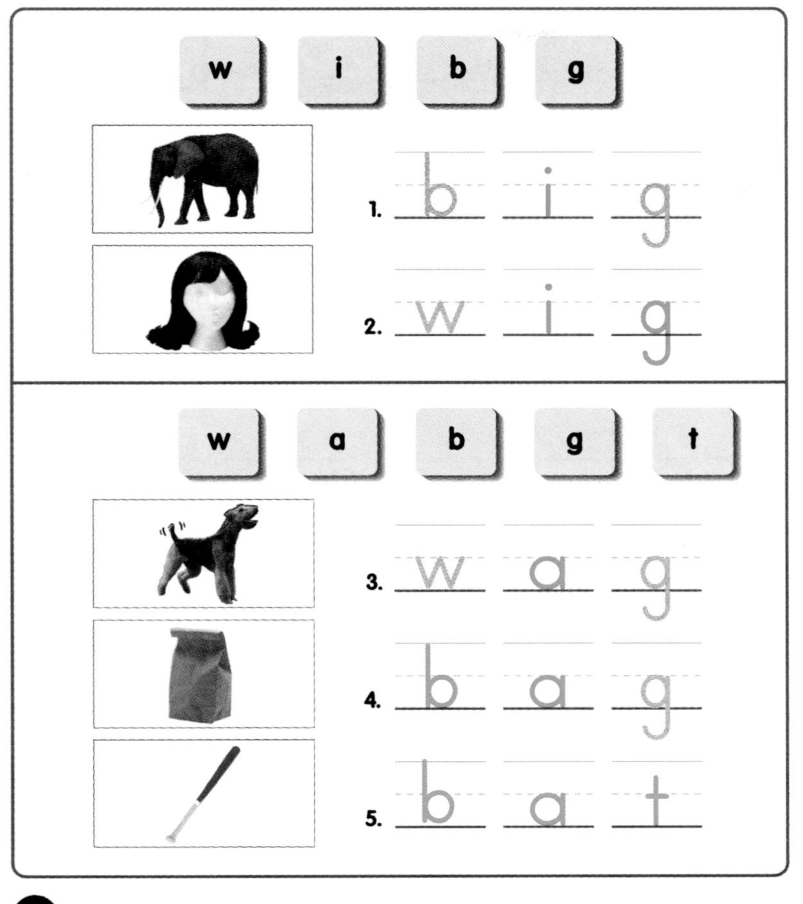

BUILD WORDS

Look at each picture. Use the letter tiles to write each picture name.

w i b g

1. b i g

2. w i g

w a b g t

3. w a g

4. b a g

5. b a t

Copyright © Scholastic Inc.

82 Build Words

Integrated Curriculum

SPELLING CONNECTION

CLIMB THE LADDER Write the word ***wig*** on the chalkboard, and model blending. Have children replace **/w/,** the first sound in ***wig,*** with **/p/** to make a new word. Ask children what letter stands for **/p/**. Then replace the letter ***w*** in ***wig*** with the letter ***p,*** and blend the new word. Continue by changing one letter at a time to build new words. You may wish to use this sequence of words:

pig
pin
pan
can

MATH CONNECTION

WINTER FUN Ask children to name the season whose name begins with **/w/.** (***winter***) Then help children generate a list of fun winter activities such as skiing, ice skating, or making snowmen. Ask children to raise their hands every time you name a winter activity that they enjoy. Record the number counted beside each activity on the list. Then help the class to create a graph showing their favorite winter activities.

CHALLENGE

Give partners or small groups of children the following beginning and ending words of a three-word ladder. Have children change one letter in each to complete the ladder.

- **lot** (hot) **hop**
- **mat** (man) **can**
- **pan** (pin) **win**

Note: The words in parentheses are possible answers.

ESL English contains some sounds and letters that may be completely unfamiliar to children acquiring English. For example, Spanish does not contain the letter ***w***. To provide additional support, have children search through stories for words that begin with ***w***. Then have children use the pronunciation feature in WiggleWorks Plus to hear how each word is said. Ask children to repeat each word before searching for a new one.

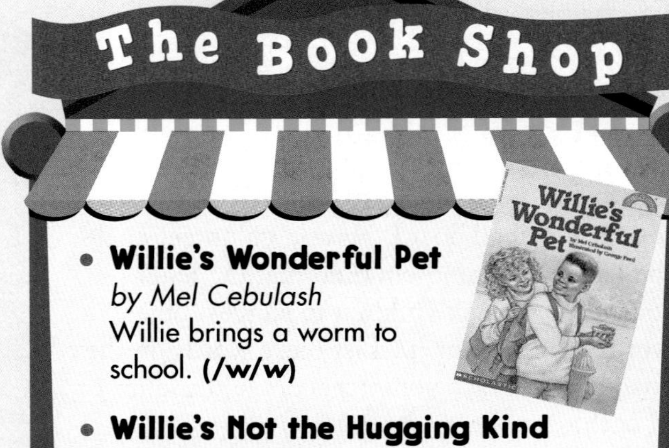

The Book Shop

- **Willie's Wonderful Pet**
  by Mel Cebulash
  Willie brings a worm to school. **(/w/w)**
- **Willie's Not the Hugging Kind**
  by Joyce Barrett
  A little boy doesn't want hugs. **(/w/w)**

Lesson 50

pages 83-84

Recognize and Write /j/j

QUICKCHECK ✔

Can children:

- ✔ **orally segment words?**
- ✔ **write capital and small *Jj*?**
- ✔ **recognize /j/?**
- ✔ **identify the letter that stands for /j/?**

If YES go to Read and Write.

TEACH

Develop Phonemic Awareness

Oral Segmentation Explain that you are going to say a word. Ask children to say the first sound in the word and then the rest of the word. For example, if you say ***fan,*** children will say **/f/...*an*.** Continue with these words:

- sock
- fun
- men
- lick
- jam
- jump
- jet
- jog

Write the Letter Explain to children that the letter ***j*** stands for /j/, the sound they hear at the beginning of ***jam***. Write ***Jj*** on the chalkboard. Point out the capital and small forms of the letter. Then model for children how to write the letter.

Have children write both forms of the letter on the chalkboard or in the air with their fingers.

READ AND WRITE

Connect Sound-Symbol Write the word ***jog*** on the chalkboard. Have a volunteer circle the letter ***j***. Remind children that the letter ***j*** stands for /j/. Ask children to suggest other words that contain /j/. List these words on the chalkboard. Have volunteers circle the letter ***j*** in each.

Complete Activity Pages Read aloud the directions on pages 83–84. Review each picture name with children.

HANDWRITING

Name

Jar begins with the j sound.

J j

J J J

j j j

Jam is in the jar.

Supporting All Learners

VISUAL/AUDITORY LEARNERS

FIND THE WORD Read aloud "John Jacob Jingleheimer Schmidt" on pages 22–23 in the *Big Book of Rhymes and Rhythms, 1A*. Have children identify all the words with /j/ in the rhyme. Write these words on the chalkboard. Invite volunteers to add other /j/ words to the list.

AUDITORY LEARNERS

LISTEN FOR /J/ Have children listen to "Jack Be Nimble" and "John Jacob Jingleheimer Schmidt" on the Sounds of Phonics Audiocassette (Teamwork). Ask children to identify all the words they hear that begin with /j/ in each song. Write the words that children name and then read the list, emphasizing the initial /j/ in each.

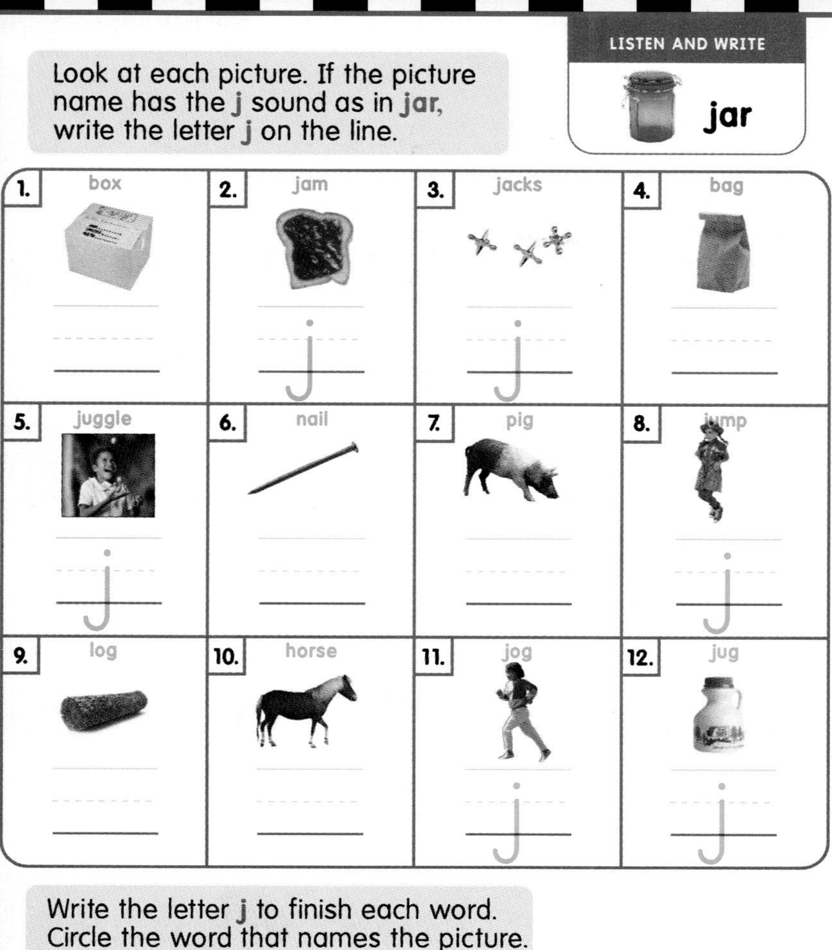

LISTEN AND WRITE

jar

Look at each picture. If the picture name has the j sound as in jar, write the letter j on the line.

1. box
2. jam
3. jacks
4. bag
5. juggle
6. nail
7. pig
8. jump
9. log
10. horse
11. jog
12. jug

Copyright © Scholastic Inc.

Write the letter j to finish each word. Circle the word that names the picture.

13. jog
14. job

84 Recognize /j/j

Integrated Curriculum

SPELLING CONNECTION

LIST WORDS Have children write the following word parts on sheets of paper: ___***am*** and ___***ob***. Have volunteers add the letter ***j*** to each. Then have children write words that rhyme with ***jam*** under that word, and words that rhyme with ***job*** under that word.

SCIENCE CONNECTION

POTATO ART Cut a potato in half, and carve a backwards ***j*** out of the flat side, so that a ***j*** will print. Have children press the potato ***j*** in paint and print it on paper. Children can add letters to each painted ***j*** to make words.

Phonics Connection

Literacy Place: *Problem Patrol*
Teacher's SourceBook, pp. T250–251;
Literacy-at-Work Book, pp. 57–58

My Book: *Jumping*

Big Book of Rhymes and Rhythms, 1A: "John Jacob Jingleheimer Schmidt," pp. 22-23

Chapter Book: *A Lot of Hats*, Chapter 9

EXTRA HELP

Using the red magnetic letters from the Phonics and Word Building Kit, spell the word ***jam***. Have children say the sound of each letter in the word in sequence. Then ask children to blend the word aloud. Review any sound-letter relationships that cause difficulty, and model blending when necessary. Then scramble the letters and ask children to reform it with the magnetic letters. Continue with the words ***jog*** and ***Jim***.

In Spanish **/j/** does not exist. The letter ***j*** is sometimes pronounced as **/h/**. Therefore, children might need additional practice discriminating and pronouncing **/j/**. Display pictures of the following: **jet, jump, jar, gate,** and **hat**. Read aloud each picture name. Then have children select the picture names that begin with **/j/**.

Lesson 51

page 85

Blend Words With /j/j

QUICKCHECK ✔

Can children:

✔ orally blend word parts?

✔ identify /j/?

✔ blend words with /j/j?

If YES go to Read and Write.

TEACH

Develop Phonemic Awareness

Oral Blending Say the following word parts, and ask children to blend them. Provide feedback and modeling when necessary.

/j/...am	/j/...og	/j/...ob
/j/...ump	/j/...ug	/j/...acks

Connect Sound-Symbol Remind children that the letter *j* stands for /j/ as in *jam*. Write *jam* on the chalkboard, and have a volunteer circle the letter *j*. Model how to blend the word.

I can put the letters *j, a,* and *m* together to make the word *jam.* Let's say the word slowly as I move my finger under the letters. Listen to how I string together the sound that each letter stands for to make the word: *jaaaammmm, jaamm, jam.*

Then help children blend *jog* and *job*.

READ AND WRITE

Blend Words Have volunteers read the following words and sentence. Model blending when necessary.

- jam job Jim
- log cat cab
- Jan can jog.

Complete Activity Page Read aloud the directions on page 85. You may wish to review the art.

BLEND WORDS

Name

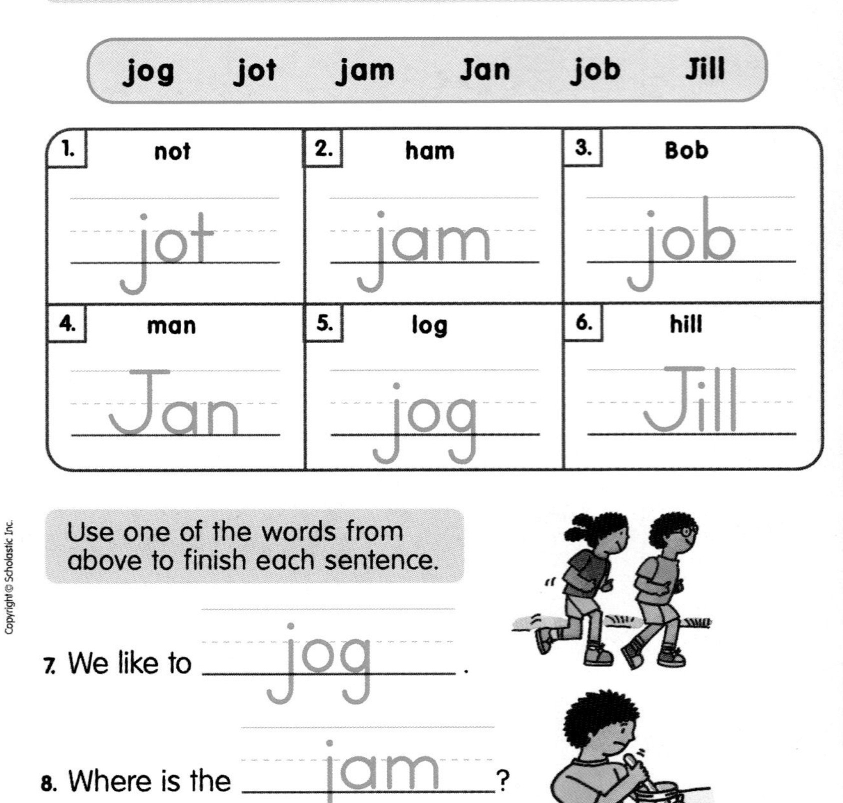

Read each word. Find the word in the box that rhymes with it. Then write the word on the line.

jog jot jam Jan job Jill

1. not	2. ham	3. Bob
jot	jam	job
4. man	**5. log**	**6. hill**
Jan	jog	Jill

Use one of the words from above to finish each sentence.

7. We like to jog.

8. Where is the jam?

Blend and Write Words With /j/j 85

Supporting All Learners

KINESTHETIC LEARNERS

HOT POTATO GAME Have children sit in a circle and play a hot potato game. To play, children pass a potato or an object such as a chalkboard eraser around the circle. Each time a child passes the potato or other small object, he or she must name a word that begins with /j/. Repeat the game with words and names that begin with /w/.

The Book Shop

- **Norma Jean, Jumping Bean**
  by Joanna Cole
  Norma Jean, a kangaroo, naturally likes to jump.
  (/j/j)

POETRY

Read the poem. Finish each sentence with words from the poem.

A Bat and a Cat

A bat in a bib
Said to a cat in a wig,
"I see a pan of ham!"

The cat in the wig
Said to the bat in the bib,
"I see a pot of jam!"

"Can we?" said the cat.
"Why not?" said the bat.

So they sat and sat.
Mmmm, Mmmm, Mmmm!

1. "I see a pan of ham," said the bat.
2. "I see a pot of jam," said the cat.

Supporting All Learners

KINESTHETIC LEARNERS

LETTER-SOUND ACTS Invite children to think of words and names that begin with **/j/, /w/,** and **/b/**. Write the words on a chart. Children can work together to think of scenes to act out, using as many of the words as possible. To get started, they may act out the following examples: ***Jen jogs and bats a ball; Ben and Will put jam on a bun;*** and ***Jim washes his wig in a big tub.***

TACTILE LEARNERS

SAND WRITING Partners can take turns writing and reading words with **/j/, /w/,** and **/b/** in sand. You may wish to have children keep a list of the words they make.

CHALLENGE

Have children generate a list of words that begin with **/j/**. Record these words on the chalkboard. Then have children create a story using as many **/j/** words on the chalkboard as they can. Write the dictated story on chart paper for group and individual reading. Return to the story, rereading it in subsequent lessons.

Write and Read Words With /b/b, /w/w, /j/j

QUICKCHECK ✔

Can children:

✔ **spell words with /b/b, /w/w, and /j/j?**

✔ **read a poem?**

If YES go to Read and Write.

TEACH

Link to Spelling Review with children the following sound-spelling relationships: /**b**/*b*, /**w**/*w*, and /**j**/*j*. Say one of these sounds, and ask a volunteer to write the spelling that stands for the sound on the chalkboard. For example, the letter ***b*** stands for /**b**/. Continue with all the sounds. You may also wish to review /**f**/*f*, /**n**/*n*, and /**k**/*c*.

Phonemic Awareness Oral Segmentation

Say ***job***, and have children orally segment the word. (/**j**/ /**o**/ /**b**/) Ask them how many sounds the word has. (3) Draw three boxes on the chalkboard, and have a volunteer write the spelling that stands for each sound in the word ***job***. Continue with the words ***wig*** and ***bad***.

j	o	b

READ AND WRITE

Dictate Dictate the following words and sentence. Have children write the words and sentence on a sheet of paper. When they are finished, write the words and sentence on the chalkboard, and have children correct their papers.

- **win** **bit** **jam**
- **can't** **they** **who**
- **He can jog up the hill.**

Complete Activity Page Read aloud the directions on page 86. Have children complete the page independently.

Lesson 53

pages 87-88

Read and Review Words With /b/b, /w/w, /j/j

QUICKCHECK ✔

Can children:

✔ **read words with /b/*b*, /w/*w*, and /j/*j*?**

✔ **recognize high-frequency words?**

If YES go to Read and Write.

TEACH

Review Sound-Spellings Review the sound-spelling relationships from the past few lessons, including /**b**/*b*, /**w**/*w*, and /**j**/*j*. Say one of these sounds, and have a volunteer write on the chalkboard the spelling that stands for the sound. For example, the letter ***b*** stands for /**b**/. Then display pictures of objects whose names begin with one of these sounds. Have children write the letter that each picture name begins with.

Review High-Frequency Words Review the high-frequency words ***why, can't we, they,*** and ***for***. Write on the chalkboard sentences with each high-frequency word. Say one word, and have volunteers circle the word in the sentences. Ask children to use each word in a sentence.

READ AND WRITE

Build Words Distribute the following letter cards: ***a, b, f, h, i, j, m, n, t, w***. Allow children time to build as many words as possible using the letter cards. Children can write their words on a separate sheet of paper.

Build Sentences Display the following words on note cards: ***I, we, they, the, a, had, like, and, to, job, win, big***. Have children make sentences using the words. For example: ***They had a big job***. Have children work in small groups to complete the activity.

Complete Activity Pages Read aloud the directions on pages 87–88. Review the art with children.

Name

Check each word as you read it to a partner.
Circle any words you need to practice.

I can read!

☐ bat	☐ band	☐ bad
☐ bit	☐ bill	☐ big
☐ cab	☐ will	☐ wig
☐ win	☐ jog	☐ bib
☐ jam	☐ job	☐ cob

Lookout Words!

☐ why	☐ can't	☐ they	☐ we	☐ for
☐ them	☐ where	☐ like	☐ my	☐ to

Supporting All Learners

VISUAL LEARNERS

CLASS DICTIONARY Add words that begin with ***b, w,*** and ***j*** to the class dictionary, or encourage children to add words that begin with these letters to their individual dictionaries. Children can draw, select illustrations from libraries of computer art, or cut out pictures from magazines to help clarify word meanings.

KINESTHETIC LEARNERS

CATS AND BATS Have children play a game of Cats and Bats by forming two groups, one group as the "cats" and the other group as the "bats." Then draw a tic-tac-toe grid on the chalkboard. The "cats" and "bats" take turns writing a ***c*** or ***b*** in the square of their choice. After writing a letter, the group must say a word that begins with that letter. If they can not think of a word, the opposite team gets a chance to. If that team cannot think of a word, the letter gets erased from the square. The first team that gets three ***c***'s or ***b***'s in a row, wins.

TEST YOURSELF

Fill in the bubble next to the sentence that tells about each picture.

1.	● Bill will jog. ○ Bill will pat the cat.
2.	○ Pam likes to mop. ● Pam likes to bat.
3.	○ He likes the cat. ● He likes the bib.
4.	● They sit in the cab. ○ They sat on the cot.
5.	○ Where is the pan? ● Where is the jam?
6.	○ She can jog to the cab. ● The man can see the wig.

Make sure you fill in the bubble neatly.

EXTRA HELP

Encourage children who are experiencing difficulty with the standardized assessment format to work with more confident learners until they become comfortable with the procedure.

CHALLENGE

Children can write a letter to either the cat or the bat in "A Bat and a Cat." Have children think of questions they would like to ask the bat or the cat before writing.

Display the following picture cards in random order, and have children identify their names, offering help as needed: **ball, bat, boat, box, jar, jump, watch, wig, window**. As children repeat the picture names, have them sort the cards into three groups according to the beginning sounds: **/b/** as in ***ball***, **/j/** as in ***jar***, and **/w/** as in ***watch***.

Integrated Curriculum

WRITING CONNECTION

CAT AND BAT CHIT-CHAT Have children write what else the cat and the bat might say to each other. Children can draw pictures of the cat and the bat, then write what they might say in speech balloons.

WIGGLEWORKS TECHNOLOGY

IN YOUR OWN WORDS Children can use the Record Tool on the computer to tape themselves reading the lists of words. They can then play them back to decide which words they know best, which cause them to hesitate, and which they do not know.

For additional practice with **/b/b, /w/w,** and **/j/j,** have children read the following My Books: *Big and Bigger, Who Walks?,* and *Jumping*. For information on using the My Books on the computer, see the WiggleWorks Plus My Books Teaching Plan.

I CAN READ! OPTIONS

The I Can Read! page can be used for one or all of the following:

- paired reading
- individual assessment
- choral reading
- homework practice
- program placement

pages 89-90

Read Words in Context

TEACH

Assemble the Story Ask children to remove pages 89–90. Have children fold the pages in half to form the Take-Home Book.

Preview the Story Preview *Pigs at Bat,* a story about a big pig who can hit a baseball hard but cannot run fast. Ask a volunteer to read the title. Invite children to browse through the first two pages of the story and to comment on anything they notice. Suggest that they point out any unfamiliar words. Read these words aloud as children blend them. Then have children predict what they think the story might be about.

READ AND WRITE

Read the Story Read the story aloud, or have volunteers take turns reading aloud a page at a time. Discuss items of interest on each page, and encourage children to help each other with any blending difficulties. The following prompts may help children while reading:

- **What letter sounds do you know in the word?**
- **What do the pictures tell you about the story?**

Reflect and Respond Have children share their reactions to *Pigs at Bat*. What did they like most? What do they think of Bob the Pig? Encourage children to think of some funny words or pictures to add to the story.

Develop Fluency Reread the story as a choral reading, or have partners reread the story. Reread the story on subsequent days to develop fluency and increase reading rate.

Ask children how they figure out unfamiliar words and have them model blending. Encourage children to look for words with **/b/*b*, /w/*w*,** and **/j/*j*** and to apply what they have learned about these sound-spelling relationships to decode the words.

Supporting All Learners

KINESTHETIC LEARNERS

READERS THEATER Children can role-play Bob and the other pigs on his team. Provide tips for role-playing the other pigs, such as looking high and far into the air when Bob hits the ball, applauding wildly, urging on Bob as he runs, and watching Bob get thrown out before he reaches first base.

VISUAL LEARNERS

HOW MANY? Children can keep count of how many words with **/b/*b*, /w/*w*,** and **/j/*j*** they find in the Phonics Readers. Suggest that they read Phonics Readers #19, #20, and #21.

Reflect on Reading

ASSESS COMPREHENSION

To assess their understanding of the story, ask children questions such as:

- **Who is Bob?** ***(Bob is a big pig who is at bat.)***
- **What can Bob do?** ***(He can bat the ball hard and far.)***
- **What can't Bob do?** ***(He can't run fast.)***

HOME-SCHOOL CONNECTION

Send home *Pigs at Bat*. Encourage children to read the story to a family member. You may also suggest that children make a game out of searching for words they can read at home. Parents and other family members may provide praise for decoding efforts as well as feedback to help children know when they have read a word correctly.

WRITING CONNECTION

Have children write sentences using words with **/b/b, /w/w,** and **/j/j.** Children can also illustrate their sentences. Then children can exchange sentences and pictures. Have children identify the words with **/b/b, /w/w,** and **/j/j** in the sentences they were given.

CHALLENGE

Challenge children to use words with **/b/b, /w/w,** and **/j/j** to add sentences to *Pigs at Bat* or to write another simple story.

EXTRA HELP

Have children read *Pigs at Bat* aloud as a choral reading. Then partners can read the story to each other by reading alternate sentences.

Phonics Connection

Phonics Readers:
#19, *Who Has a Bill?*
#20, *Where Is It?*
#21, *Jim*

Lesson 55

pages 91-92

Recognize and Write /z/z

QUICKCHECK ✔

Can children:

✔ replace sounds in words?

✔ write capital and small Zz?

✔ recognize /z/?

✔ identify the letter that stands for /z/?

If YES go to Read and Write.

TEACH

Develop Phonemic Awareness

Phonemic Manipulation Explain to children that you will read a list of words. Children will make new words by replacing the first sound in each word with /z/. For example, if you say the word ***lip,*** children are to replace /l/ with /z/ and say the word ***zip***. Tell children that some of the words will be nonsense, or made-up, words. Use the following words:

- pig
- hip
- funny
- bag
- hero
- silly
- boom
- cone
- too

Write the Letter Explain that z stands for /z/, the sound at the beginning of ***zip***. Write Zz on the chalkboard. Point out the capital and small forms of the letter. Model how to write the letter.

Have children write both forms of the letter in the air with their fingers. Have volunteers practice writing the letter on the chalkboard.

READ AND WRITE

Connect Sound-Symbol Write ***zip*** on the chalkboard. Have a volunteer circle the z. Remind children that z stands for /z/. Ask children to suggest other words that contain /z/. List them on the chalkboard. Have volunteers circle the letter z in each.

Complete Activity Pages Read aloud the directions on pages 91–92. Review each picture name with children.

HANDWRITING

Name

Zebra begins with the z sound.

Zz

Z Z Z

z z z

Where is the zebra?

Supporting All Learners

AUDITORY LEARNERS

LISTEN FOR /z/z Read aloud "Zip, Zoom" on page 24 in the *Big Book of Rhymes and Rhythms, 1A*. Have children find all the words with /z/ in the rhyme. Write these words on the chalkboard. Invite volunteers to add other /z/ words to the list.

VISUAL LEARNERS

WORD WALL On a large note card, write the letter z and the key word ***zebra***. Display the card on the Word Wall. Have children suggest words with /z/z, write them on small note cards, and add them to the wall. Remind children to look for words that begin with z during their reading. Add these words to the wall.

TACTILE/AUDITORY LEARNERS

MAGNETIC LETTERS Provide children with magnetic letters from the Phonics and Word Building Kit for all the sounds learned so far. Invite children to build words, including some words with z. Then have children say each word as they trace its letters.

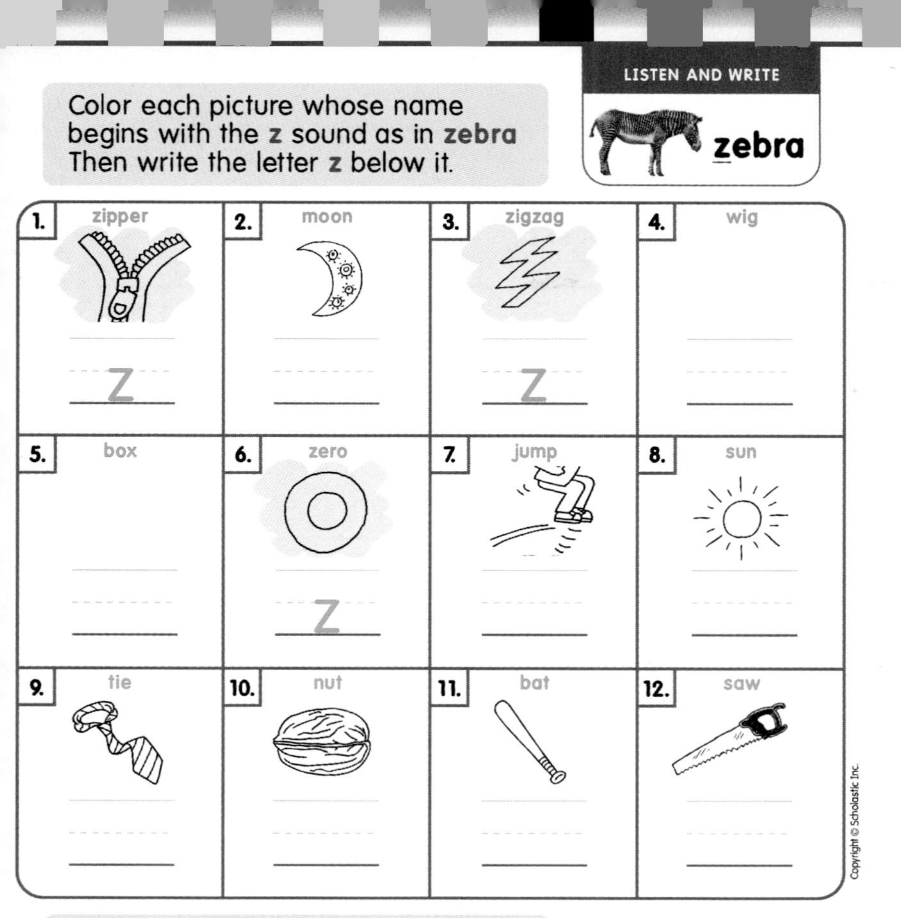

Write the letter z to finish each word.
Read the words to a friend.

13. z ip 14. z ap

Integrated Curriculum

SPELLING CONNECTION

Z* GAME** Write the following word parts on index cards, one word part per card: ***__ip, __ap, __ig, __ag. Place the cards in a bag. Have children pick one card each. Ask children to write the word part on a sheet of paper and then add the letter **z**. Model for children how to blend each word. Then have children take turns saying the words they made. Repeat the game by returning the word-part cards to the bag.

SOCIAL STUDIES CONNECTION

FIND /z/*z* WORDS You may wish to have children look in newspapers, magazines, and books for words with **/z/*z***. Children can collect these words on a ***Zz*** poster or in a ***Zz*** book.

Phonics Connection

Literacy Place: *Problem Patrol*
Teacher's SourceBook, pp. T252–253;
Literacy-at-Work Book, pp. 59–60

My Book: *My Noisy Zipper*

Big Book of Rhymes and Rhythms, 1A: "Zip, Zoom," p. 24

Chapter Book: *A Lot of Hats,* Chapter 9

CHALLENGE

Have children create a **/z/** story. Begin with a **z** title such as "*Zip, the Zebra.*" Write the dictated story on chart paper, and return to it in subsequent lessons.

EXTRA HELP

Have children work together to write five or more words with **z**. Then children can use these words to write or dictate silly sentences such as ***The zebra zigs and zags***.

If possible, put on a jacket that has a zipper and say "Zip the zipper" as you zip it up. Invite children to repeat the phrase as they take turns zipping the zipper. They may like to sing this song to the tune of "Row, Row, Row Your Boat": ***Zip, zip, zip the zipper/Zip and zip and zip/Zip, zip, zip the zipper/Zip the zipper up!***

Lesson 56

pages 93-94

Blend and Build Words With /z/z

QUICKCHECK ✔

Can children:

✔ **orally blend word parts?**

✔ **identify /z/?**

✔ **blend and build words with /z/z and phonograms *-an, -in*?**

If YES go to Read and Write.

TEACH

Develop Phonemic Awareness

Oral Blending Say the following word parts. Ask children to orally blend them.

/z/...ip	/z/...ap	/z/...ero
/z/...ipper	/z/...ebra	/z/...igzag

Connect Sound-Symbol Review that *z* stands for /z/ as in *zap*. Write *zap* on the chalkboard. Have a volunteer circle the *z*. Then model blending.

Introduce the Phonograms Write *-an* and *-in* on the chalkboard. Point out the sounds these phonograms stand for. Add *f* to the beginning of *-an*. Model how to blend *fan*. Add *w* to the beginning of *-in*, and model how to blend *win*. Ask children to suggest words that rhyme with *fan* and *win*. List these words on the chalkboard in separate columns.

READ AND WRITE

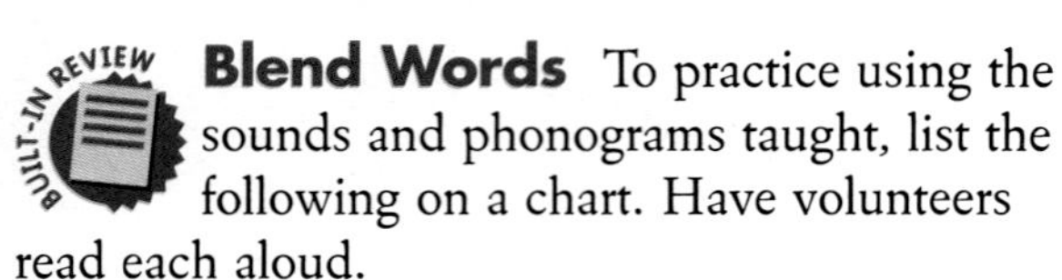

Blend Words To practice using the sounds and phonograms taught, list the following on a chart. Have volunteers read each aloud.

- zip zap zigzag
- pan can in
- The man can zip up the hill.

Complete Activity Pages Read aloud the directions on pages 93 and 94. Review the art with children.

Name

Look at each picture. Circle the word that best finishes each sentence. Then write the word on the line.

1.	He will zip it up.	hip / (zip) / zap
2.	She can zap it.	zip / (zap) / sip
3.	I see a zigzag.	zip / (zigzag) / zap

Look at the picture. Then unscramble the words to make a sentence that tells about the picture. Write the sentence on the line.

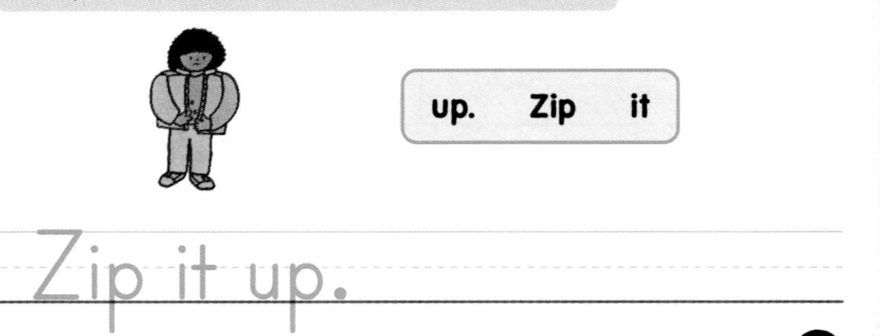

Supporting All Learners

TACTILE LEARNERS

CLAY WORDS Have children use clay to form the words parts *-an* and *-in*. Then have children use clay to form letters to add to *-an* and *-in* to make words. You may wish to suggest that children form the words ***man, can, tin, tan, pan, fin,*** and ***pin***. Blend these words for children as they repeat them aloud. Encourage children to look for the word parts *-an* and *-in* as they read.

AUDITORY LEARNERS

PHONOGRAM SONGS Children can listen to "Z Was a Zebra, " "Zum Gali Gali," "Mix a Pancake," and "Patty Cake" on the Sounds of Phonics Audiocassette (Community Involvement) from the Phonics and Word Building Kit. The first two songs feature the letter *z*; the other two feature the phonogram *-an*. Ask children to identify words with the target sound or phonogram as they listen.

KINESTHETIC LEARNERS

BUILD WORDS Children can write the word parts *-an* and *-in*, or use a pocket chart and letter cards to build a word. Have children add a letter to the beginning of each to make a word. Children can replace the initial consonant to build a second word.

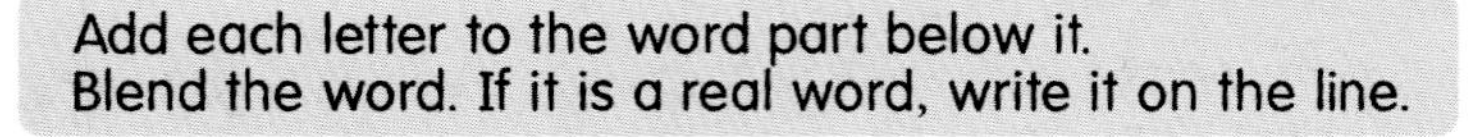
Add each letter to the word part below it.
Blend the word. If it is a real word, write it on the line.

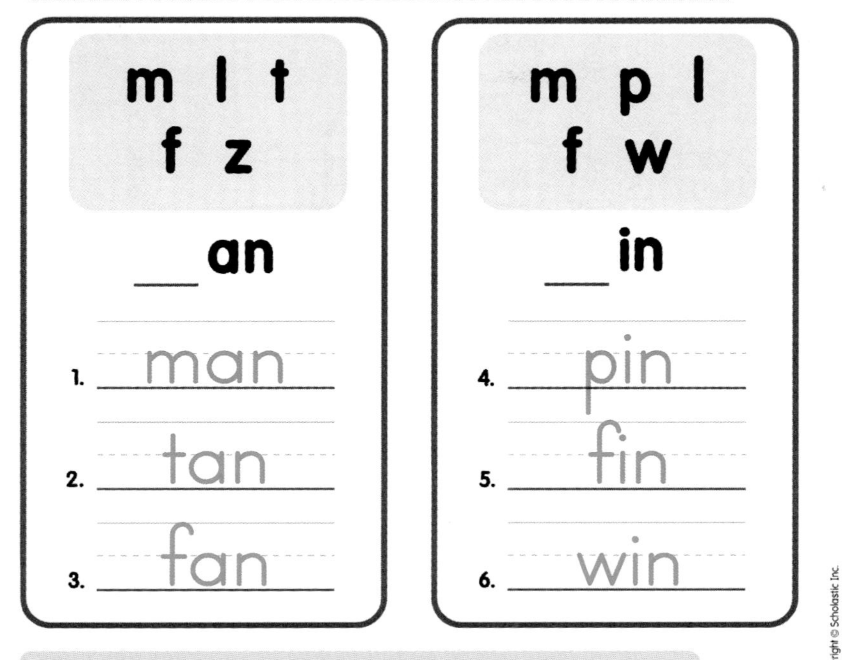

Write a sentence using one of the words you made.

7. Answers will vary.

Integrated Curriculum

WRITING CONNECTION

QUESTIONS, QUESTIONS Have children write or dictate questions using words with **/z/z** and words with ***-an*** and ***-in***. You may wish to provide an example such as ***Where can you see a zebra?*** Display the questions on a bulletin board. Have children identify words with **/z/z** and words with ***-an*** and ***-in***.

MATH CONNECTION

WORD PUZZLES Children can work together to make word puzzles with the phonograms ***-an*** and ***-in***. You can use the following puzzles as models:

can – c + t = tan
win + t – w = tin

EXTRA HELP

For children who need additional practice building words with phonograms, distribute familiar letter and phonogram cards from the Phonics and Word Building Kit. As children create words, emphasize the blending of the sounds represented by each letter and phonogram.

CHALLENGE

Invite children to make flip books featuring the phonogram ***-an***. The top word may be ***can***. Under the ***c,*** children can write ***f, m, p,*** and ***t,*** one letter per page. Children can take turns reading the words in their flip books and can make up sentences using the words.

- **My "x, y, z" Sound Box**
  by Jane Moncure
  Three children – X, Y, and Z – search for appropriate *x, y,* and *z* objects. **(/z/z)**

- **Zella, Zack, and Zodiac**
  by Bill Peet
  Zella the zebra helps Zack the Ostrich, and later he returns the favor. **(/z/z)**

Lesson 57

pages 95–96

Recognize and Write /d/*d*

QUICKCHECK ✔

Can children:

- ✔ **write capital and small *Dd*?**
- ✔ **recognize /d/?**
- ✔ **identify the letter that stands for /d/?**

If YES go to Read and Write.

TEACH

Develop Phonemic Awareness

Oddity Task Explain to children that you are going to say a list of three words. Children are to choose the word that does not end with the same sound as the other two words. For example, say the words ***red, bad,*** and ***cat***. Ask children which word does not belong. Explain that ***red*** and ***bad*** end with /d/. ***Cat*** does not. Therefore, the word ***cat*** does not "belong." Continue with these words:

• had	can	sad
• sun	did	led
• hid	mad	hen
• dad	lid	bus
• crab	glad	made

Write the Letter Explain that the letter *d* stands for **/d/,** the sound at the end of the words ***red*** and ***bad***. Write ***Dd*** on the chalkboard. Point out the capital and small forms of the letter. Model how to write the letter.

Have children write both forms of the letter in the air. You may also wish to have volunteers practice writing the letter on the chalkboard.

READ AND WRITE

Connect Sound-Symbol Write ***dot*** and ***mad*** on the chalkboard. Have a volunteer circle ***d*** in each word. Remind children that ***d*** stands for **/d/**. Ask children to suggest other words that begin or end with **/d/**. List these on the chalkboard. Have volunteers circle the ***d*** in each.

Complete Activity Pages Read aloud the directions on pages 95–96. Review each picture name with children.

Supporting All Learners

AUDITORY/VISUAL LEARNERS

WORD HUNT Read aloud "Hickory, Dickory, Dock" on page 25 in the *Big Book of Rhymes and Rhythms, 1A*. Have children find all the words with **/d/** in the rhyme. Write these words on the chalkboard. Invite volunteers to add other **/d/** words to the list.

VISUAL/KINESTHETIC LEARNERS

LETTER CARDS Distribute the letter card ***d*** to each child. Then have children draw three connected boxes on a sheet of paper. Tell children that you are going to say a list of words. All of the words contain **/d/**. Some words begin with **/d/**, and some words end with **/d/**. If children hear **/d/** at the beginning of the word, they are to place the letter card in the first box. If they hear **/d/** at the end, they are to place the card in the last box. Use the following words: ***sad, dog, dig, had, dip, dot, hid,*** and ***lid***.

LISTEN AND WRITE

Look at each picture. If the picture name begins with the **d** sound as in **dog**, write the letter **d** on the first line. If it ends in **d** as in **sad**, write **d** on the second line.

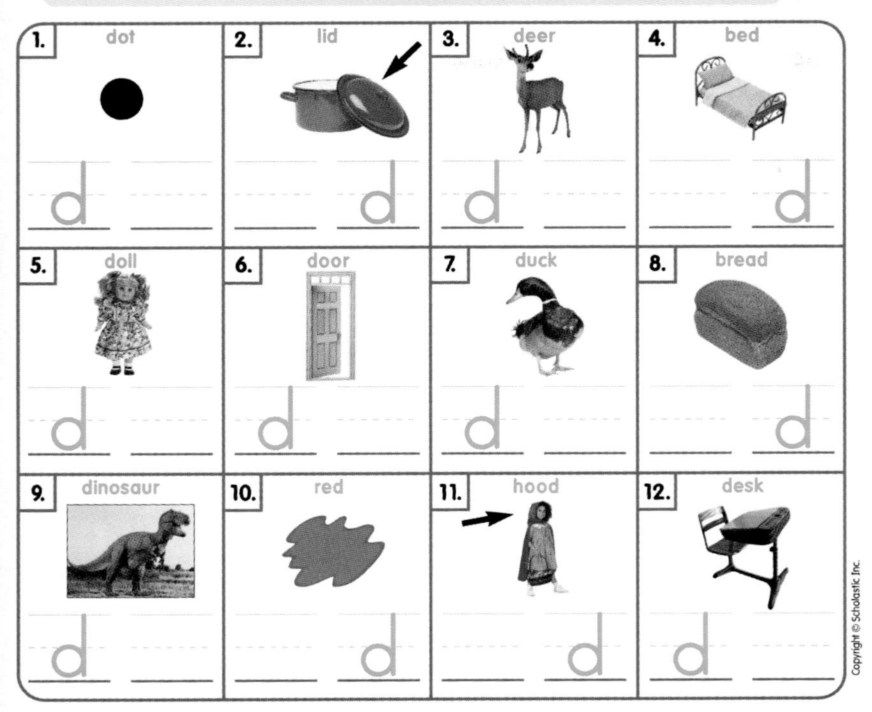

Write the letter **d** to finish each word. Circle the word that names the picture.

13. ha d

14. d ig

CHALLENGE

Have children write or dictate a list of words with **/d/d**. Then have children use these words to write or dictate a simple, short story.

EXTRA HELP

Display the pocket ABC cards and pocket chart from the Phonics and Word Building Kit. Have children whose names contain ***d*** use the ABC cards to make their names in the pocket chart. Children can use the pocket ABC cards and pocket chart to make other **/d/d** words.

Have children acquiring English look through discarded magazines for pictures whose names begin with **/d/**— for example: **dog, dad, dish, dinner, duck, door, desk,** and **dots**. Have them cut out the pictures they find and paste them on separate sheets of paper to make a book. Children can write ***Dd*** on the cover and then draw lots of dots.

Integrated Curriculum

SPELLING CONNECTION

LETTER TIME Have children write the following word parts: **_og, _ig, sa_, hi_, di_**. Then have children add the letter ***d*** to each word part to make a word. Model for children how to blend each word. Then have children replace the ***d*** with another letter to make another word. Children can write or dictate that word.

ART CONNECTION

PICTURE THIS Have children draw pictures of things with names that contain **/d/d**. Children can combine their pictures into a ***Dd*** collage or make their own posters of things with **/d/d**. Have children write or dictate the name of each picture as a label underneath that picture.

Phonics Connection

Literacy Place: *Problem Patrol*
Teacher's SourceBook, pp. T306–307;
Literacy-at-Work Book, pp. 68–69

My Book: *The Dog Didn't Do It*

Big Book of Rhymes and Rhythms, 1A: "Hickory, Dickory, Dock," p. 25.

Chapter Book: *A Lot of Hats*, Chapter 10

ASSESSMENT

Follow up on previous assessments. Compare your current observations on children's abilities to match specific sounds and letters and to blend words to previous observations. You may also ask children to talk about the ways in which they have become better readers and writers.

Lesson 58

pages 97-98

Blend and Build Words With /d/d

QUICKCHECK ✔

Can children:

✓ orally blend word parts?

✓ recognize /d/?

✓ blend and build words with /d/d?

If YES go to Read and Write.

TEACH

Develop Phonemic Awareness

Oral Blending Say the following word parts. Ask children to blend them. Provide corrective feedback and modeling when necessary.

/d/...ip	/d/...esk	/d/...og
/d/...ance	/d/...ive	/d/...eep

Connect Sound-Symbol Review that *d* stands for /d/ as in *dig*. Write the word *dig* on the chalkboard. Have a volunteer circle the *d*. Then model how to blend the word.

THINK ALOUD

I can put *d*, *i*, and *g* together to make the word *dig*. Let's say the word slowly as I move my finger under the letters.

READ AND WRITE

Blend Words To practice using the sounds taught, list the following on a chart. Have volunteers read each aloud. Model blending when necessary.

- mad dad dot
- Dad had a dog.

Build Words Distribute these letter cards to children: *a, b, c, d, i, g, m, o, s,* and *t*. If children have their own set of cards, have them locate it. Have children build as many words as possible.

Complete Activity Pages Read aloud the directions on pages 97–98. Review each picture name with children.

Name

Circle the word that names each picture. Then write the word on the line.

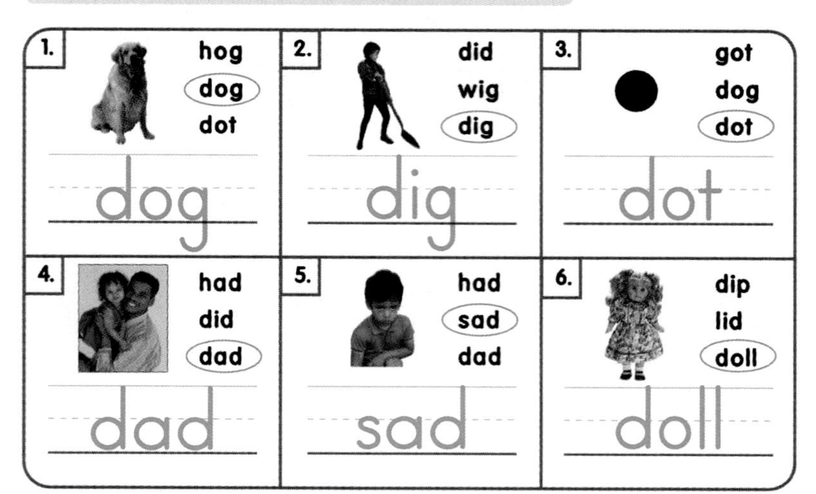

Use one of the words from above to finish each sentence.

7. My dog likes to dig.

8. My dad is not sad.

Supporting All Learners

VISUAL LEARNERS

ANIMAL MOBILE Have children draw pictures of animals with **/d/d, /j/j, /z/z, /b/b,** or **/w/w** in their names. Help children to label these pictures. Then attach each picture to a piece of string. Attach the strings to a hanger to make an animal mobile.

KINESTHETIC/VISUAL LEARNERS

WORD SORT Write the following words on index cards, and have children sort them according to whether they begin with **/d/** or end with **/d/**: ***hid, mad, sad, dot, dip, bad, dig,*** and ***dog***.

AUDITORY/VISUAL LEARNERS

CREATE A WORD Write the word ***dad*** on the chalkboard. Model blending. Then have a volunteer replace the first **/d/** in ***dad*** with **/s/** to make a new word. Ask children what letter stands for **/s/**. Continue by having children change one letter at a time to build new words.

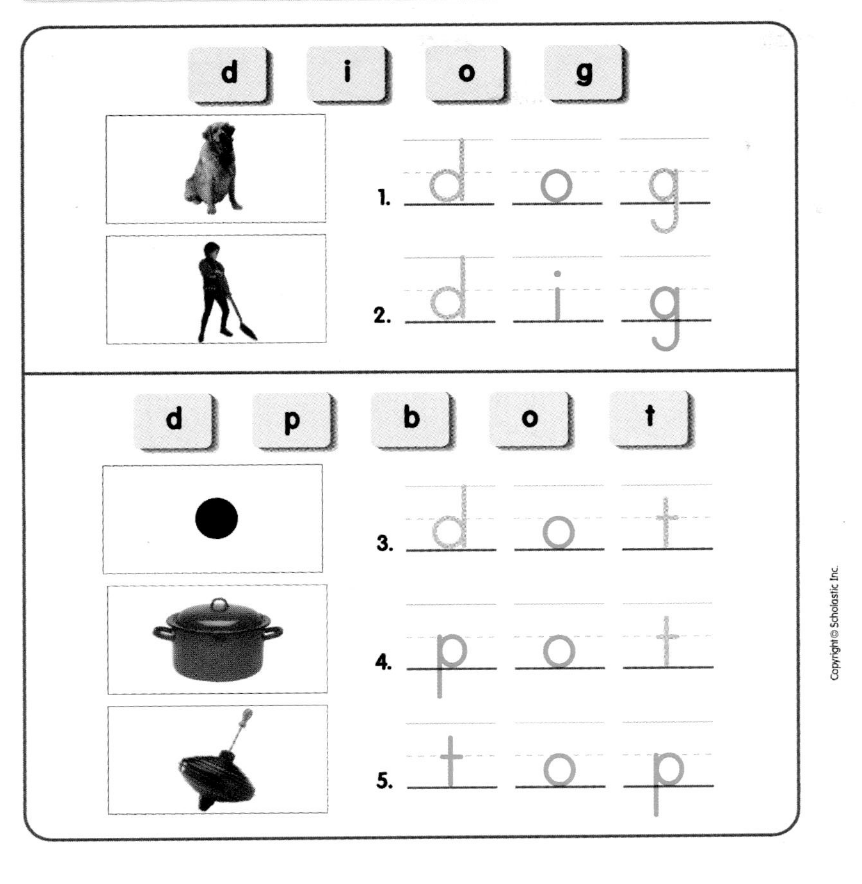

Integrated Curriculum

WRITING CONNECTION

TO DO Have children write or dictate a "to do" list using words with **/d/d**. You may wish to provide an example such as ***Get a dog for Dad***. Children can write or dictate things to do on strips of paper. Then you can display the things to do on a bulletin board. Children can revisit the list to identify words with **/d/d**.

SCIENCE CONNECTION

DOUGH LETTERS Children can mix four cups of flour, one cup of salt, and one and three-quarter cups of warm water to make dough. Then children can form ***Dd*** and words with ***Dd*** out of the dough.

CHALLENGE

Give children the first and last words of a three-word ladder. Have children change one letter in each word to complete the word ladder. Use the following words.

- **pig** **(dig)** **dog**
- **mat** **(mad)** **dad**
- **dip** **(zip)** **lip**

Note: The words in parentheses are possible answers.

EXTRA HELP

To review sounds you have taught recently, distribute the letter cards for ***d, j, z, b,*** and ***w*** from the Phonics and Word Building Kit. Say words such as ***dog, zebra, bird, jacks, walrus***, and ***desk***. Have children hold up the letter card that matches the beginning sound in each word.

- **Dad's Dinosaur Day**
  by Diane Dawson Hearn
  Mikey and his dad have fun when his dad pretends to be a dinosaur. **(/d/d)**

Lesson 59

pages 99-100

Recognize and Write /r/r

QUICKCHECK ✔

Can children:

- ✔ **orally segment words?**
- ✔ **write capital and small *Rr*?**
- ✔ **recognize /r/?**
- ✔ **identify the letter that stands for /r/?**

If YES go to Read and Write.

TEACH

Develop Phonemic Awareness

Oral Segmentation Explain to children that you are going to say words by saying their parts. Children will listen to the parts and then say the whole word. For example, say the sounds /**d**/ /**i**/ /**g**/. Guide children to say the whole word: ***dig***. Continue with the following word parts:

/r/ /a/ /g/	/r/ /o/ /b/
/r/ /a/ /t/	/r/ /a/ /n/
/r/ /i/ /p/	/r/ /o/ /t/

Write the Letter Explain that ***r*** stands for /**r**/, the sound at the beginning of ***ran***. Write ***Rr*** on the chalkboard. Point out the capital and small forms of the letter. Model how to write the letter.

Have children write both forms of the letter in the air. You may also wish to have volunteers practice writing the letter on the chalkboard.

READ AND WRITE

Connect Sound-Symbol Write the word ***ran*** on the chalkboard, and have a volunteer circle the letter ***r***. Ask children to suggest other words that contain /**r**/. List them on the chalkboard. Have volunteers circle each letter ***r***.

Write the following on the chalkboard: ***_ip, _at, _od, _ag, _ob***. Have volunteers add ***r*** to each. Model how to blend each word. Then have children say each word and write it.

Complete Activity Pages Read aloud the directions on pages 99–100. Review each picture name with children.

Supporting All Learners

AUDITORY/VISUAL LEARNERS

WORD HUNT Read aloud "Rain" on page 26 in the *Big Book of Rhymes and Rhythms, 1A*. Have children find all the words with **/r/** in the rhyme. Write these words on the chalkboard. Invite volunteers to add other **/r/** words to the list.

KINESTHETIC/AUDITORY LEARNERS

BINGO To review and reinforce recognition of the following initial sounds, play the Bingo Game in the Phonics and Word Building Kit: **/b/, /d/, /f/, /h/, /k/, /l/, /m/, /p/, /r/, /s/, /t/,** and **/w/**.

AUDITORY LEARNERS

GAME TIME Pairs of children can play a version of tic-tac-toe with words that have **/r/r**. Draw a tic-tac-toe grid on the chalkboard. Give each child the label of X or O. As children play the tic-tac-toe game, they must say a word with **/r/r** before they can place an X or O in a square. List these words on the chalkboard. When each game is finished, have children say each word in the list.

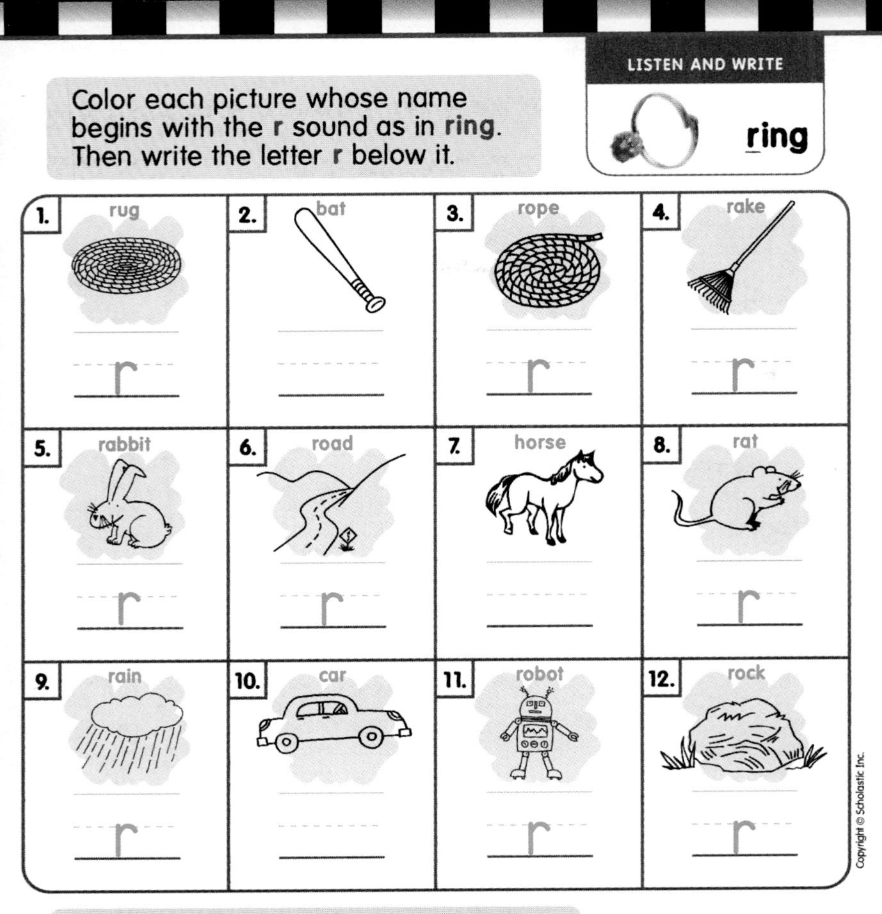

Write the letter r to finish each word.
Circle the word that names the picture.

14. r an

Integrated Curriculum

WRITING CONNECTION

Have children work together to write or dictate a list of words that begin with **/r/r**. Then children can write or dictate tongue-twister sentences such as ***The red rat ran into Ron and Rita***. Display the sentences. Challenge children to revisit them, reading them as fast as possible.

SCIENCE CONNECTION

R* WORDS** Have children write on construction paper large words with **/r/r**, one word per page. Children can then glue rice or raisins to each ***r in each word. Children can also exchange papers and identify each ***r*** word.

Phonics Connection

Literacy Place: *Problem Patrol*
Teacher's SourceBook, pp. T308–309;
Literacy-at-Work Book, pp. 70–71

My Book: *Too Much Rain*

Big Book of Rhymes and Rhythms, 1A: "Rain," p. 26.

Chapter Book: *A Lot of Hats,* Chapter 11

ESL Encourage children acquiring English to write words that begin with **/r/** on the Magnet Board in WiggleWorks Plus. Then they can click on the words and hear them read aloud.

EXTRA HELP

On a large note card, write the letter ***r*** and the key word ***ring***. Display the card on the Word Wall. Have pairs of children suggest words with **/r/r**, write them on small note cards, and add them to the wall. Remind children to look for words that begin with ***r*** during their reading. Add these words to the wall.

Lesson 60

page 101

Blend Words With /r/r

QUICKCHECK ✔

✔ orally blend word parts?

✔ identify /r/?

✔ blend words with /r/r?

If YES go to Read and Write.

TEACH

Develop Phonemic Awareness

Oral Blending Say the following word parts, and ask children to blend them. Provide corrective feedback and modeling when necessary.

/r/ /a/ /t/	/r/ /e/ /d/	/r/ /o/ /b/
/r/ /i/ /p/	/r/ /u/ /n/	/r/ /u/ /g/

Connect Sound-Symbol Review that *r* stands for /r/ as in ***rip***. Write ***rip*** on the chalkboard. Have a volunteer circle the *r*. Then model how to blend the word.

I can put the letters *r*, *i*, and *p* together to make the word *rip*. Let's say the word slowly as I move my finger under the letters. Listen to how I string together the sound that each letter stands for to make the word: *rrrrriiiip, rriip, rip*.

Help children blend the words *rat* and *rob*.

READ AND WRITE

BUILT-IN REVIEW

Blend Words To practice using the sound taught and to review previously taught sounds, list the following words and sentence on a chart. Have volunteers read each aloud. Model blending when necessary.

- **rat** **red** **ran**
- **rag** **rip** **rot**
- **The rat ran and ran.**

Complete Activity Page Read aloud the directions on page 101. Review the art with children.

BLEND WORDS

Name

Look at each picture. Circle the word that best finishes each sentence. Then write the word on the line.

1.	The cat likes the rat	cat / rat / ran
2.	They ran to the top.	ran / can / rat
3.	Will she rip it up?	hip / rap / rip
4.	Where is the rag?	rat / rib / rag
5.	She got rid of it.	rip / rid / ran

Supporting All Learners

KINESTHETIC/VISUAL LEARNERS

ACT IT OUT Write words such as ***run, red, rest, rip, rat, roof, rabbit, ring,*** and ***rug*** on sheets of paper, one word per sheet. Have children take turns picking a sheet of paper. Help them to read the word, then act it out while others try to guess it. Children acting out the word can point to things, pantomime, and draw pictures if they wish.

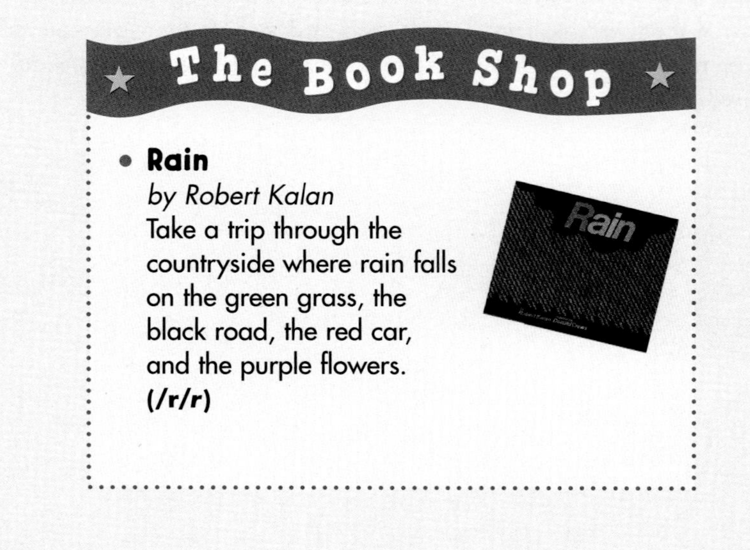

The Book Shop

- **Rain**
  by Robert Kalan
  Take a trip through the countryside where rain falls on the green grass, the black road, the red car, and the purple flowers.
  (/r/r)

POETRY

Read the poem. Write a sentence telling about the rat.

THE RAT RAN

Zip, zip, zip!
The rat ran.

It ran to the mat
And it hid in the hat.

Zip, zip, zip!
The rat ran.

It ran to the can
And it hid in the pan.

Why can't the rat
Sit for a bit?

Zip, zip, zip!

Answers will vary.

Supporting All Learners

AUDITORY LEARNERS

CLIMB THE LADDER! Draw a large ladder on chart paper. Invite children to "climb" the ladder by naming words that begin with **/r/*r***. Write the words on the ladder, one word per rung, beginning with the bottom rung, and working toward the top. Remind children to look for words that begin with ***r*** when they read. Make a second word ladder on which to add the words children find.

VISUAL LEARNERS

LETTER TIME Write words that begin with **/z/*z*, /d/*d*,** and **/r/*r*** on cards, one word per card. Then distribute the letter cards ***z, d,*** and ***r*** to children. Hold up each word card, one at a time. Have children hold up the letter card that corresponds to the first letter in the word you hold up. Then ask children what other words begin with this sound and letter.

EXTRA HELP

Help children work together to dictate sentences about their neighborhood or town using words with **/z/*z*, /d/*d*,** and **/r/*r***.

Write and Read Words with /z/z, /d/d, /r/r

QUICKCHECK ✔

Can children:

✔ **spell words with /z/*z*, /d/*d*, /r/*r*?**

✔ **read a poem?**

If YES go to Read and Write.

TEACH

Link to Spelling Review with children the following sound-spelling relationships: /**z**/***z***, /**d**/***d***, /**r**/***r***. Say one of these sounds. Have a volunteer write on the chalkboard the spelling that stands for the sound. For example, ***r*** stands for /**r**/. Continue with all the sounds. Also review /**b**/***b***, /**w**/***w***, and /**j**/***j***.

Phonemic Awareness Oral Segmentation
Then say ***dot***. Have children orally segment the word. (/**d**/ /**o**/ /**t**/) Ask them how many sounds the word contains. *(3)* Draw three connected boxes on the chalkboard. Have a volunteer write the spelling that stands for each sound in ***dot*** in the appropriate box. Continue with ***zip*** and ***rag***.

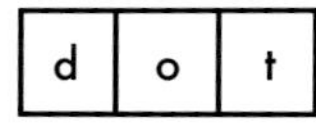

d	o	t

READ AND WRITE

Dictate Dictate the following words and sentence. Have children write the words and sentence on a sheet of paper. When children are finished, write the words and sentence on the chalkboard, and have children make any necessary corrections on their papers.

- **zap** **rat** **dip**
- **the** **why** **we**
- **The dog ran in a zigzag.**

Complete Activity Page Read aloud the directions on page 102. Children can read the poem independently or with a partner. Have children circle words that contains /**r**/.

Lesson 62

pages 103-104

Read and Review Words With /z/z, /d/d, /r/r

QUICKCHECK ✔

Can children:

✔ read words with /z/z, /d/d, /r/r?

✔ recognize high-frequency words?

If YES go to Read and Write.

TEACH

Review Sound-Spellings Review with children the sound-spelling relationships from the past few lessons. These include /z/z, /d/d, and /r/r. Say one of these sounds. Have a volunteer write the spelling that stands for the sound on the chalkboard. For example, ***d*** stands for /d/. Continue with all the sounds. Then display pictures of objects whose names begin with one of these sounds. Have children write the letter that each picture name begins with.

Review High-Frequency Words Review the high-frequency words ***why, can't we, they,*** and ***for***. Write sentences on the chalkboard containing each. Say one word. Have volunteers circle the word in the sentences. Have children generate additional sentences for each word.

READ AND WRITE

Build Words Distribute these letter cards: ***a, b, d , g, h, i, n, p, r, s, t,*** and ***z***. If children have their own set of cards, have them locate it. Have children build as many words as possible using the letter cards. They can record their words on a separate sheet of paper.

Build Sentences Write the following words on note cards: ***I, we, can, they, the, rat, and, zip, it, rip, up,*** and ***dig***. Display the cards, and have children make sentences using the words. To get them started, suggest ***I can rip it up***.

Complete Activity Pages Read aloud the directions on pages 103–104. Review the art with children.

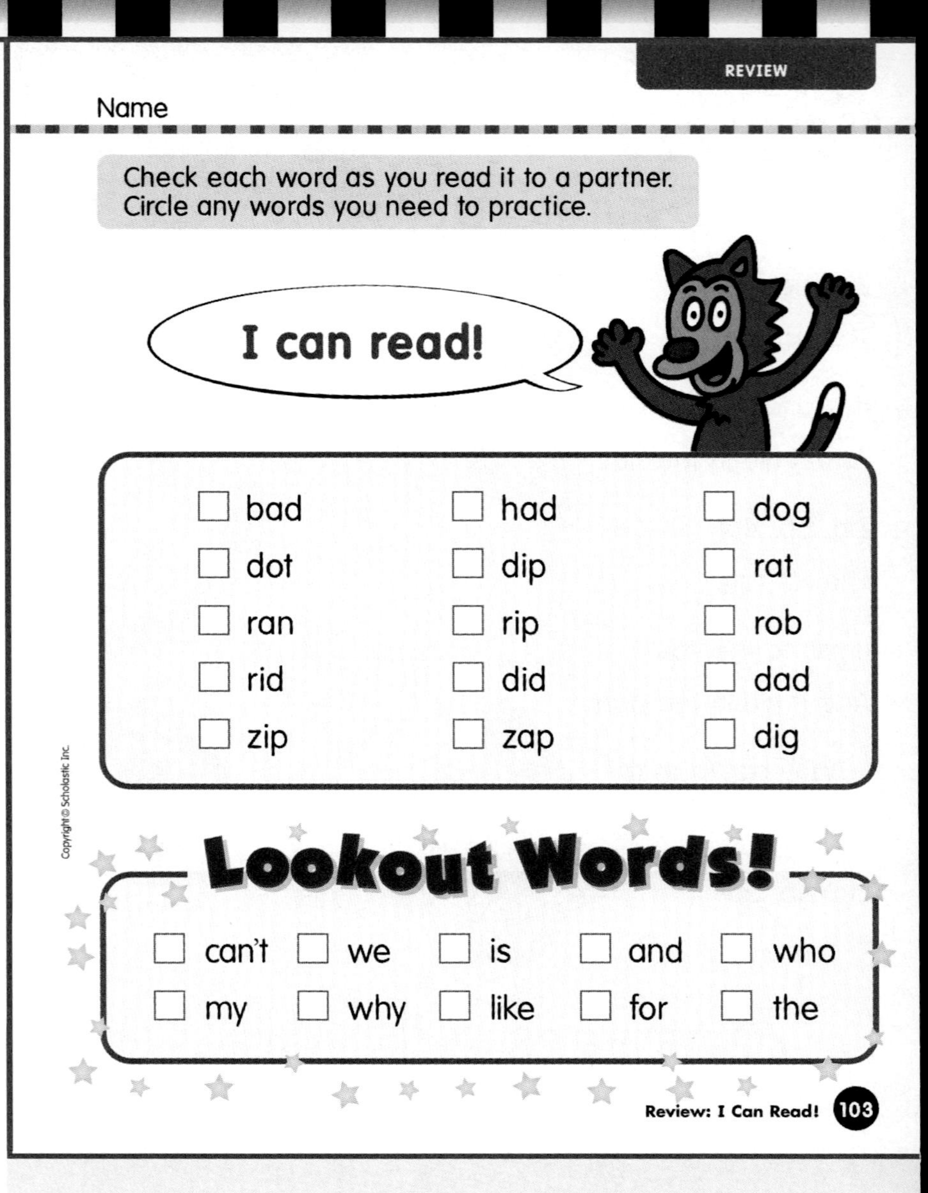

Supporting All Learners

AUDITORY/VISUAL LEARNERS

LETTER TIME To help children listen for and identify **/d/** and **/r/**, play "Hickory Dickory Dock" or "Hey Diddle Diddle" (for **/d/**) and "Ride a Cock Horse" or "Row Your Boat" (for **/r/**) on the Sounds of Phonics Audiocassettes from the Phonics and Word Building Kit. Each time children raise their hands to signal their having heard **/d/** or **/r/**, write the letter ***d*** or ***r*** on the chalkboard.

KINESTHETIC/VISUAL LEARNERS

GAME TIME Have children write six words that begin with **/z/z,** six words that begin with **/d/d,** and six words that begin with **/r/r** on index cards, one word per card. Shuffle the cards, and lay them out, facedown. Then have children take turns turning over two cards and reading the word on each. If the words begin with the same letter, the child keeps the pair of cards, if the words do not begin with the same letter, the child turns the cards facedown again.

TEST YOURSELF

Fill in the bubble next to the sentence that tells about each picture.

1.	● The dog is bad. ○ The pig is big.
2.	○ He had a dog. ● He can zip it up.
3.	○ Did he sit a bit? ● Did he rip the bib?
4.	○ I can see the rat. ● I can see the cat.
5.	● She ran to Dad. ○ She had a big hat.
6.	○ Dan can see the pad. ● Dan can see the pan.

Read each sentence carefully.

Integrated Curriculum

WRITING CONNECTION

WRITING EXTENSION Have children generate a list of **/z/z, /d/d,** and **/r/r** words. Record these words on the chalkboard. Then have children create a story using as many of these words as they can. You may wish to begin with a title such as "Zip and Dot Run a Race." Write the dictated story on chart paper for group and individual reading. Return to the story, rereading it in subsequent lessons.

TECHNOLOGY CONNECTION

WIGGLEWORKS Children can create their sentences using the computer. They can then use the Paint Tools to illustrate their work. For information and writing activities, see the WiggleWorks Plus Teaching Plan.

Also, for additional practice with **/z/z, /d/d,** and **/r/r,** have children read the following My Books: *My Noisy Zipper, The Dog Didn't Do It,* and *Too Much Rain*. For information on using the My Books on the computer, see the WiggleWorks Plus My Books Teaching Plan.

I CAN READ! OPTIONS

The I Can Read! page can be used for one or all of the following:

- paired reading
- individual assessment
- choral reading
- homework practice
- program placement

CHALLENGE

To present a challenge, have children sort the words in the I Can Read! list. Write the words on index cards, and invite children to think of as many ways as possible to sort them.

EXTRA HELP

Use the following picture cards from the Phonemic Awareness Kit to help reteach **/d/, /z/,** and **/r/: dig, dog, duck, zebra, zipper, red, ring, rope,** and **run**. Hold up each card, say the picture name, and ask children to tell the sound it begins with. Then have a volunteer write the letter that stands for the sound on the chalkboard.

pages 105-106

Read Words in Context

TEACH

Assemble the Story Ask children to remove pages 105–106. Have children fold the pages in half to form the Take-Home Book.

Preview the Story Preview *Rob's Cab,* a story about characters trying to fit into a small cab. Have a volunteer read aloud the title. Invite children to browse through the first two pages of the story and to comment on anything they notice. Suggest that they point out any unfamiliar words. Read these words aloud as children repeat them. Then have children predict what the selection might be about.

READ AND WRITE

Read the Story Read the story aloud, or have volunteers take turns reading aloud a page at a time. Discuss anything of interest. Encourage children to help each other with any blending difficulties. The following prompts may help children who need extra support:

- **What letter/sounds do you know in each word?**
- **What do the pictures tell you about the story?**

Reflect and Respond Have children share their reactions to *Rob's Cab.* What was funny about the story? Could this story really happen? Why or why not?

Develop Fluency Reread the story as a choral reading, or have partners reread the story independently. Children can reread the story to develop fluency and increase reading rate.

Continue to review challenging sound-spelling relationships for children needing additional support.

Supporting All Learners

VISUAL/TACTILE LEARNERS

PICTURE THIS Have children draw pictures of Rob's cab on another busy day. Children can write or dictate captions for their drawings. Then children can take turns sharing their drawings and captions.

KINESTHETIC LEARNERS

WORD SCRAMBLE Use the pocket ABC cards and pocket chart from the Phonics and Word Building Kit to build the following scrambled words: ***der, izp, ropd, gdo, nur, pir, beraz, prid, gid, mozo, atr,*** and ***perzip***. Have children unscramble the letters to make words that begin with **/r/r, /d/d,** and **/z/z.**

KINESTHETIC LEARNERS

ACT OUT AN INTERVIEW Have pairs of children take turns acting out the parts of an interviewer and Rob from *Rob's Cab*. The interviewer should ask "Rob" questions about his job driving a cab. For example, what does Rob like and not like about the job? You may wish to tape-record children's interviews.

2

Dan and his dad ran to the cab.
Can they fit in the cab?

Zip, zip, zip!
The cab can zip up the hill.

3

Nan ran to the cab.
She had a fan,
a fat pig,
a big hat,
and a lid for a pan.
Can she fit in the cab?

Reflect On Reading

ASSESS COMPREHENSION

ONGOING ASSESSMENT To assess their understanding of the story, ask children questions such as:

- **Who is Rob? (*the cab driver*)**
- **What can fit in Rob's cab? (*A lot can fit, including two people, a dog, a pig, a fan, a hat, and a lid for a pan.*)**
- **Where did Nan's things fit? (*on top of the cab*)**

HOME-SCHOOL CONNECTION

HOME/SCHOOL CONNECTION Send home *Rob's Cab*. Encourage children to read the story to a family member. You may also wish to suggest that children play I Spy at home, in a supermarket, or while traveling by bus, car, or subway. Family members may spy an object that begins with ***d, r,*** or ***z,*** tell the sound they hear at the beginning of the object's name, and challenge one another to name the object.

WRITING CONNECTION

ROB'S BUS Have children write or dictate a story about Rob's bus. How would the story be different from *Rob's Cab*? Would all the passengers fit in the bus? Where would the bus go? Have children think about such questions, then work together to write or dictate the story.

CHALLENGE

Have children find words with **/r/r, /d/d,** or **/z/z** in the classroom or school. Children can count how many words they found and record the results of their counts in a chart such as the one shown.

4			
3			
2			
1			
	Words With **/r/r**	Words With **/d/d**	Words With **/z/z**

Phonics Connection

Phonics Readers:
#22, *Zzzzz*
#23, *Dig!*
#24, *Drip, Drop, Drip!*

EXTRA HELP

Children can draw a picture of Rob and Rob's cab in their town or neighborhood. Tell children to be sure to include special landmarks, stores, or signs in their pictures.

Lesson 64

pages 107-108

Recognize and Write /e/e

QUICKCHECK ✔

Can children:

✔ **orally segment words?**

✔ **write capital and small *Ee*?**

✔ **recognize /e/?**

✔ **identify the letter that stands for /e/?**

If YES go to Read and Write.

TEACH

Develop Phonemic Awareness

Oral Segmentation Explain to children that you are going to say a word. Children will tell you how many sounds they hear. For example, have children listen to ***rip***. Then say the word sound by sound: /r/ /i/ /p/. Guide children to understand that the word ***rip*** contains three separate sounds. Continue with the following words:

- get
- wet
- let
- set
- net
- bed
- ten
- led
- pet
- fed
- jet
- men
- pen
- red
- hen

Write the Letter Explain that ***e*** stands for /e/, the sound in the middle of the word ***bed***. Write ***Ee*** on the chalkboard. Point out the capital and small forms of the letter. Then model how to write the letter.

Have children write both forms of the letter in the air with their fingers. Volunteers can practice writing the letter on the chalkboard.

READ AND WRITE

Connect Sound-Symbol Write the word ***pet*** on the chalkboard, and have a volunteer circle the letter ***e***. Remind children that the letter ***e*** stands for /e/. Ask children to suggest other words that contain /e/. List these words on the chalkboard. Have volunteers circle the letter ***e*** in each one.

Complete Activity Pages Read aloud the directions on pages 107–108. Review each picture name with children.

Supporting All Learners

AUDITORY/VISUAL LEARNERS

WORD SEARCH Read aloud "Open Them, Shut Them" on page 27 in the *Big Book of Rhymes and Rhythms, 1A*. Have children find all the words with **/e/** in the rhyme. Write these words on the chalkboard. Invite volunteers to add other **/e/** words to the list.

VISUAL/KINESTHETIC LEARNERS

EGG CARTON EXCHANGE Give small groups of children an empty egg carton and twelve slips of paper. Have them write twelve words, each on a separate slip. Ask them to include words that contain **e**. Have children place one slip of paper in each egg carton cup. Groups can then exchange their egg cartons and practice reading each other's words.

VISUAL/AUDITORY LEARNERS

OUR PETS Have children draw and cut out pictures of their favorite pets or pets that they would like to have. Children can write or dictate names for their imaginary pets using words with **/e/**. Children can also write or dictate sentences about the pets. For example: ***My dog Jet likes to get wet.*** Display the pictures, names, and sentences on a bulletin board.

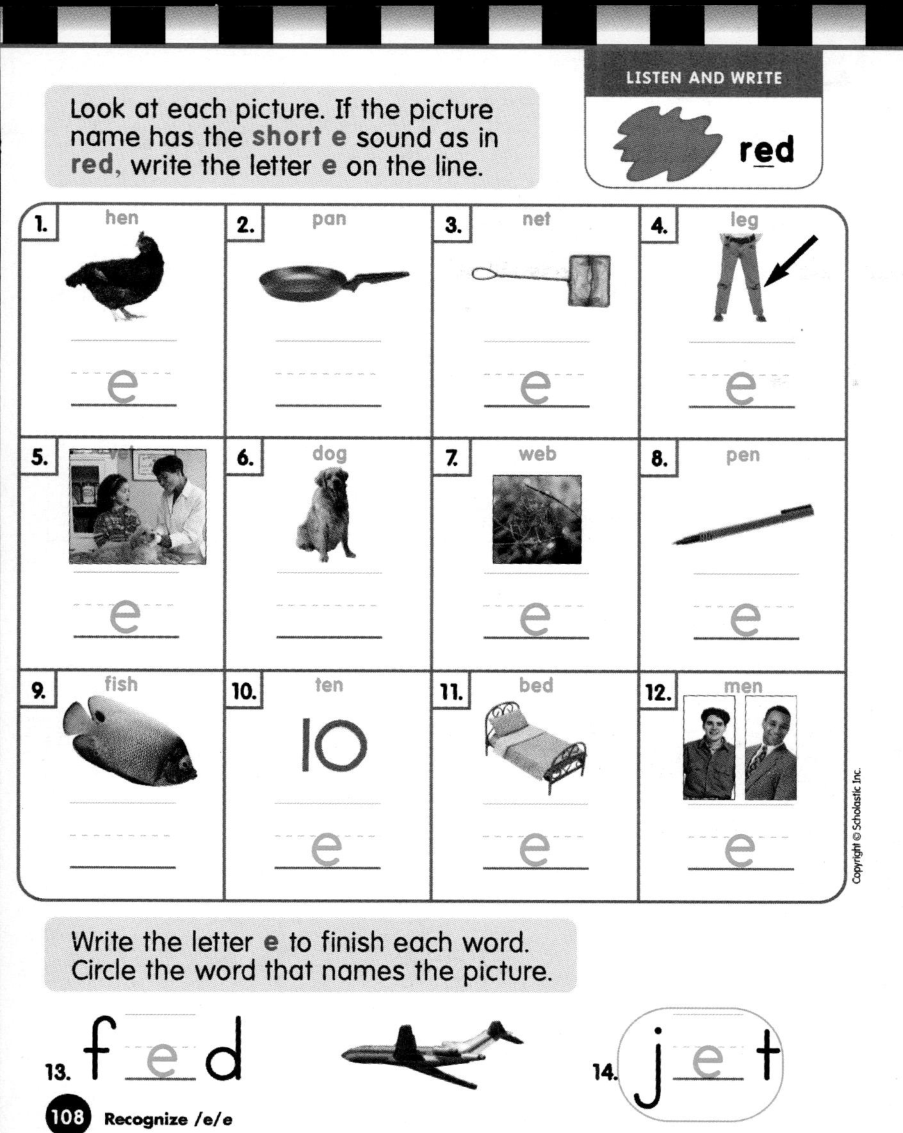
LISTEN AND WRITE

red

Look at each picture. If the picture name has the **short e** sound as in **red**, write the letter **e** on the line.

1. hen	2. pan	3. net	4. leg
e		e	e
5. vet	**6. dog**	**7. web**	**8. pen**
e		e	e
9. fish	**10. ten**	**11. bed**	**12. men**
	e	e	e

Write the letter **e** to finish each word. Circle the word that names the picture.

13. f e d 14. j e t

108 Recognize /e/e

Copyright © Scholastic Inc.

Integrated Curriculum

SPELLING CONNECTION

ADD AN *Ee* Write the following word parts on the chalkboard: ***p_n, m_t, b_d, j _ t,*** and ***h_n***. Have volunteers add the letter ***e*** to each one. Model for children how to blend each word. Have children say each word and write it on a sheet of paper. Then have children dictate sentences using these words.

SOCIAL STUDIES CONNECTION

COUNT TO TEN Have children collect ten ***/e/e*** words from books, magazines, and newspapers. Children can then share and compare their lists.

Phonics Connection

Literacy Place: *Team Spirit*
Teacher's SourceBook, pp. T56–57;
Literacy-at-Work Book, pp. 8–9

My Book: *Getting Wet*

Big Book of Rhymes and Rhythms, 1A: "Open Them, Shut Them," p. 27.

Chapter Book: *Fun With Zip and Zap*, Chapter 1

EXTRA HELP

Use one of the puppets in the Phonemic Awareness Kit to reinforce how to write the letter ***e***, to review that ***/e/*** is written as ***e***, and to model blending the word ***pet***.

Give each child acquiring English a self-sticking note with ***e*** written on it. Have children trace the ***e*** and then take turns attaching it to pictures such as the following that you have displayed randomly: **the number 10, jet, bed, a blotch of red, net, hen,** and **egg**. Then invite children to name all the pictures and tell what vowel sound they hear in each picture name.

Lesson 65

page 109

Blend Words With /e/e

QUICKCHECK ✔

Can children:

✓ orally blend word parts?

✓ identify /e/?

✓ blend words with /e/e?

If YES go to Read and Write.

TEACH

Develop Phonemic Awareness

Oral Blending Say the following word parts. Ask children to blend them. Provide corrective feedback and modeling when necessary.

/m/ /e/ /t/ /f/ /e/ /d/ /l/ /e/ /d/
/t/ /e/ /n/ /j/ /e/ /t/ /p/ /e/ /n/

Connect Sound-Symbol Review that *e* stands for /e/ as in ***bed***. Write ***bed*** on the chalkboard, and have a volunteer circle the ***e***. Model for children how to blend the word.

THINK ALOUD

I can put the letters b, e, and d together to make the word bed. Let's say the word slowly as I move my finger under the letters. Listen to how I string together the sound that each letter stands for to make the word: beeeeed, beed, bed.

Help children blend the words ***set*** and ***men***.

READ AND WRITE

BUILT-IN REVIEW

Blend Words To practice using the sound taught and to review previously taught sounds, list the following words and sentence on a chart. Have volunteers read each aloud. Model blending when necessary.

- **pen** **red** **wet**
- **jet** **men** **fell**
- **He fed the red hen.**

Complete Activity Page Read aloud the directions on page 109. Review the art with children.

Name

Look at each picture. Circle the word that best finishes each sentence. Then write the word on the line.

1.	He will pet the dog.	(pet) pit pot
2.	She is wet.	win (wet) met
3.	He will get the pan.	jet got (get)
4.	They met at the cab.	mat net (met)
5.	The dog likes to beg.	big (beg) bag
6.	Bill led the dog.	let lid (led)

Supporting All Learners

KINESTHETIC/VISUAL LEARNERS

PORTFOLIO

WORD INDEX Invite children to go on a scavenger hunt to collect things or pictures of things whose names contain the short ***e*** sound, such as ***pen, pet, hen, men, bed,*** and ***jet***. Display the collected items or pictures, and have children write or dictate the name of each on an index card.

The Book Shop

- **An Extraordinary Egg**
  by Leo Lionni
  A frog finds a beautiful pebble that hatches. **(/e/e)**
- **Emma's Pet**
  by David McPhail
  Imagination turns homework into construction activities. **(/e/e)**

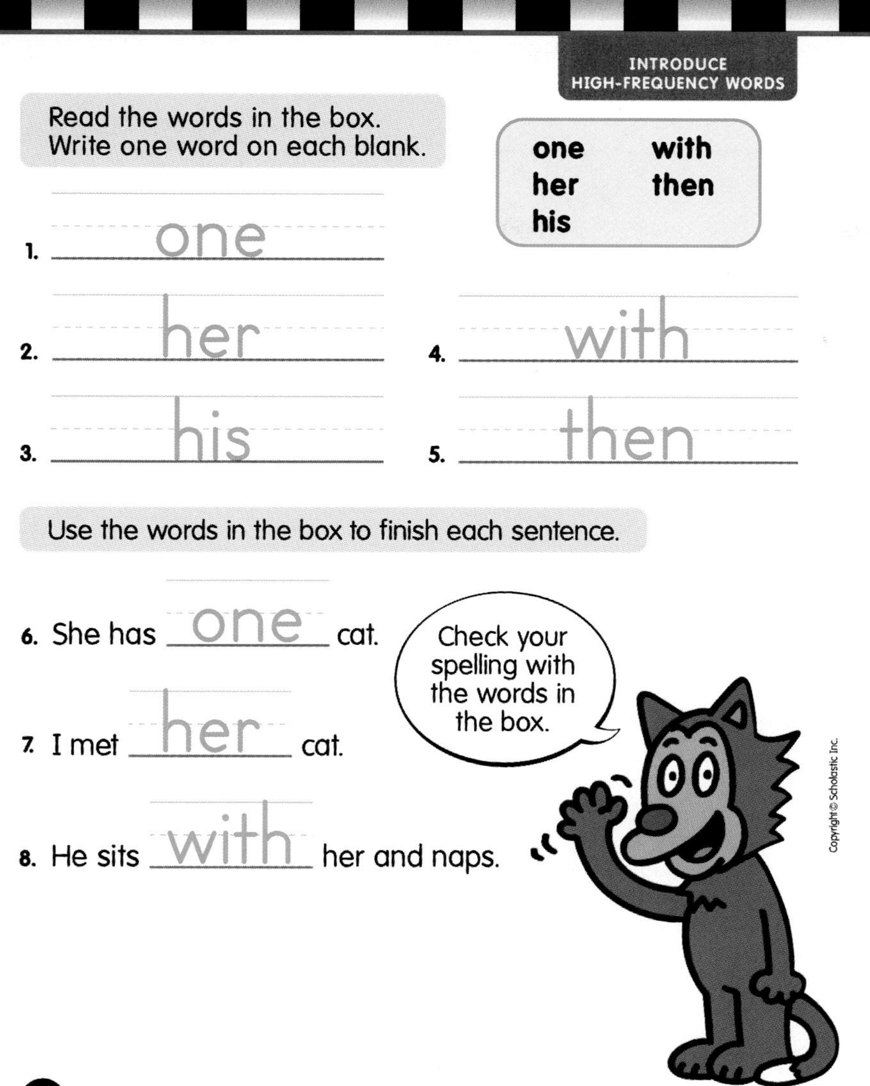
INTRODUCE HIGH-FREQUENCY WORDS

Read the words in the box.
Write one word on each blank.

one	with
her	then
his	

1. one
2. her
3. his
4. with
5. then

Use the words in the box to finish each sentence.

6. She has one cat.
7. I met her cat.
8. He sits with her and naps.

Supporting All Learners

AUDITORY/VISUAL LEARNERS

CLASS LIST Have children list the names of everyone in the class. Write the list on chart paper. Have volunteers point to a name and use it in a sentence with one or more high-frequency words.

EXTRA HELP

For children who need additional support reading the high-frequency words, use Phonics Reader #19, *Who Has a Bill?* to review the word ***with*** in context.

Relate new words to known words through the use of charts or other graphic organizers. Make a tree diagram with ***he*** at the top. On one side, list names of boys in the room. On the other, write ***his,*** and give examples of things boys in the class have, such as ***his red pen***. Then do the same with the words ***she*** and ***her***.

page 110

Recognize High-Frequency Words

QUICKCHECK ✔

Can children:

- ✔ **recognize and write the high-frequency words *one, her, his, with, then*?**
- ✔ **complete sentences using the high-frequency words?**

If YES go to Read and Write.

TEACH

Introduce the High-Frequency Words

Write the high-frequency words ***one, her, his, with,*** and ***then*** in sentences on the chalkboard. Read them aloud, underline the words, and ask children if they recognize them. You may wish to use these sentences:

1. **One cat sat with Ben.**
2. **His name is Sam.**
3. **One dog sat with Ben.**
4. **Her name is Pam.**
5. **Then the cat and dog ran.**

Ask volunteers to dictate sentences using the high-frequency words.

READ AND WRITE

Practice Write each high-frequency word on a note card. Read each aloud. Then do the following:

- **Mix the cards.**
- **Display one card at a time, and ask children to state each word aloud.**
- **Have children spell each word aloud, clapping on each letter.**
- **Ask children to write each word in the air as they state aloud each letter. Then have them write each word on a sheet of paper.**

Complete Activity Page Read aloud the directions on page 110. In the top section, children will write each word one time. In the bottom section, children will complete each sentence using one of the words in the box.

Lesson 67

pages 111-112

Recognize and Write /g/g

QUICKCHECK ✔

Can children:

✔ replace sounds in words?

✔ write capital and small ***Gg***?

✔ recognize /g/?

✔ identify the letter that stands for /g/?

If YES go to Read and Write.

TEACH

Develop Phonemic Awareness

Phonemic Manipulation Explain to children that you are going to read a list of words. Children are to replace the last sound in each word with /g/. For example, read the word ***bet***. Ask children to identify the last sound, /t/. Guide them to replace /t/ with /g/ and to say the new word ***beg***. Continue with the following words:

• win	• hot	• job
• bat	• dip	• bit
• led	• bud	• rat
• pin	• lot	• tan

Write the Letter Explain to children that the letter **g** stands for /g/, the sound they hear in the words *get* and *dog*. Write *Gg* on the chalkboard. Point out the capital and small forms of the letter. Then model for children how to write the letter.

Have children write both forms of the letter in the air with their fingers. You may also wish to have volunteers practice writing the letter on the chalkboard.

READ AND WRITE

Connect Sound-Symbol Write the words *got* and *dog* on the chalkboard. Have a volunteer circle the letter **g** in each word. Ask children to suggest other words that begin or end with /g/. List them on the chalkboard.

Complete Activity Pages Read aloud the directions on pages 111–112. Review each picture name with children.

HANDWRITING

Name

Goat begins with the g sound.

Gg

G G G

g g g

Get the big goat.

Write the Letter *Gg* 111

Supporting All Learners

AUDITORY/VISUAL LEARNERS

READ ABOUT *Gg* Read aloud "Gobble, Gobble" on page 28 in the *Big Book of Rhymes and Rhythms, 1A*. Have children find all the words with **/g/** in the rhyme. Write these words on the chalkboard. Invite volunteers to add other **/g/** words to the list.

KINESTHETIC/AUDITORY LEARNERS

GET TO THE LETTER Select four or five letter sounds, and write each letter on a large sheet of paper, one letter per page. Place the papers in a large play area, making sure each is at least two feet away from the next. Say a word that begins with one of the letter sounds, and guide children to run, jump, or hop to the appropriate letter. Then have children say a word with that letter.

LISTEN AND WRITE

Look at each picture. If the picture name begins with the **g** sound as in **goat**, write the letter **g** on the first line. If it ends in **g** as in **big**, write **g** on the second line.

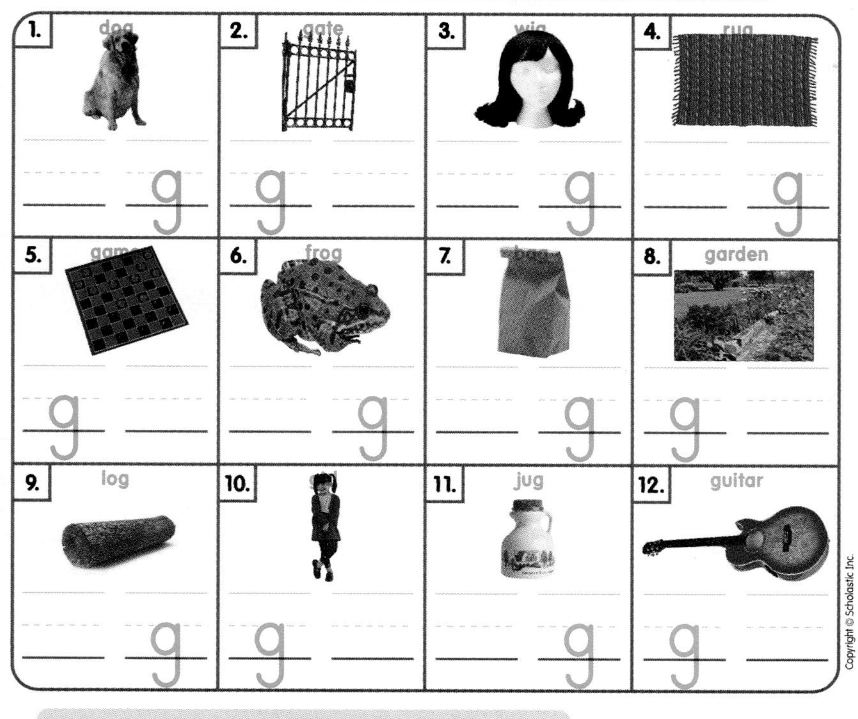

Write the letter **g** to finish each word. Circle the word that names the picture.

13. pi g

14. g et

Integrated Curriculum

SPELLING CONNECTION

MAKE A WORD Write the following word parts on index cards, one word part per card: ***_et, le_, ra_, lo_,*** and ***di_***. Place the word parts in a bag. Then have children take turns picking a word part and holding it up. Volunteers can hold up the letter card for ***g*** next to the word part. Have children say each new word that is made with ***g*** and write it on a sheet of paper.

SCIENCE CONNECTION

GOOD FOR GREEN Have children collect or draw pictures of things in nature that are green. Children can use the pictures to make a "Good for Green" poster. Have children write or dictate labels for the pictures on the poster. Then children can identify each word with ***/g/g***.

CHALLENGE

On a large note card, have children write the letter ***g*** and the key word ***goat***. Display the card on the Word Wall. Have children suggest words with ***/g/g***, write them on small note cards, and add them to the wall. Remind children to look for words that begin or end with ***g*** during their reading. Add these words to the wall.

EXTRA HELP

Help children find the words with **/g/** in the rhyme "Gobble, Gobble" on page 28 in the *Big Book of Rhymes and Rhythms, 1A*.

Phonics Connection

Literacy Place: *Team Spirit*
Teacher's SourceBook, pp. T58–59;
Literacy-at-Work Book, pp. 10–11

My Book: *Get It, Max!*

Big Book of Rhymes and Rhythms, 1A: "Gobble, Gobble," p. 28.

Chapter Book: *Fun With Zip and Zap,* Chapter 2

Lesson 68

pages 113-114

Blend and Build Words With /g/g

QUICKCHECK ✔

Can children:

- ✓ **orally blend word parts?**
- ✓ **identify /g/?**
- ✓ **blend and build words with /g/g and phonograms *-en, -et*?**

If YES go to Read and Write.

TEACH

Develop Phonemic Awareness

Oral Blending Say these word parts. Ask children to blend them. Offer corrective feedback.

/g/...et	**/g/...ot**	**/g/...ame**
/f/...og	**/r/...ag**	**/l/...eg**

Connect Sound-Symbol Review that g stands for /g/ as in *got*. Write *got* on the chalkboard. Circle g. Model how to blend the word. Then help children blend the words *gas* and *leg*.

Introduce the Phonograms Write *-en* and *-et* on the chalkboard. Point out the sounds the phonograms stand for. Add ***m*** to the beginning of ***-en***. Model how to blend the word. Have children repeat ***men*** as you blend the word again. Then add ***l*** to the beginning of ***-et***. Model how to blend ***let***. Then ask children to suggest words that rhyme with ***men*** and ***let***. List them on the chalkboard in separate columns. Have children underline ***-en*** or ***-et*** in each word.

READ AND WRITE

BUILT-IN REVIEW

Blend Words To practice using the sounds and phonograms taught, list the following on a chart. Have volunteers read each aloud. Model blending.

- **get** **men** **ten**
- **pet** **set** **sell**
- **Did the hen get wet?**

Complete Activity Pages Read aloud the directions on pages 113–114. Review each picture name with children.

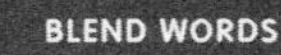

Name

Circle the word that names each picture. Then write the word on the line.

1. big / **(bag)** / beg — bag
2. **(leg)** / log / led — leg
3. dig / **(dog)** / dad — dog
4. end / **(egg)** / beg — egg
5. big / **(dig)** / dot — dig
6. wag / win / **(wig)** — wig

Use one of the words from above to finish each sentence.

7. My pen is in the bag.
8. Is the egg in the nest?

Supporting All Learners

KINESTHETIC LEARNERS

WORD SORT Write the following words on note cards: ***net, jet, ten, met, let, pen, set, get, hen, wet, bet,*** and ***pet***. Have children sort the words into two piles: words that contain ***-en*** and words that contain ***-et***. Then challenge children to think of other ways to sort the cards.

VISUAL LEARNERS

WORD COUNT Have children count how many word cards they have from the Word Sort activity. Then have children count how many ***-en*** words they have and how many ***-et*** words they have. Challenge children to use two ***-en*** words or two ***-et*** words in a sentence.

KINESTHETIC LEARNERS

BUILD WORDS Have the children use ABC cards and the pocket chart to make the word parts ***-en*** and ***-et***. Have children add a letter to the beginning of each phonogram to make a new word. Have children replace the initial consonant in the first word to build a second word.

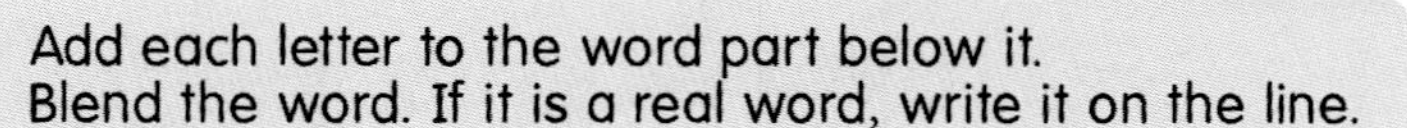

Add each letter to the word part below it.
Blend the word. If it is a real word, write it on the line.

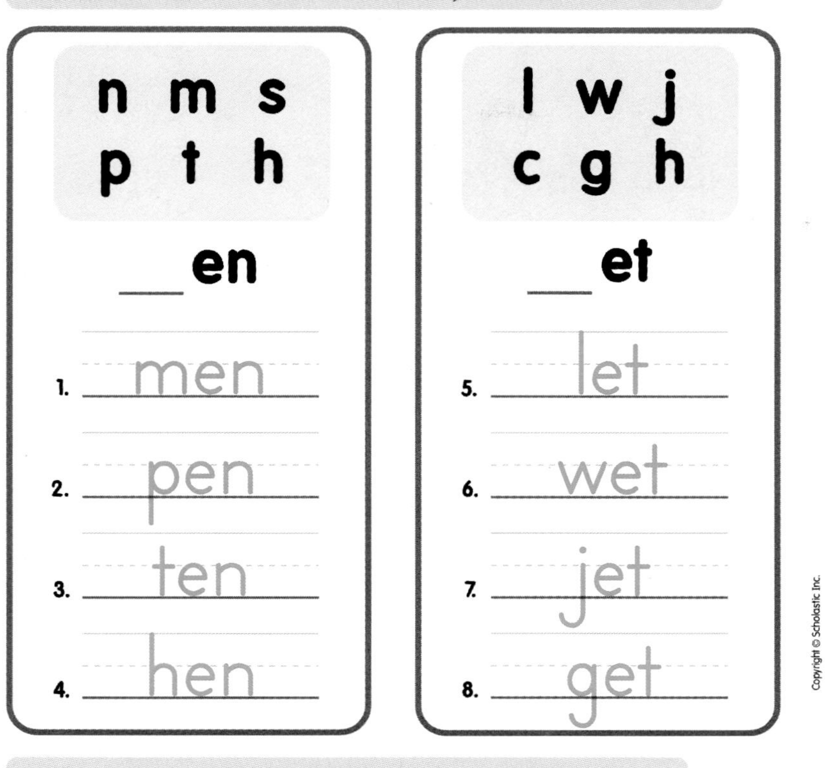

n m s p t h	l w j c g h
__en	__et
1. men	5. let
2. pen	6. wet
3. ten	7. jet
4. hen	8. get

Write a sentence using one of the words you made.

9. Answers will vary.

Integrated Curriculum

WRITING CONNECTION

STORY TIME Have children generate a list of words that begin with **/g/g**. You may wish to start the list with words such as ***game, get, leg,*** or ***rag***. Record the list of words on the chalkboard. Have children work together to create a story using as many of the **/g/g** words as possible. Write the dictated story on chart paper for group and individual reading. Have children draw illustrations for the story. Return to the story, rereading it in subsequent days.

ART CONNECTION

***Gg* POSTER** Have children draw pictures of things with names that have **/g/g** on a large sheet of paper. You may wish to suggest things such as goat, leg, game, or garden. Help children label their drawings. When the poster is finished, display it. Have volunteers take turns identifying the drawings and picture names. Have other volunteers identify the **/g/g** in each picture name. Invite children to revisit the poster in subsequent days to review the pictures, labels, and **/g/g**.

EXTRA HELP

For children needing additional phonemic awareness training, see Scholastic Phonemic Awareness Kit. The oral blending exercises will help children orally string together sounds to form words. This is necessary for children to be able to decode, or sound out, words while reading.

CHALLENGE

Have children write silly **/g/g** sentences. You may wish to provide examples such as ***Ben can get a green goat***. After children finish their sentences, they can circle words with **/g/g,** then exchange sentences with partners to look for more **/g/g** words.

- **Go, Dog, Go!**
  by Phil Eastman
  In this playful book, dogs have all kinds of fun! **(/g/g)**
- **The Golly Sisters Go West**
  by Betsy Byars
  Sisters have comic adventures in the West. **(/g/g)**

Lesson 69

pages 115-116

Recognize and Write /ks/x

QUICKCHECK ✔

Can children:

✔ orally segment words?

✔ write capital and small Xx?

✔ recognize /ks/?

✔ identify the letter that stands for /ks/?

If YES go to Read and Write.

TEACH

Develop Phonemic Awareness

Oral Segmentation Explain to children that you are going to say words in parts. Children will listen to the parts and then say the whole word. For example, say the sounds /g/ /e/ /t/. Guide children to say the whole word *get*. Continue with the following word parts:

/a/ /ks/	**/o/ /ks/**
/f/ /o/ /ks/	**/m/ /i/ /ks/**
/b/ /o/ /ks/	**/f/ /i/ /ks/**
/w/ /a/ /ks/	**/s/ /i/ /ks/**

Write the Letter Explain to children that the letter *x* stands for /**ks**/, the sounds at the end of the word *fox*. Write *Xx* on the chalkboard. Point out the capital and small forms of the letter. Then model for children how to write the letter.

Have children write both forms of the letter in the air with their fingers. You may also wish to have volunteers practice writing the letter on the chalkboard.

READ AND WRITE

Connect Sound-Symbol Write the word *wax* on the chalkboard, and have a volunteer circle the letter *x*. Remind children that the letter *x* stands for /**ks**/. Ask children to suggest other words that contain /**ks**/. List these words on the chalkboard. Have volunteers circle the letter *x* in each.

Complete Activity Pages Read aloud the directions on pages 115–116. Review each picture name with children.

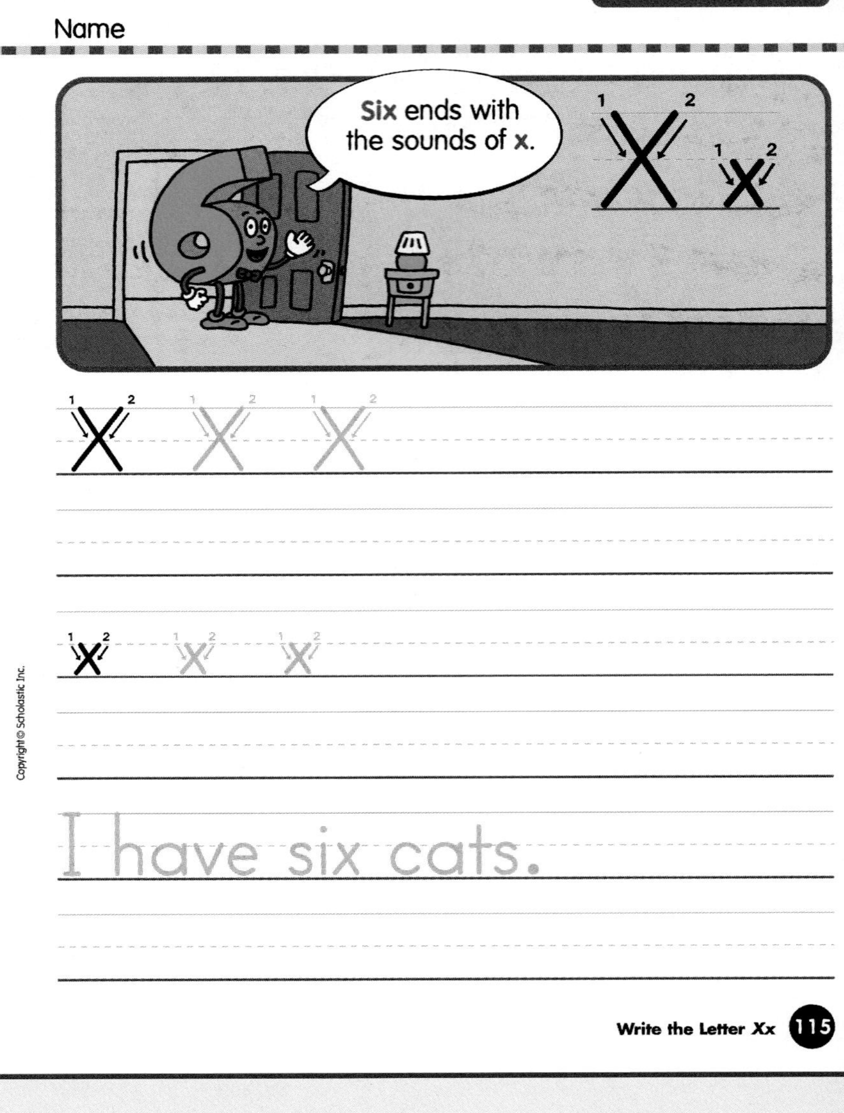

Supporting All Learners

AUDITORY/VISUAL LEARNERS

LETTER TIME Read aloud "Six Little Fishies" on page 29 in the *Big Book of Rhymes and Rhythms, 1A*. Have children find all the words with **/ks/** in the rhyme. Write these words on the chalkboard. Invite volunteers to add other **/ks/** words to the list.

KINESTHETIC/AUDITORY/VISUAL LEARNERS

GO FISH! GAME Make cards for the following sets of words: ***box, fix, fox, mix; wig, big, hog, dog; bad, did, dad, hid; cab, rob, bib, rib***; and ***fan, can, pin, win***. Invite partners to play a variation of Go Fish in which the goal is to get four word cards with the same ending sound. In this variation, children start with four cards. As in the usual game, children ask each other for cards and "go fish" from the remaining cards. When they have four of a kind, they must read the words and put the set down.

KINESTHETIC/VISUAL/AUDITORY LEARNERS

ALL ABOUT *Xx* Place the ABC cards for ***a, b, f, i, m, o, s,*** and ***x*** in the pocket chart. Make the word ***ax***. Model how you blend short ***a*** and ***x*** to make ***ax***. Then ask children to come to the pocket chart and make other **/ks/*x*** words using the ABC cards. Each time children make a word, orally blend it with them.

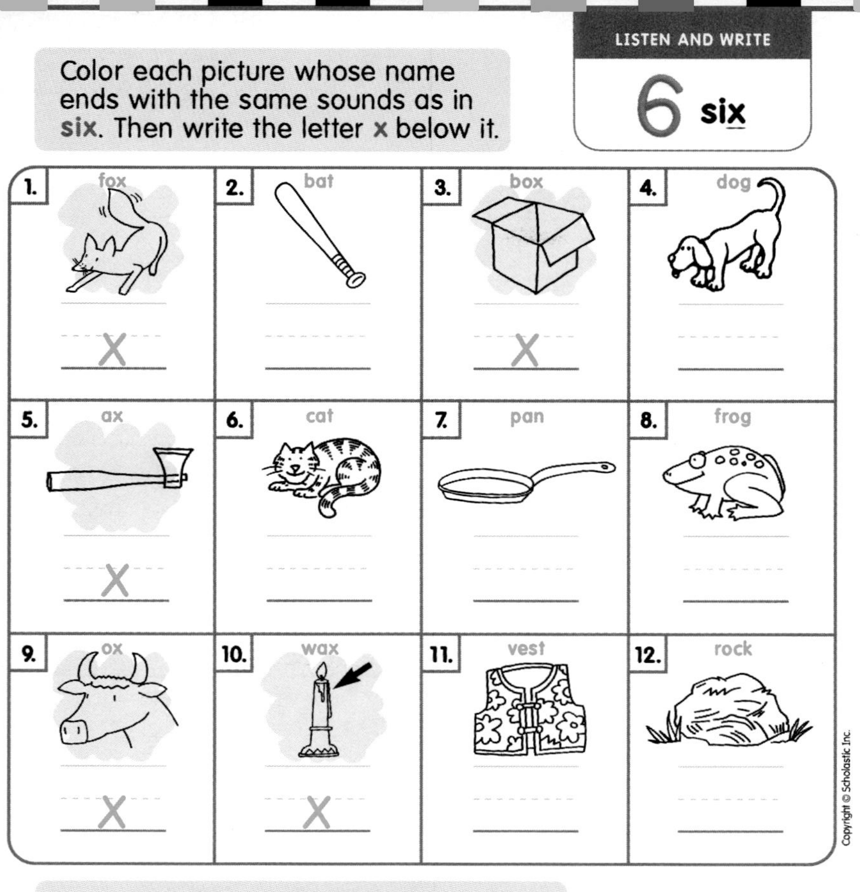
LISTEN AND WRITE

6 six

Color each picture whose name ends with the same sounds as in six. Then write the letter x below it.

1. fox	2. bat	3. box	4. dog
x		x	
5. ax	6. cat	7. pan	8. frog
x			
9. ox	10. wax	11. vest	12. rock
x	x		

Write the letter x to finish each word. Circle the word that names the picture.

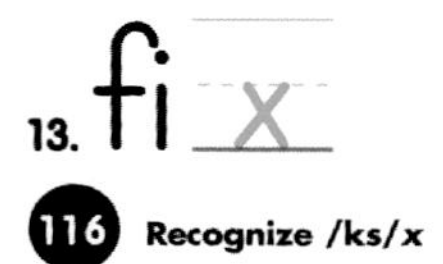
13. fi x

14. mi x

116 Recognize /ks/x

Integrated Curriculum

SPELLING CONNECTION

Xx* MARKS THE SPOT** Write these sentences on strips of paper, one sentence per strip: ***The fo__ ran from the dog; I put my hat in the bo__; We mi__ flour and water in a bowl; She can fi__ the car; Cut down the tree with an a__. Have children pick a sentence, copy it, add the letter ***x*** to the incomplete word, and illustrate it. Then have children take turns reading their sentences aloud. Display the sentences, and invite children to identify each word with **/ks/*x***.

SOCIAL STUDIES CONNECTION

WORD SEARCH Have children visit the school or local library to look through books, magazines, and newspapers for words with **/ks/*x***. Children can write the words they find on a poster, then illustrate some of the words.

Phonics Connection

Literacy Place: *Team Spirit*
Teacher's SourceBook, pp. T60–61;
Literacy-at-Work Book, pp. 12–13

My Book: *Fox and His Car*

Big Book of Rhymes and Rhythms, 1A: "Six Little Fishies," p. 29

Chapter Book: *Fun With Zip and Zap*, Chapter 2

ESL Invite children acquiring English to make a big ***X*** by crossing their arms in front of them. For each word they hear you say that ends with **/ks/**, they can make the ***X;*** otherwise, they just let their arms relax. Say words such as these: ***mix, ox, leg, fox, six, get, ax, fix, net, pen, box, dog,*** and ***wax***.

EXTRA HELP

Draw a large car out of construction paper, and call it an "X-mobile." Invite children to write words with ***x*** on self-sticking notes and to stick them on the car. Orally blend each word children write.

Lesson 70

page 117

Blend Words With /ks/x

QUICKCHECK ✔

Can children:

- ✓ **orally blend word parts?**
- ✓ **identify /ks/?**
- ✓ **blend words with /ks/x?**

If YES go to Read and Write.

TEACH

Develop Phonemic Awareness

Oral Blending Say the following word parts. Ask children to blend them. Provide corrective feedback and modeling when necessary.

o.../ks/	**a.../ks/**	**fo.../ks/**
mi.../ks/	**bo.../ks/**	**si.../ks/**

Connect Sound-Symbol Review with children that ***x*** stands for **/ks/** as in ***fix***. Write ***fix*** on the chalkboard, and have a volunteer circle the ***x***. Then model how to blend the word.

THINK ALOUD

I can put the letters *f, i,* and *x* together to make the word *fix*. Let's say the word slowly as I move my finger under the letters.

Help children blend the words ***six*** and ***box***.

READ AND WRITE

BUILT-IN REVIEW

Blend Words To practice using the sound taught and to review previously taught sounds, list the following words and sentence on a chart. Have volunteers read each aloud. Model blending when necessary.

- **fix** **fox** **box**
- **ox** **mix** **six**
- **The mix is in the box.**

Complete Activity Page Read aloud the directions on page 117. Review each picture name with children.

Name

Circle the word that names each picture.
Then write the word on the line.

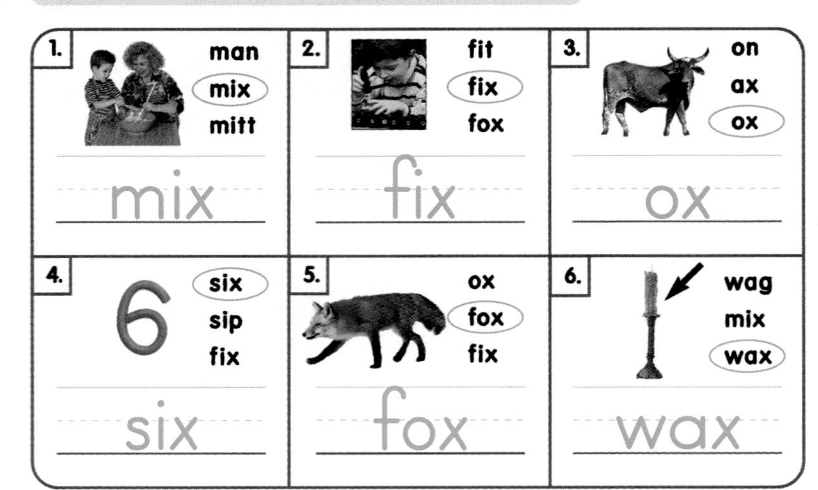

Write the word that names each picture.

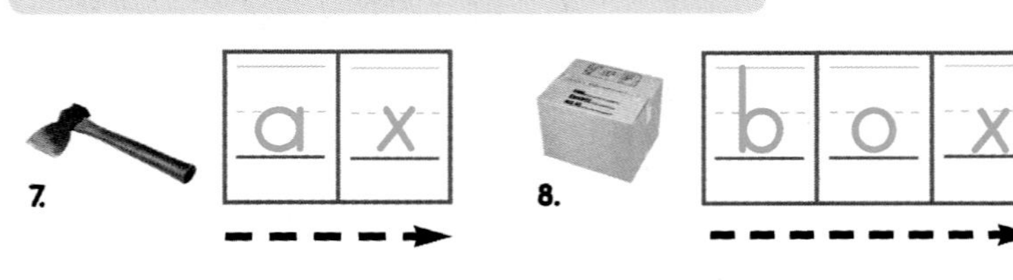

Supporting All Learners

TACTILE LEARNERS

TIC-TAC-TOE Draw a large tic-tac-toe grid on chart paper. Have children make ***x***'s and ***o***'s out of clay. Then have children play tic-tac-toe with a twist. Each time a child places an ***x*** on the grid, that child must say a word with ***x;*** each time a child places an ***o*** on the grid, that child must say a word with ***o***. The first child to place three ***x***'s or ***o***'s on the grid and to say three ***x*** or ***o*** words, wins the game.

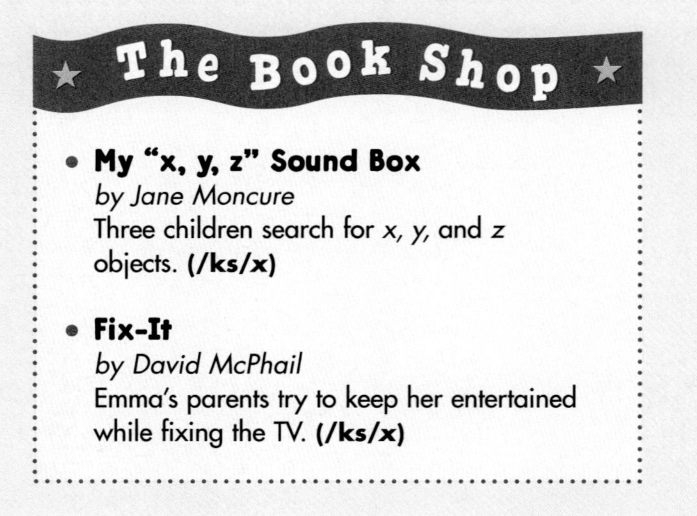

The Book Shop

- **My "x, y, z" Sound Box**
  by Jane Moncure
  Three children search for *x, y,* and *z* objects. **(/ks/x)**
- **Fix-It**
  by David McPhail
  Emma's parents try to keep her entertained while fixing the TV. **(/ks/x)**

POETRY

Read the poem. Write a sentence about one of the items in the box.

IN THE BOX

In the box I had
six cats,
six dogs,
six hats,
and
six logs,

six wigs,
six nets,
six pigs,
and
six jets.

In the box I had
a lot!

Answers will vary.

Supporting All Learners

AUDITORY LEARNERS

LEARNING CENTER For additional practice, use Center 5, "Rhyme Time," Activity 2 from *Quick-and-Easy Learning Centers: Phonics*. To help children match rhyming words, recognize spelling patterns, and emphasize sound-spelling relationships, adapt the activity for use with this poem.

CHALLENGE

Have children generate a list of words that begin or end with **/e/e, /g/g,** and **/ks/x**. Record these words on the chalkboard. Then have children create a story using as many of these words as they can. Write the dictated story on chart paper for group and individual reading. Return to the story, rereading it in subsequent lessons.

Write and Read Words With /e/e, /g/g, /ks/x

QUICKCHECK ✔

Can children:

✔ **spell words with /e/e, /g/g, /ks/x?**

✔ **read a poem?**

If YES go to Read and Write.

TEACH

Link to Spelling Review the following sound-spelling relationships: /*e*/*e*, /*g*/*g*, /**ks**/*x*. Say one of these sounds. Have a volunteer write the spelling that stands for the sound on the chalkboard. For example, ***x*** stands for /**ks**/. Continue with all the sounds. You may also wish to review /**d**/*d*, /**r**/*r*, and /**z**/*z*.

Phonemic Awareness Oral Segmentation
Say *get*. Have children orally segment the word. (/*g*/ /*e*/ /*t*/) Ask them how many sounds it contains. *(3)* Draw three connected boxes on the chalkboard. Have a volunteer write the spelling that stands for each sound in the word ***get*** in the appropriate box. Continue with ***leg*** and *fox*.

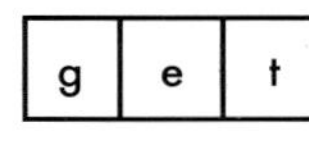

READ AND WRITE

Dictate Dictate the following words and sentence. Have children write them. When children are finished, write the words and sentence on the chalkboard. Have children make any necessary corrections.

- **fix** **box** **let**
- **with** **one** **can't**
- **The fox got wet.**

Complete Activity Page Read aloud the directions on page 118. Children can read the poem independently or with partners.

Lesson 72

pages 119-120

Read and Review Words With /e/e, /g/g, /ks/x

QUICKCHECK ✔

Can children:

✔ **read words with /e/e, /g/g, /ks/x?**

✔ **recognize high-frequency words?**

If YES go to Read and Write.

TEACH

Review Sound-Spellings Review the sound-spelling relationships from the past few lessons. These include /e/e, /g/g, and /ks/x. Say one of these sounds. Have a volunteer write the spelling that stands for the sound(s) on the chalkboard. For example, *e* stands for /e/. Continue with all the sounds. Then display pictures of objects whose names begin or end with one of these sounds. Have children write the letter that each picture name begins or ends with as the picture is displayed.

Review High-Frequency Words Review the high-frequency words *one, her, his, with,* and *then*. Write sentences on the chalkboard containing each high-frequency word. Say one word, and have volunteers circle it. Have children generate additional sentences for each word.

READ AND WRITE

Build Words Distribute these letter cards: *b, d, e, f, g, h, i, n, o, t,* and *x*. If children have their own set of cards, have them locate it. Have children build as many words as possible using the letter cards. Suggest that they record their words on a separate sheet of paper.

Build Sentences Write the following words on note cards: *I, his, her, fix, box, got, the, pen, in, is,* and *get*. Display the cards. Have children make sentences using the words. To get them started, suggest the sentence *The pen is in the box*.

Complete Activity Pages Read aloud the directions on pages 119–120. Review the art with children.

REVIEW

Name

Check each word as you read it to a partner. Circle any words you need to practice.

I can read!

☐ get	☐ fix	☐ bag
☐ leg	☐ beg	☐ men
☐ hen	☐ pen	☐ ten
☐ led	☐ fed	☐ bed
☐ red	☐ box	☐ six

Lookout Words!

☐ one	☐ her	☐ his	☐ with	☐ then
☐ they	☐ said	☐ to	☐ she	☐ up

Supporting All Learners

VISUAL LEARNERS

WORD WALL Invite children to find as many I Can Read! words on the Word Wall as they can. Ask children to add to the Word Wall any I Can Read! words that do not already appear.

KINESTHETIC LEARNERS

WORD PUZZLES Write +, –, = on cards, one sign per card. Have children use these cards and the pocket ABC cards and pocket chart from the Phonics and Word Building Kit to make word puzzles with ***e, g,*** and ***x***. Children can write the puzzles they make. Use the following puzzle as a model:

f o x – x + g = fog

TEST YOURSELF

Fill in the bubble next to the sentence that tells about each picture.

1.	○ See the six men hop. ● Six men sit.
2.	○ Ned will get the box. ● The fox sits in a box.
3.	○ She had to mix it. ● She can fix the pen.
4.	● Ben got into his bed. ○ Ben led the men to the pen.
5.	○ The man had ten hens. ● The hen sat on her eggs.
6.	○ He ran with his pet. ● He ran with his net.

You're becoming a great reader!

Integrated Curriculum

WRITING CONNECTION

ONCE UPON A BOX Have children use words with ***/e/e, /g/g,*** and ***/ks/x*** as well as the high-frequency words to write or dictate sentences about things they would like to find in a box or put in a box. Record the sentences on chart paper. Have children revisit the sentences to identify words with ***/e/e, /g/g, /ks/x,*** and the high-frequency words.

TECHNOLOGY CONNECTION

For additional practice with ***/e/e, /g/g, /ks/x,*** have children read the following My Books: *Getting Wet; Get It, Max; and Fox and His Car*. For information on using the My Books on the computer, see the WiggleWorks Plus My Books Teaching Plans. You may also wish to have children use words from I Can Read! to write sentences or stories on the computer using the WiggleWorks writing area. See the WiggleWorks Plus Teaching Plan for ideas.

I CAN READ! OPTIONS

The I Can Read! page can be used for one or all of the following:

- **paired reading**
- **individual assessment**
- **choral reading**
- **homework practice**
- **program placement**

CHALLENGE

Children can make a poster of animals with names that have ***/e/e, /g/g,*** or ***/ks/x***. Have children draw or cut out pictures of these animals. Then children can write or dictate a sentence about each animal.

To help children acquiring English participate with greater confidence in the Review activities, go over the key letter sounds, decodable words, and high-frequency words beforehand. For the Build Words and Build Sentences activities, you may wish to pair these children with those whose primary language is English, then let the same partners work together to complete the pages.

EXTRA HELP

Help children look for words with ***/e/e, /g/g,*** and ***/ks/x*** on signs, labels, and books in the school or neighborhood library.

Lesson 73

pages 121-122

Read Words in Context

TEACH

Assemble the Story Ask children to remove pages 121–122. Have children fold the pages in half to form the Take-Home Book.

Preview the Story Preview *Big Ben and His Ox,* a story about teamwork. Have a volunteer read aloud the title. Invite children to browse through the first two pages of the story and to comment on anything they notice. Suggest that children point out any unfamiliar words. Read these words aloud. Model how to blend them. Then have children predict what they think the story might be about.

READ AND WRITE

Read the Story Read the story aloud, or have volunteers take turns reading aloud a page at a time. Discuss with children anything of interest on each page. Ask them to help each other with any blending difficulties. The following prompts may help children who need extra support while reading:

- **What letter sounds do you know in the words?**
- **Are there any word parts you know?**
- **What do the pictures tell you about the story?**

Reflect and Respond Have children share their reactions to the story. What did they like best about the story? What parts do they think really could—or could not—happen, and why? Encourage children to write another page that could be added to the book.

Develop Fluency You may wish to reread the story as a choral reading, or have partners reread the story independently. Provide time for children to reread the story on subsequent days to develop fluency and increase reading rate.

Supporting All Learners

KINESTHETIC/VISUAL LEARNERS

PICK A CARD Shuffle a small set of letter cards, including ***e, g,*** and ***x***. Turn the pile facedown. Have children take turns picking cards and calling out the letter on each. If children pick a letter card for ***e, g,*** or ***x,*** they should say a word with **/e/e, /g/g,** or **/ks/x**. Write the words children say on the chalkboard or on chart paper.

KINESTHETIC LEARNERS

ACT IT OUT Invite children to act out the story of *Big Ben and His Ox*. Have children read the story together, then take turns acting out the parts of Ben and Jen. You may wish to have children make props and act out the story for the class.

TACTILE LEARNERS

BEN AND JEN Children can use clay to mold the figures of Ben and Jen. Then have children mold the letters of Ben's and Jen's names out of clay. After children spell out each name with the clay letters, children can display each character with the appropriate name.

CHALLENGE

Have children collect and count words with **/e/e, /g/g,** and **/ks/x.** Children can record the results of their counts in a chart such as the one shown.

3			
2			
1			
	Words With **/e/e**	Words With **/g/g**	Words With **/ks/x**

EXTRA HELP

Read aloud *Big Ben and His Ox* with children. Then read it again, asking children to stop you whenever you read a word with **/e/, /g/,** or **/ks/**. Write the words children identify on the chalkboard. Have volunteers underline the ***e, g,*** or ***x*** in each word.

Reflect On Reading

ASSESS COMPREHENSION

To assess their understanding of the story, ask children questions such as:

- **Who is Big Ben?** ***(a big man with a big ax)***
- **What does Big Ben want to do?** ***(He wants to get ten logs up the hill.)***
- **How does Jen help Ben?** ***(Jen, the ox, helps Ben to pull the logs up the hill.)***

HOME-SCHOOL CONNECTION

Send home *Big Ben and His Ox.* Encourage children to read the story to a family member.

WRITING CONNECTION

JEN AND BEN Have children write or dictate sentences about another adventure Ben and Jen might have together or another problem they might solve. How else could Jen help Ben?

Phonics Connection

Phonics Readers:
#25, Let's Grow Them
#26, Max's Pet

Lesson 74

pages 123-124

Recognize and Write /k/k

QUICKCHECK ✔

Can children:

- ✔ **substitute sounds in words?**
- ✔ **write capital and small *Kk*?**
- ✔ **recognize /k/?**
- ✔ **identify the letter that stands for /k/?**

If YES go to Read and Write.

TEACH

Develop Phonemic Awareness

Phonemic Manipulation Explain to children that you are going to read a list of words. Children are to replace the last sound in each word with **/k/**. For example, read the word ***sit***. Ask children to identify the last sound, **/t/**. Guide them to replace **/t/** with **/k/** and to say the new word they form: ***sick***. Continue with the following words.

• pin	• lip	• bag
• pat	• sat	• dog
• rot	• but	• log

Write the Letter Explain that ***k*** stands for **/k/**, the sound they hear at the beginning of ***king*** or at the end of ***book***. Ask children what other letter can stand for **/k/**. ***(c)*** Then write ***Kk*** on the chalkboard. Point out the capital and small forms of the letter. Then model how to write the letter.

Have children write both forms of the letter in the air with their fingers. You may also wish to have volunteers practice writing the letter.

READ AND WRITE

Connect Sound-Symbol Write ***kick*** on the chalkboard, and have a volunteer circle ***ck***. Tell children that ***ck*** also stand for **/k/**. Ask children to suggest other words that begin or end with **/k/**. List these words on the chalkboard. Have volunteers circle ***k*** or ***ck*** in each.

Complete Activity Pages Read aloud the directions on pages 123–124. Review each picture name with children.

HANDWRITING

Name

Kick begins and ends with the k sound.

K k

K K K

k k k

Kick the ball.

Write the Letter *Kk* 123

Supporting All Learners

AUDITORY/VISUAL LEARNERS

HURRY FOR *Kk*! Read aloud "One, Two" on pages 30–31 in the *Big Book of Rhymes and Rhythms, 1A*. Have children find all the words with **/k/** in the rhyme. Write these words on a "Hurray for Kk" poster or bulletin board. Invite volunteers to add other **/k/** words to the list.

KINESTHETIC/VISUAL LEARNERS

PICTURE SORT Distribute the following picture cards from the Phonemic Awareness Kit: **book, kite, king, duck, sock**. Have children sort the cards according to whether they hear **/k/** at the beginning or at the end of each picture name. Then give children the **cake** picture card, and ask them why it may go in either pile.

VISUAL/AUDITORY LEARNERS

FROM WORD PARTS TO WORDS Have children write the following word parts: ***pi__, li__, ba__, lo__, du__***. Children can add the letters ***ck*** to each. Model how to blend each word. Then have children say each word.

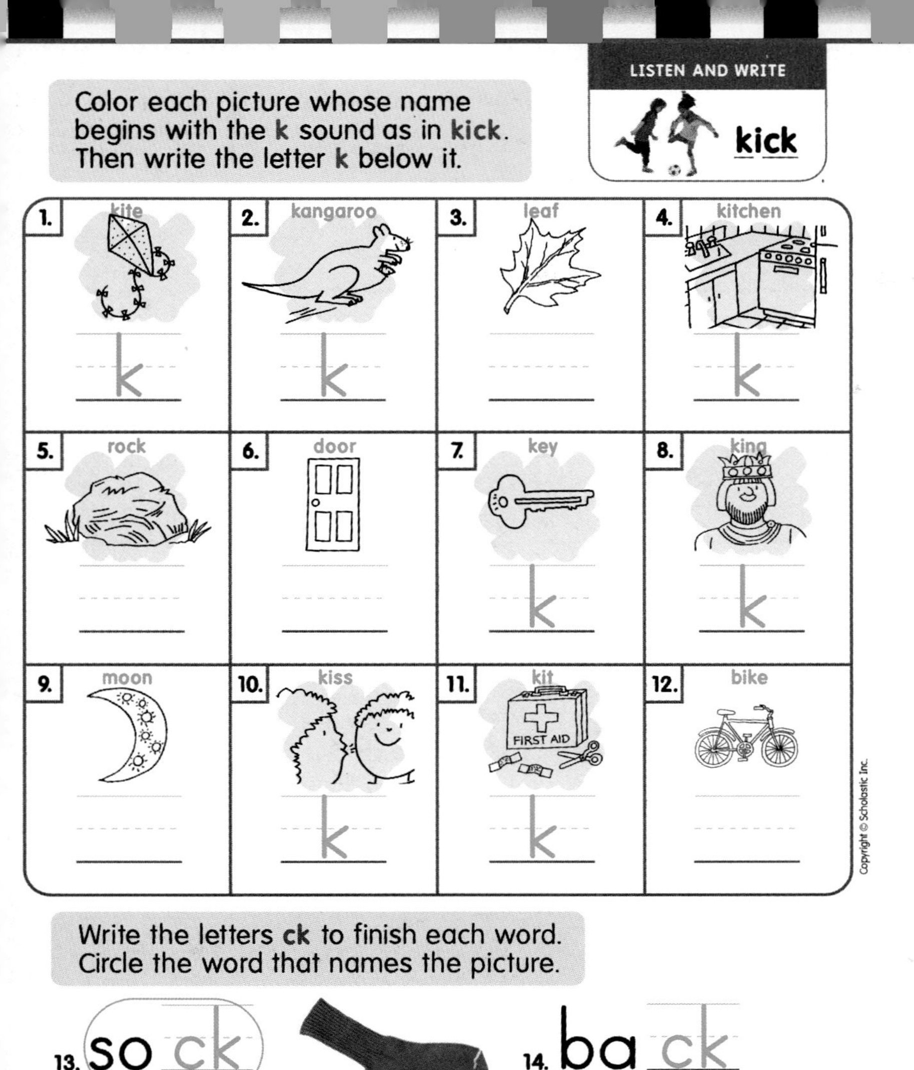

Color each picture whose name begins with the k sound as in kick. Then write the letter k below it.

LISTEN AND WRITE

kick

1. kite — k
2. kangaroo — k
3. leaf
4. kitchen — k
5. rock
6. door
7. key — k
8. king — k
9. moon
10. kiss — k
11. kit — k
12. bike

Write the letters ck to finish each word. Circle the word that names the picture.

13. so ck

14. ba ck

124 Recognize /k/k, ck

Copyright © Scholastic Inc.

Integrated Curriculum

WRITING CONNECTION

RHYME TIME Have children write or dictate a list of words with **/k/k, ck**. Then write a silly rhyming sentence such as the following on the chalkboard: ***The duck went to the dock, looking for a rock.*** Read it aloud. Then have children change the sentence by replacing one or more of the **/k/k, ck** words, or make up new sentences using the words from the list.

SCIENCE CONNECTION

K* ANIMALS** Have children write or talk about kittens, ducks, and kangaroos. Then have children draw pictures of the animals they talked about. Children can write or dictate the name of each animal and what they know about it. Give children an opportunity to share their ***k-animals with each other.

EXTRA HELP

Display the ABC card for ***Kk***. Have children whose names contain ***Kk*** write their names. Then have children suggest other **/k/** words to add to the card. Display the card for future reference when reading and writing. For additional practice, have children use Center 1, What's in a Name? from *Quick-and-Easy Learning Centers: Phonics*. The reproducible masters for Name Rhymes, Name Scramble, and Name Twisters focus on rhymes and oddity tasks.

CHALLENGE

Have children write or dictate **/k/k, ck** tongue twisters. You may wish to display the following **/k/** tongue twister as a model: ***Black ducks pack rocks.*** Have children share their tongue twisters, and challenge each other to say them ten times fast.

Phonics Connection

Literacy Place: *Team Spirit*
Teacher's SourceBook, pp. T102–103; Literacy-at-Work Book, pp. 20–21

My Book: *Hen and Duck*

Big Book of Rhymes and Rhythms, 1A: "One, Two," pp. 30–31

Chapter Book: *Fun With Zip and Zap*, Chapter 4

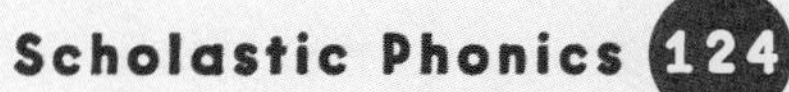

Lesson 75

pages 125-126

Blend and Build Words With /k/k

QUICKCHECK ✔

Can children:

✔ **orally blend word parts?**

✔ **identify the sound the letters *k* and *ck* stand for?**

✔ **blend and build words with /k/k?**

If YES go to Read and Write.

TEACH

Develop Phonemic Awareness

Oral Blending Say these word parts. Ask children to blend them. Provide corrective feedback.

/k/...it	/k/...ite	/k/...angaroo
/k/...eep	/k/...ick	/k/...ind

Connect Sound-Symbol Review that *k* and *ck* stand for /k/ as in ***kick***. Write ***kick*** on the chalkboard. Have a volunteer circle the *k* at the beginning and the *ck* at the end. Then model how to blend the word.

I can put *k*, *i* and *ck* together to make *kick*. Let's say it slowly as I move my finger under the letters.

READ AND WRITE

Blend Words To practice using the sounds taught, have volunteers read each word aloud. Model blending as needed.

- **kit** **kick** **sick**
- **pick** **lock** **sack**
- **I got my sock back.**

Build Words Distribute these letter cards: *a, b, c, i, k, l, s, o, p*. If children have their own set of cards, have them locate it. Provide time for children to build as many words as possible using the letter cards.

Complete Activity Pages Read aloud the directions on pages 125–126. Review the art with children.

Name

Look at each picture. Circle the word that best finishes each sentence. Then write the word on the line.

	Sentence	Choices
1.	She will kick it.	sock, pick, (kick)
2.	He is sick.	sack, (sick), sock
3.	Jack can pack the bag.	sack, pick, (pack)
4.	Kim ran to the rock.	dock, back, (rock)
5.	He will put the kit back.	pack, (back), box
6.	Dad will lock it up.	pick, lick, (lock)

Supporting All Learners

KINESTHETIC/VISUAL LEARNERS

WORD WALL On a large note card, write **k** and **ck** and the key word ***kick***. Display the card on the Word Wall. Have children suggest words with **/k/**, write them on small note cards, and add them to the wall. Remind children to look for words that begin with **k** or end with **ck** during their reading. Add these words to the wall.

AUDITORY LEARNERS

I HEAR *Kk*! Have children write ***Kk*** on sheets of paper, one sheet per child. Then read aloud a story or stories such as those highlighted under The Book Shop. Have children hold up their ***Kk's*** whenever they hear a word with **/k/**. Ask a volunteer to point out the word with ***/k/k, ck***. Then children can repeat the word.

VISUAL LEARNERS

LADDER CLIMB Write the first and last words of a three-word ladder on the chalkboard. Children can change one letter or letter pair in each word to complete the word ladder. Have children write the new word on the chalkboard. Use the following words.

- **sick** **(sock)** **rock**
- **pig** **(pick)** **lick**

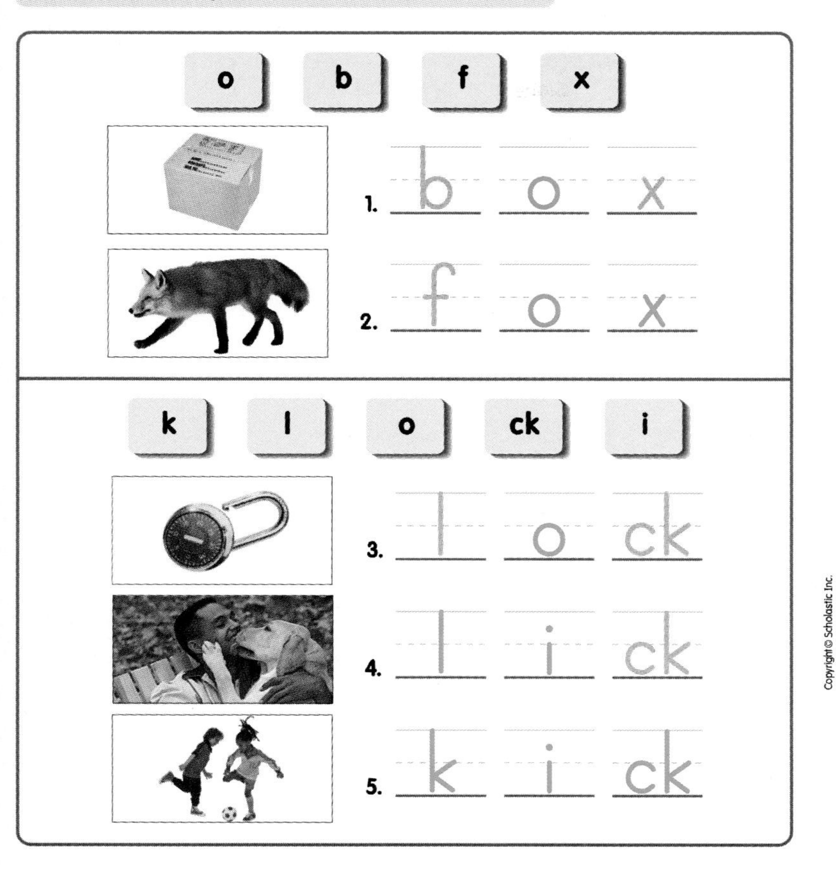
Look at each picture. Use the letter tiles to write each picture name.

o b f x

1. b o x
2. f o x

k l o ck i

3. l o ck
4. l i ck
5. k i ck

Integrated Curriculum

SPELLING CONNECTION

BUILD WORDS Write ***kit*** on the chalkboard, and model blending. Then ask a volunteer to blend the word. Tell children to replace **/k/** with **/b/** to make a new word. Blend the new word. Continue by changing one letter at a time to build more new words. For example:

bit
big
bag
rag

SOCIAL STUDIES CONNECTION

FIND /k/*k*, *ck* WORDS You may wish to have children look in newspapers and magazines for words with **/k/*k*, *ck***. Children can collect these words on a ***Kk*** poster or in a ***Kk*** book.

EXTRA HELP

For children needing additional phonemic awareness training, see the Scholastic Phonemic Awareness Kit. The oral segmentation exercises will help children to break apart words sound by sound. This is necessary for children to be able to encode, or spell, words while writing.

Display pictures of objects whose names begin and end with **/k/**. These pictures may include the following **rock, kick, kite, kangaroo, clock,** and **sock**. Prompt children acquiring English to take turns identifying each picture name and dictating sentences with these words. Write the sentences on the chalkboard, and help children to read them chorally.

- **Kitten Can**
  by Bruce McMillan
  This book shows the things a kitten can do. **(/k/*k*)**

- **Cookie's Week**
  by Cindy Ward
  Cookie creates mischief for every day of the week. **(/k/*k*)**

Lesson 76

pages 127-128

Recognize and Write /u/u

QUICKCHECK ✔

Can children:

✔ **write capital and small *Uu*?**

✔ **recognize /u/?**

✔ **identify the letter that stands for /u/?**

If YES go to Read and Write.

TEACH

Develop Phonemic Awareness

Oddity Task Explain to children that you will say three words and hold up a picture of each. (If available, use the picture cards in the Phonemic Awareness Kit. If not, paste pictures to index cards.) Two of the words have the same middle sound; the other does not. Children are to choose the word that does not belong. For example, you may hold up the picture cards for **cup, bus,** and **pig,** and say the words. Point out that ***cup*** and ***bus*** have the same middle sound; ***pig*** does not. Therefore, ***pig*** does not "belong." Continue with the following word lists.

- **nut** **mop** **run**
- **bun** **wig** **tub**
- **hat** **sun** **duck**

Write the Letter Explain that ***u*** stands for **/u/,** the sound in the middle of ***nut*** and ***run***. Write ***Uu*** on the chalkboard. Point out the capital and small forms of the letter. Then model how to write the letter. Have children write both forms of the letter in the air. You may also wish to have volunteers practice writing the letter on the chalkboard.

READ AND WRITE

Connect Sound-Symbol Write ***dug*** on the chalkboard. Have a volunteer circle the ***u***. Remind children that ***u*** stands for **/u/**. Ask children to suggest other words that contain **/u/**. List them on the chalkboard. Have volunteers circle the ***u*** in each.

Complete Activity Pages Read aloud the directions on pages 127–128. Review each picture name with children.

HANDWRITING

Name

Umbrella begins with the **short** u sound.

Uu

U U U

u u u

Up goes the umbrella.

Supporting All Learners

VISUAL/AUDITORY LEARNERS

***Uu* TOO** Read aloud "See a Penny" on page 32 in the *Big Book of Rhymes and Rhythms, 1A*. Have children find all the words with **/u/** in the rhyme. Write these words on the chalkboard. Invite volunteers to add other **/u/** words to the list.

KINESTHETIC LEARNERS

SPIN IT GAME To practice reading words with short ***u***, children may enjoy playing "Spin It!" on pages 13–15 in *Quick-and-Easy Learning Games: Phonics*.

AUDITORY/VISUAL LEARNERS

WHERE ARE THE *Uu*'s? Have children listen to "Twinkle, Twinkle Little Star" and "Wheels on the Bus" on the Sounds of Phonics Audiocassette (Managing Information). Invite children to identify all the words they hear that contain **/u/** in each song. Write the words that children name on a chart, and then read the list emphasizing **/u/** in each.

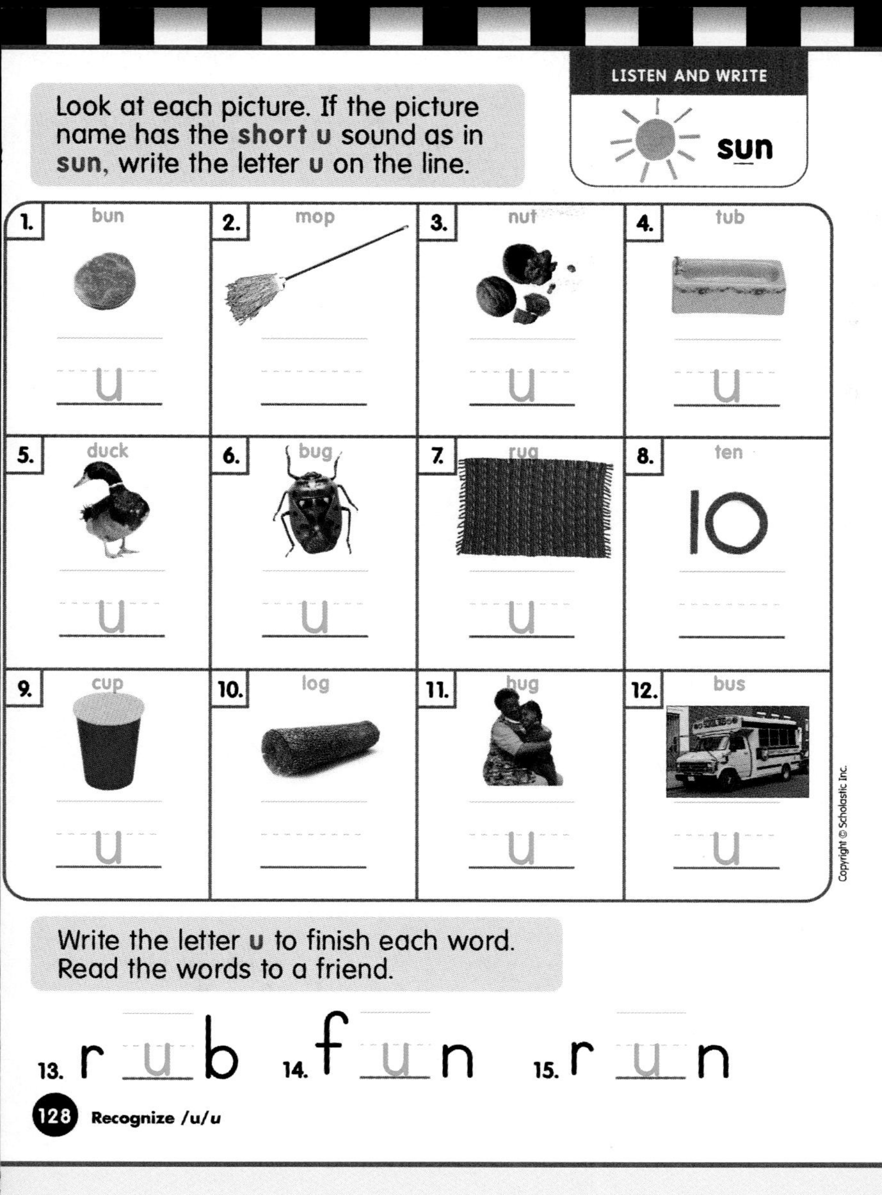
Look at each picture. If the picture name has the **short u** sound as in **sun**, write the letter **u** on the line.

LISTEN AND WRITE

sun

1. bun u
2. mop
3. nut u
4. tub u
5. duck u
6. bug u
7. rug u
8. ten
9. cup u
10. log
11. hug u
12. bus u

Copyright © Scholastic Inc.

Write the letter **u** to finish each word. Read the words to a friend.

13. r u b 14. f u n 15. r u n

128 Recognize /u/u

Integrated Curriculum

SPELLING CONNECTION

FUN WITH WORDS Write the following word parts on the chalkboard: ***n_t, h_g, c_t, s_n, b_t***. Have children add the letter ***u*** to each. Model for children how to blend each word. Then have children say each word and write it on a sheet of paper. You may wish to have children suggest other words they can make by changing one or more letters in each word .

SOCIAL STUDIES CONNECTION

READ ALL ABOUT IT! Have children look through books, magazines, and newspapers for words with ***u***. Children can write these words on the chalkboard, read them aloud, then underline the ***u*** in each word.

CHALLENGE

Play word dominoes by making dominoes out of tagboard or heavy paper. Instead of dots, each section of the domino tile can contain word parts—consonants, vowels, and phonograms. Make at least 42 domino tiles. Have children play dominoes using the tiles. They must turn all the tiles facedown and then draw one tile from the pile. Using the letters on the tile and on any tiles already displayed, they must try to make a word. Children should continue until all the tiles have been used.

EXTRA HELP

Have children suggest **/u/*u*** words. List the words on the chalkboard or on chart paper. Then have children underline each ***u*** in the words.

Phonics Connection

Literacy Place: *Team Spirit*
Teacher's SourceBook, pp. T152–153;
Literacy-at-Work Book, pp. 33–34

My Book: *With Nuts*

Big Book of Rhymes and Rhythms, 1A: "See a Penny," p. 32

Chapter Book: *Fun With Zip and Zap*, Chapter 5

Lesson 77

page 129

Blend Words With /u/u

QUICKCHECK ✔

Can children:

✔ **orally blend word parts?**

✔ **identify /u/**

✔ **blend words with /u/u?**

If YES go to Read and Write.

TEACH

Develop Phonemic Awareness

Oral Blending Say the following word parts, and ask children to blend them. Provide corrective feedback and modeling when necessary.

/s/ /u/ /n/ /f/ /u/ /n/ /r/ /u/ /g/
/t/ /u/ /b/ /h/ /u/ /t/ /d/ /u/ /k/

Connect Sound-Symbol Review that ***u*** stands for /u/ as in ***sun***. Write ***sun*** on the chalkboard. Have a volunteer circle the ***u***. Then model how to blend the word.

THINK ALOUD

I can put the letters *s, u,* and *n* together to make the word *sun.* Let's say the word slowly as I move my finger under the letters. Listen to how I string together the sound that each letter stands for to make the word: *ssssuuuunnnn, ssuunn, sun.*

Help children blend the words ***fun*** and ***cut***.

READ AND WRITE

BUILT-IN REVIEW

Blend Words To practice using the sound taught and to review previously taught sounds, list the following words and sentence. Have volunteers read each aloud. Model blending.

- us bus tug
- but nut cut
- We had fun on the bus.

Complete Activity Page Read aloud the directions on page 129. Review each picture name with children.

Name

Circle the word that names each picture.
Then write the word on the line.

Write the word that names each picture.

Supporting All Learners

TACTILE LEARNERS

NUT WORDS Have children write large words with **/u/u** on construction paper, one word per page. Children can then glue nuts or nut shells to each ***u*** in the words.

The Book Shop

- **Hunches in Bunches**
  by Dr. Seuss
  A boy has many hunches. **(/u/u)**
- **Fun/No Fun**
  by James Stevenson
  The author/artist lists what was fun and not in his childhood. **(/u/u)**

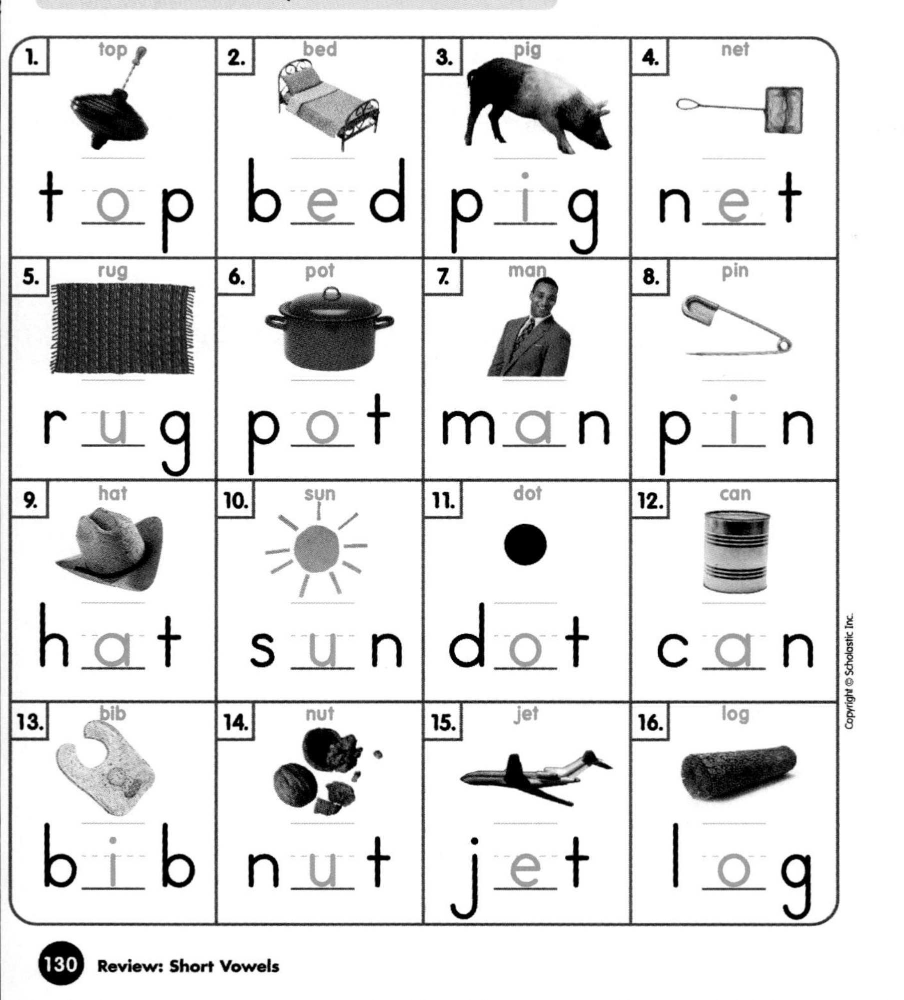
REVIEW

Look at each picture. Write the missing letter to finish the picture name.

1. top — t o p
2. bed — b e d
3. pig — p i g
4. net — n e t
5. rug — r u g
6. pot — p o t
7. man — m a n
8. pin — p i n
9. hat — h a t
10. sun — s u n
11. dot — d o t
12. can — c a n
13. bib — b i b
14. nut — n u t
15. jet — j e t
16. log — l o g

Supporting All Learners

KINESTHETIC LEARNERS

FROG HOP GAME To practice reading words with short vowel sounds, children may enjoy playing "Frog Hop" on pages 16–20 in *Quick-and-Easy Learning Games: Phonics*. (To adapt the game to children's needs, eliminate the game cards containing blends that children have not yet studied, such as ***st***.)

VISUAL LEARNERS

SHORT VOWEL WORDS Have children draw pictures of things with names that have ***/a/a, /e/e, /i/i, /o/o,*** or ***/u/u***. You may wish to suggest things such as *net, bat, pin, hut,* and *frog*. Help children label each picture. Children can display their drawings on a poster, bulletin board, or in a book.

EXTRA HELP

For children who need extra support, use only the game cards from "Frog Hop." (Eliminate those cards containing blends that children have not yet studied, such as ***st***.) Partners can be guided or prompted to say each picture name, to listen for the sounds that make it up, and to determine the short vowel sound that completes each word. Children may also sort the game cards according to short vowel sounds.

Review Short Vowels

QUICKCHECK ✔

Can children:

- ✔ **review sounds previously taught?**
- ✔ **blend words with short vowel *u*?**
- ✔ **write words with short vowels?**

If YES go to Read and Write.

TEACH

Link to Spelling Review with children the following sound-spelling relationships: /***a***/***a***, /***e***/***e***, /***i***/***i***, /***o***/***o***, /***u***/***u***. Say one of these sounds. Have a volunteer write on the chalkboard the spelling that stands for the sound. For example, ***e*** stands for /***e***/. Continue with all the sounds.

Phonemic Awareness Oral Segmentation
Say ***but***. Have children orally segment the words. (/**b**/ /**u**/ /**t**/) Ask them to say the word slowly and to tap lightly on their desks for each sound the word contains. Draw three connected boxes on the chalkboard. Have a volunteer write the spelling that stands for each sound in ***but*** in the appropriate box.

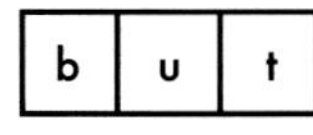

Repeat and continue the activity by changing ***u*** to ***a*** to form ***bat; a*** to ***i*** to form ***bit***, and ***i*** to ***e*** to form ***bet***. Blend to emphasize the vowel sound.

READ AND WRITE

Practice Distribute three counters and one copy of the Segmentation Reproducible Master on page T29 to each child. Explain that you are going to read aloud a word. Children will count how many sounds they hear in the word, placing one counter on each box on the reproducible master. Use these groups of words in order: ***at, it, pit; an, in, win; on, an, fan; cut, but, bat***. Ask children which sound changes or is added.

Complete Activity Page Read aloud the directions on page 130. Review each picture name with children.

Lesson 79

pages 131-132

Recognize and Blend Words With /th/ *th*

QUICKCHECK ✔

Can children:

- ✔ **substitute sounds in words?**
- ✔ **listen for /th/?**
- ✔ **identify the sound the letters *th* stand for?**
- ✔ **blend words with /th/*th*?**

If YES go to Read and Write.

TEACH

Develop Phonemic Awareness

Phonemic Manipulation Explain to children that they are going to play a consonant riddle game. Say a word. Children are to think of a word that rhymes and begins with /**th**/.

- **What rhymes with *pen* and starts with /th/?**
- **What rhymes with *tin* and starts with /th/?**
- **What rhymes with *sick* and starts with /th/?**

Connect Sound-Symbol Write ***tin*** and ***thin*** on the chalkboard as you read each aloud. Ask children what sounds they hear at the beginning of each word and how the words are different. Point out that the /**th**/ sound at the beginning of ***thin*** is spelled by two letters together, ***t*** and ***h***. Explain that ***th*** can stand for the beginning sound in ***thin*** and ***that***.

READ AND WRITE

BUILT-IN REVIEW

Blend Words List the following words and sentences. Have volunteers read each aloud. Model how to blend the words.

- **that** **this** **then**
- **them** **thin** **than**
- **That man is thin.**
- **Get that dog for them.**

Complete Activity Pages Read aloud the directions on pages 131–132. Review each picture name with children.

LISTEN AND WRITE

Name

Color each picture whose name begins with the **th** sound as in **thumb**. Then write the letters **th** below it.

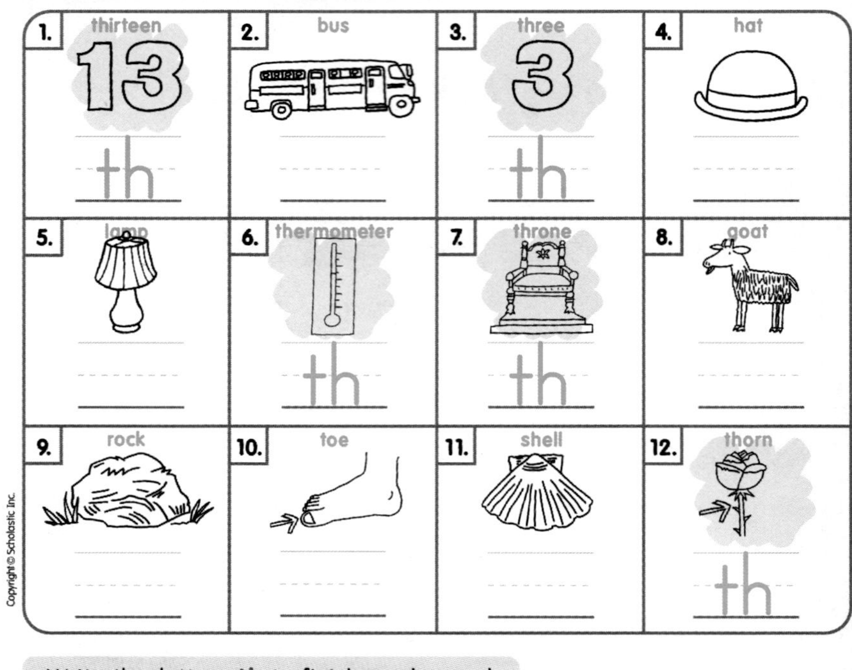

Write the letters **th** to finish each word. Read the words to a friend.

13. thin 14. thick 15. then

Supporting All Learners

AUDITORY/VISUAL LEARNERS

WORD HUNT Read aloud "Little Bo Peep" on page 33 in the *Big Book of Rhymes and Rhythms, 1A*. Have children find all the words with **/th/** in the rhyme. Write these words on the chalkboard. Invite volunteers to add other **/th/** words to the list.

VISUAL LEARNERS

WORD COUNT Have children read the ***th*** words on the chalkboard and count how many words they have collected.

KINESTHETIC LEARNERS

WORD CARDS Have children write the words from the Connect Sound-Symbol exercise on note cards, one word per card. Children can use the note cards as flash cards to practice reading the words. Suggest that they create cards for other ***th*** words.

BLEND WORDS

Read each word. Then find the word in the box that rhymes with it. Write the word on the line.

thin	this
then	that
thick	than

1. pan than
2. pin thin
3. cat that
4. sick thick
5. men then
6. miss this

Use one of the words from above to finish each sentence.

7. Pick up that thick log.
8. Pick up that thin stick.

Copyright © Scholastic Inc.

132 Blend and Write Words With /th/*th*

ESL Because the blend ***th*** does not appear in Spanish and is pronounced differently in other languages, provide opportunities for children acquiring English to record their own pronunciations of words such as ***then, them, the,*** and ***that***. Children can play the recording back to compare the pronunciations with the pronunciations of native English speakers.

CHALLENGE

Partners can play a word-building game with the magnetic letters from the Phonics and Word Building Kit. To begin, make the word ***the*** with the magnetic letters. Then have children take turns changing letters to make other words with **/th/*th***. Challenge children to see how many words they can make.

Integrated Curriculum

SPELLING CONNECTION

Write the following word parts on the chalkboard: ***_an, _at, _en***. Have children write the word parts on paper. Then children can add ***th*** to each. Ask children to suggest other words that contain ***th***. List these words on the chalkboard. Have children add these words to their papers. Have children circle the letters ***th*** in each word. Then have volunteers dictate sentences using words with ***th***. Write these sentences on the chalkboard.

SOCIAL STUDIES CONNECTION

WORD SEARCH Have children look through books, magazines, and newspapers for words with ***th***. List the words on the chalkboard or on chart paper. You may also wish to record the words on the Word Wall. Then children can use the words to dictate sentences about their school or town.

Phonics Connection

Literacy Place: *Team Spirit*
Teacher's SourceBook, pp. T154–155;
Literacy-at-Work Book, pp. 35–36

My Book: *The Bakers*

Big Book of Rhymes and Rhythms, 1A: "Little Bo Peep," p. 33

Chapter Book: *Fun With Zip and Zap,* Chapter 6

Lesson 80

page 133

Blend Words With Phonograms

QUICKCHECK ✔

Can children:

✔ **blend words with phonograms *-un, -ut?***

✔ **build words with *-un, -ut?***

If YES go to Read and Write.

TEACH

Introduce the Phonograms Write the phonograms *-un* and *-ut* on the chalkboard. Point out the sounds these phonograms stand for. Add the letter *s* to the beginning of *-un,* and model for children how to blend the word formed. Have children repeat the word ***sun*** aloud as you blend the word again. Then add ***c*** to the beginning of *-ut,* and model for children how to blend the word formed.

Ask children to suggest words that rhyme with ***bun*** and ***cut***. List these words on the chalkboard in separate columns. Have volunteers underline the phonogram *-un* or *-ut* in each one. Be sure to include the words ***fun, run, but, hut,*** and ***nut***. Blend these words for children as they repeat them aloud. Encourage children to look for the word parts *-un* and *-ut* as they read.

READ AND WRITE

Build Words Write the word part *-un* on the chalkboard. (If available, use a pocket chart and letter cards.) Have children add a letter to the beginning of the phonogram to make a new word. Continue by having children replace the initial consonant or blend in each word to build a new word. For example, children may build ***run*** from ***fun,*** then ***bun*** from ***fun***.

Continue the activity by building words with the word part *-ut*.

Complete Activity Page Read aloud the directions on page 133. Children can complete the page independently.

PHONOGRAMS

Name

Add each letter to the word part below it.
Blend the word. If it is a real word, write it on the line.

b m f l r s c	b c d h n w l
__un	__ut
1. bun	5. but
2. fun	6. cut
3. run	7. hut
4. sun	8. nut

Write a sentence using one of the words you made.

9. Answers will vary.

Write Words With Phonograms *-un, -ut* 133

Supporting All Learners

KINESTHETIC LEARNERS

FLIP BOOK Make or help children make flip books. The top word can be ***but***. Under the ***b***, put ***c, h,*** and ***n,*** one letter per page. Have children take turns reading the words in their flip books. Children can make up sentences using these words.

The Book Shop

- **That's Mine and That's Yours**
  by Angie and Chris Sage
  An older sibling has a day of give-and-take with a baby.
- **When Will the Snow Trees Grow?**
  by Ben Schechter
  A boy wants to know how things will grow.

POETRY

Read the poem. Finish each sentence using a word from the poem.

That Bug!

That big bug is on my back!
That big bug is on this rock!

That big bug is on this pack!
That big bug is on my sock!

Ug! Ug! Ug! Bug.
Go back to the rug!

1. The bug is on my sock.

2. The bug must go to the rug.

Supporting All Learners

KINESTHETIC LEARNERS

CHARADES Write words such as ***sick, pack, back, duck, fox, thin,*** and ***fix*** on note cards. Invite children to play charades by acting out the words, one at a time. Volunteers can guess each word and spell it.

VISUAL LEARNERS

SIGN SEARCH Have children look for words with **/k/*k*, *ck*, /u/*u*,** and **/th/*th*** on signs, labels, and books in your school or neighborhood library.

CHALLENGE

Have children generate a list of words that begin or end with **/k/**. Record these words on the chalkboard. Then have children create a story using as many **/k/** words on the chalkboard as they can. Write the dictated story on chart paper for group and individual reading. Return to the story, rereading it in subsequent lessons.

page 134

Write and Read Words With /k/*k*, *ck*; /u/*u*; /th/*th*

QUICKCHECK ✔

Can children:

- ✔ **spell words with /k/*k*, *ck*; /u/*u*; /th/*th*?**
- ✔ **read a poem?**

If YES go to Read and Write.

TEACH

Link to Spelling Review these sound-spelling relationships: **/k/*k*, *ck*; /u/*u*;** and **/th/*th*.** Say one of these sounds. Have a volunteer write the spelling or spellings that stand for the sound on the chalkboard. For example, ***u*** stands for **/u/**. Continue with all the sounds.

Phonemic Awareness Oral Segmentation
Say ***duck***. Have children orally segment it. (**/d/ /u/ /k/**) Ask them how many sounds the word contains. *(3)* Have a volunteer point out the sound that is shown by two letters. Draw three connected boxes on the chalkboard, and have a volunteer write the spelling that stands for each sound in ***duck*** in the appropriate box. Continue with ***sock*** and ***than***.

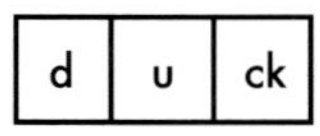

READ AND WRITE

Dictate Dictate the following words and sentence. Have children write them. When children are done, write the words and sentence on the chalkboard. Have children make any necessary corrections.

- **sick** **rug** **then**
- **one** **his** **for**
- **Pick up that rock.**

Complete Activity Page Read aloud the directions on page 134. Children can read the poem independently.

Lesson 82

pages 135-136

Read and Review Words With /k/k, ck; /u/u; /th/th

QUICKCHECK ✔

Can children:

✔ **read words with /k/k, ck; /u/u; /th/th?**

✔ **recognize high-frequency words?**

If YES go to Read and Write.

TEACH

Review Sound-Spellings Review the sound-spelling relationships from the past few lessons. These include /**k**/*k*, *ck*; /**u**/*u*; and /**th**/*th*. Say one of these sounds. Have a volunteer write the spelling or spellings that stand for the sound on the chalkboard. Continue with all the sounds. Then display pictures of objects whose names contain one of these sounds. Have children write the spelling that each picture name contains.

Review High-Frequency Words Review the high-frequency words ***one, her, his, with,*** and ***then***. Write on the chalkboard sentences containing each high-frequency word. State aloud one word, and have volunteers circle the word in the sentences.

READ AND WRITE

Build Words Distribute the following letter cards to children: ***a, b, c, e, h, i, k, n, o, s, t, u***. Provide time for children to build as many words as possible using the letter cards. Suggest that they record their words on a separate sheet of paper.

Build Sentences Write the following words on note cards: ***his, her, cat, dog, sick, sock, got, had, a, that, hat, big***. Display the cards, and have children make sentences using the words. To get them started, suggest the sentence ***Her dog got sick***.

Complete Activity Pages Read aloud the directions on pages 135–136. Review the art with children.

REVIEW

Name

Check each word as you read it to a partner.
Circle any words you need to practice.

I can read!

☐ back	☐ pack	☐ cup
☐ rock	☐ cut	☐ bus
☐ rub	☐ fun	☐ bug
☐ rug	☐ thick	☐ with
☐ thin	☐ sick	☐ that

Lookout Words!

☐ his	☐ said	☐ see	☐ she	☐ one
☐ my	☐ the	☐ can't	☐ why	☐ her

Supporting All Learners

VISUAL/KINESTHETIC LEARNERS

TREASURE HUNT Assign each child a letter of the alphabet or a phonogram. Give each child a sack or bag for collecting treasures that can be named with words that have the sound(s) of the assigned letter or phonogram. For example, a child with the letter ***t*** may collect **tape**, the number **10**, a picture of a **toad**, and a book featuring a **tiger** on its cover.

KINESTHETIC LEARNERS

MATCH MAKING Give some children picture cards from the Bingo game; give other children corresponding letter cards. Invite children to find their match.

AUDITORY LEARNERS

TRADE LISTS Children can write or dictate their own lists of words with **/k/*k*, *ck*, /u/*u*,** and **/th/*th*.** Have children include at least one word that does not belong in each list. Then children can trade lists. Have children find the words that do not belong in each list.

TEST YOURSELF

Fill in the bubble next to the sentence that tells about each picture.

1.	● We dug in the mud. ○ We can run in the mud.
2.	○ He has a big backpack. ● He sits on top of the rock.
3.	○ The dog can nap with the cat. ● The dog licks the cat.
4.	● That is his rug. ○ We sat on the thick rug.
5.	○ The bug likes the mud. ● The bug is in the sun.
6.	● Six pigs had fun in the pen. ○ Lock the cats in the pen.

Remember to look at the picture before you read the two sentences.

Integrated Curriculum

WRITING CONNECTION

LETTER TIME Help children to write a list of words with ***/k/k, ck; /u/u;*** and ***/th/th***. Have children write or dictate letters to each other using words from the list. Then children can exchange the letters and find the words with ***/k/k, ck; /u/u;*** and ***/th/th***.

TECHNOLOGY CONNECTION

MY BOOKS For additional practice with ***/k/k, ck; /u/u;*** and ***/th/th,*** have children read the following My Books: *Our Cat's Kittens, Hen and Duck, With Nuts,* and *The Bakers*. For information on using the My Books on the computer, see the WiggleWorks Plus My Books Teaching Plan.

I CAN READ! OPTIONS

The I Can Read! page can be used for one or all of the following:

- **paired reading**
- **individual assessment**
- **choral reading**
- **homework practice**
- **program placement**

EXTRA HELP

For children needing additional phonemic awareness training, see the Scholastic Phonemic Awareness Kit. The oral blending exercises will help children orally string together sounds to form words. This is necessary for children to be able to decode, or sound out, words while reading.

CHALLENGE

Have children collect and count words with ***/k/k, ck; /u/u;*** and ***/th/th***. Children can record the results of their counts in a chart such as the one shown.

3			
2			
1			
	Words With ***k, ck***	Words With ***u***	Words With ***th***

Lesson 83

pages 137-138

Read Words in Context

TEACH

Assemble the Story Ask children to remove pages 137–138. Have children fold the pages in half to form the Take-Home Book.

Preview the Story Preview *Jack,* a story about a typical day in the life of a six-year-old boy. Have a volunteer read the title. Invite children to browse through the first two pages of the story and to comment on anything they notice. Suggest that they point out any unfamiliar words. Read these words aloud. Model how to blend them. Then have children predict what they think the selection might be about.

READ AND WRITE

Read the Story Read the story aloud, or have volunteers take turns reading aloud a page at a time. Discuss anything of interest on each page. Encourage children to help each other with any blending difficulties. The following prompts may help children who need extra support while reading:

- **What letter sounds or word parts do you know in the words?**
- **What do the pictures tell you about the story?**

Reflect and Respond Have children share their reactions to *Jack*. Does it remind them of other stories they have read? What did they like best about the story? Encourage children to think of more details about Jack's day that they would like to add to the story.

Develop Fluency You may wish to reread the story as a choral reading or have partners reread the story independently. Provide time for children to reread the story on subsequent days to develop fluency and increase reading rate.

Supporting All Learners

VISUAL LEARNERS

WORD COUNT Have children count the number of words with ***/k/k, ck; /u/u;*** and ***/th/th*** in the My Books.

AUDITORY/VISUAL LEARNERS

RHYME, DRAW, AND WRITE Have children think of a rhyme that they can illustrate, such as ***a man with a fan, a fox in a box, a thin pin,*** or ***a pack on a back***. Have children illustrate their ideas and label them.

VISUAL LEARNERS

CIRCLE, UNDERLINE, BOX Copy *Jack* onto chart paper. Then have children circle words with ***/k/k, ck;*** underline words with ***/u/u;*** and draw a box around words with ***/th/th***.

Reflect On Reading

ASSESS COMPREHENSION

To assess their understanding of the story, ask children questions such as:

- **Who is Jack? (*He is a boy who goes to school on the bus, likes to play, and has a cat, a mom, and a dad.*)**
- **What does Jack like to do? (*He likes to run, dig, kick, play with his cat, and have fun in the tub.*)**
- **How does the story begin and end? (*It begins with Jack getting up in the morning, and it ends with Jack going to bed at night.*)**

HOME-SCHOOL CONNECTION

Send home *Jack*. Encourage children to read the story to a family member.

WRITING CONNECTION

DAYTIME FUN Have children write or dictate sentences about what they do during the day; or have children write about Jack's day and what they like best about it. Children can illustrate their sentences. You may wish to display the sentences on a bulletin board.

EXTRA HELP

Have partners or small groups of children read *Jack* aloud together or with you. Then children can take turns acting out the actions of Jack on each page of the story. Suggest that children read each page as volunteers act out the actions on that page.

CHALLENGE

Children can make word puzzles using words with **/k/k, ck; /u/u;** and **/th/th;** for example, ***Jack − J + p = pack***. When children have completed their word puzzles, cover the answers. Challenge children to figure out the answers to each other's puzzles.

ESL To build background for the story, encourage children to tell, draw, or act out some of the things they do from morning to night during a typical school day—before school, at school, after school, and at bedtime. Children who draw or pantomime an action might want to let others guess what that action is. Alternatively, as a response to the story, encourage children to role-play Jack's actions in sequence and relate these actions to their own experiences.

Phonics Connection

Phonics Readers:
#28, *Nick's Trick*
#29, *No Fun for Gus*
#30, *The Pet Bath*

Lesson 84

pages 139-140

Blend Words With /z/z

QUICKCHECK ✔

Can children:

✓ **distinguish between /z/ and other ending sounds?**

✓ **identify /z/?**

✓ **blend words with /z/s?**

If YES go to Read and Write.

TEACH

Develop Phonemic Awareness

Oddity Task Explain to children that they are going to listen to words, some of which end with /z/ as in the word *peas*. If children hear /z/ at the end of a word, they are to say /z/. If children say /z/ after a word that does not end with /z/, ask them to listen again as you repeat the word:

bag	bags	has	kick
is	fix	trees	hop
bugs	as	was	lid
jam	brothers	letters	toys

Connect Sound-Symbol Write the words *pan* and *pans* on the chalkboard as you read each aloud. Ask children what sound they hear at the end of each word and how the words are different. Point out that /z/ at the end of *pans* is spelled with the letter *s*.

READ AND WRITE

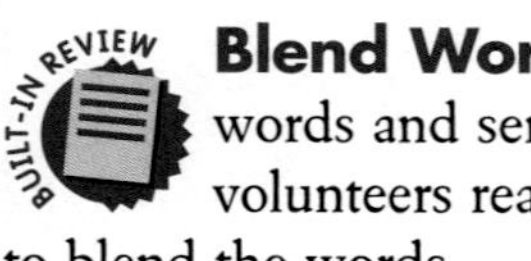

Blend Words List the following words and sentences on a chart. Have volunteers read each aloud. Model how to blend the words.

- has　　is　　was
- runs　　dogs　　fans
- His cat was sad.
- She has six big dogs.

Complete Activity Pages Read aloud the directions for pages 139–140. Review the art with children.

LISTEN AND WRITE

Name

Look at each picture. If the picture name ends with the z sound as in **peas**, color the picture.

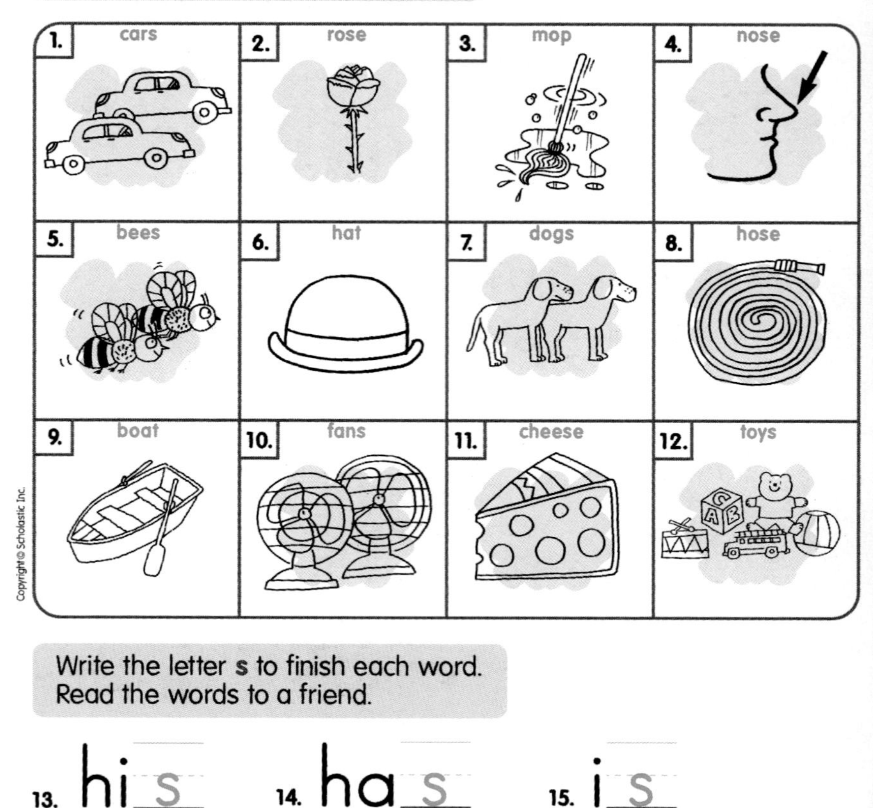

Write the letter **s** to finish each word. Read the words to a friend.

13. hi s　　14. ha s　　15. i s

Supporting All Learners

AUDITORY/VISUAL LEARNERS

/z/ SEARCH Read aloud "B-I-N-G-O!" on page 34 in the *Big Book of Rhymes and Rhythms, 1A*. Have children find the words with **/z/** in the rhyme. Write these words on the chalkboard. Invite volunteers to add other **/z/** words to the list.

TACTILE LEARNERS

PEAS PLEASE! Have children write large words with **/z/s** on construction paper or tagboard. Then have children glue dried peas on the **/z/s** in each word. Children can say the word, tracing the ***s***.

VISUAL LEARNERS

ADD A LETTER Have children write the following word parts on the chalkboard: ***can__, dog__, rag__, run__, pig__***. Have children add ***s*** to each one. Model how to blend each word. Then ask children to suggest other words that end with **/z/**. List these words on the chalkboard. Have volunteers circle the ***s*** in each.

Look at each picture. Circle the word that best finishes each sentence. Then write the word on the line.

	Sentence	Choices
1.	She likes the fans.	fuzz, fun, **(fans)**
2.	He has a big hat.	his, **(has)**, hats
3.	His cat gets fed.	Her, **(His)**, Is
4.	She runs in the sun.	rubs, **(runs)**, run
5.	She sees the cans.	cat, can, **(cans)**
6.	The buns are hot.	bus, bugs, **(buns)**

Integrated Curriculum

WRITING CONNECTION

ANSWERS PLEASE Have children work together to write a list of words that end with **/z/s**. Then children can write or dictate rhyming questions such as ***Do bees like peas?*** Display the sentences. Have children revisit them, read them aloud, and point out the words with **/z/s**.

SOCIAL STUDIES CONNECTION

SCHOOL TOUR Walk around the school with children. Look for words with **/z/s** on signs, posters, and bulletin boards. You may also wish to visit the school library to look for words with **/z/s**. Have children write the words.

Phonics Connection

Literacy Place: *Team Spirit*
Teacher's SourceBook, pp. T208–209;
Literacy-at-Work Book, pp. 44–45

My Book: *Making Friends*

Big Book of Rhymes and Rhythms, 1A: "B-I-N-G-O!" p. 34.

Chapter Book: *Fun With Zip and Zap,* Chapter 7

EXTRA HELP

Review **/z/** in ***zebra, zipper,*** and ***zigzag***. To create a bridge between children's understanding of the initial and the final consonant sound, consider using the word ***zoos***. Ask children what sound they hear at both the beginning and end of the word. Then ask what sound they hear at the end of ***peas*** and ***bees***. Note that this is the same sound they hear at the beginning of ***zebra, zipper,*** and ***zigzag***.

CHALLENGE

Children can make a chart of words with and without **/z/** added. You may wish to make a chart such as the following. Model how to fill in the chart by filling in the first space or spaces. Have children add words to the chart as they come across them in their reading.

Words Without /z/	Words With /z/
dog	dogs
run	runs

Lesson 85

pages 141-142

Recognize and Write /y/y

QUICKCHECK ✔

Can children:

✓ **orally segment words?**

✓ **write capital and small Yy?**

✓ **recognize /y/?**

✓ **identify the letter that stands for /y/?**

If YES go to Read and Write.

TEACH

Develop Phonemic Awareness

Oral Segmentation Explain to children that you will say a word sound by sound. They will listen to the parts and then say the whole word. For example, say the sounds /y/ /i/ /p/. Guide children to say the word *yip*. Continue with the following word parts.

/y/ /e/ /t/ /y/ /a/ /m/

/y/ /e/ /s/ /y/ /a/ /p/

Write the Letter Explain to children that the letter *y* stands for /y/, the sound they just heard at the beginning of ***yet, yam, yes,*** and ***yap***. Write *Yy* on the chalkboard. Point out the capital and small forms of the letter. Then model for children how to write the letter.

Have children write both forms of *Yy* in the air with their fingers. They may also practice writing the letter on the chalkboard.

READ AND WRITE

Connect Sound-Symbol Write the word ***yet*** on the chalkboard. Have a volunteer circle the letter ***y***. Remind children that the letter ***y*** stands for /y/.

Write the following word parts on the chalkboard: ***_es, _ip, _ap,*** and ***_ell***. Have volunteers add the letter ***y*** to each one. Have children blend each word, and write it on a sheet of paper.

Complete Activity Pages Read aloud the directions for pages 141–142. Review each picture name with children.

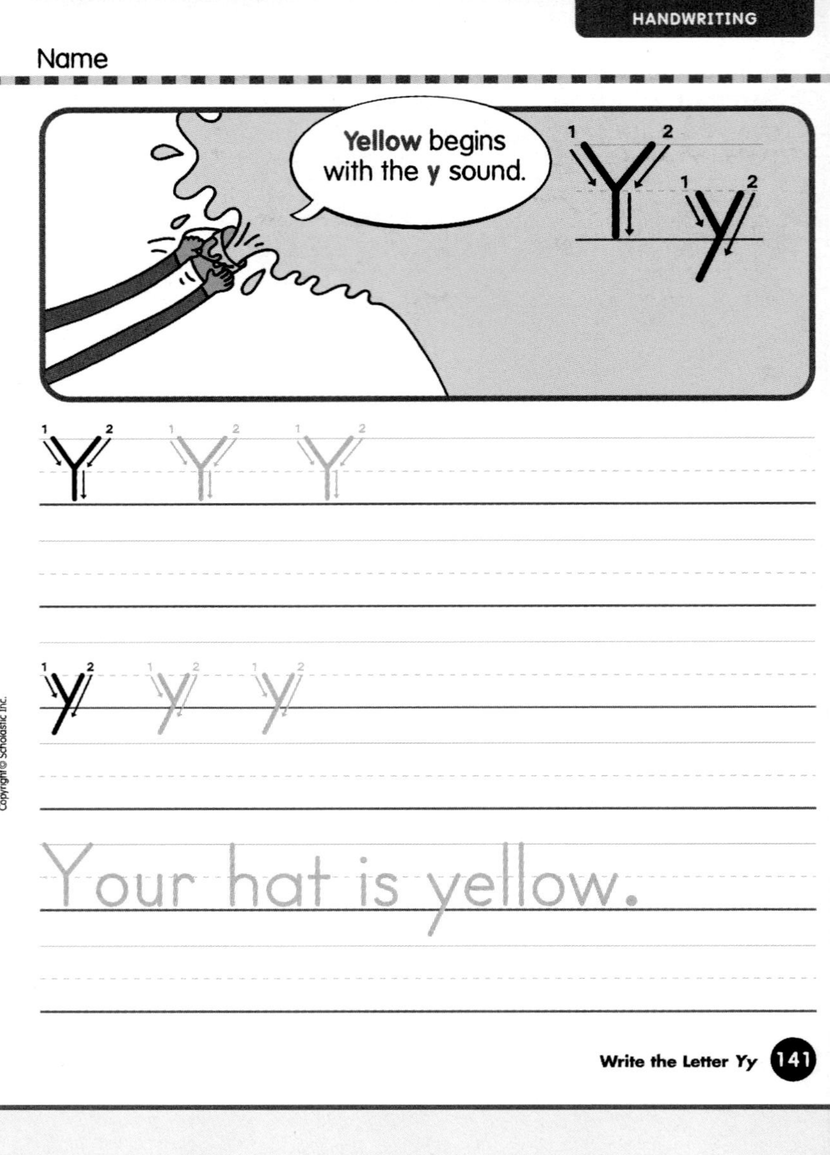

Supporting All Learners

AUDITORY/VISUAL LEARNERS

WHERE IS *Yy*? Read aloud "Yankee Doodle" on page 35 in the *Big Book of Rhymes and Rhythms, 1A*. Have children find all the words with **/y/** in the rhyme. Write these words on the chalkboard. Invite volunteers to add other **/y/** words to the list.

AUDITORY LEARNERS

PHONICS KIT Have children listen to "Stop! Look! Listen!" on the Sounds of Phonics Audiocassette (Community Involvement). Invite children to identify all the words they hear that contain **/y/**. Write the words that children name, and then read the list, emphasizing **/y/** in each.

TACTILE/VISUAL LEARNERS

ABC CARD Display the ABC card for **Yy** from the Phonics and Word Building Kit. Children whose names contain ***y*** can write their names on the ABC card. It might be necessary to point out the many sounds that ***y*** can stand for. Then have children add other words that contain ***y*** to the card. Display the card for future reference when reading and writing.

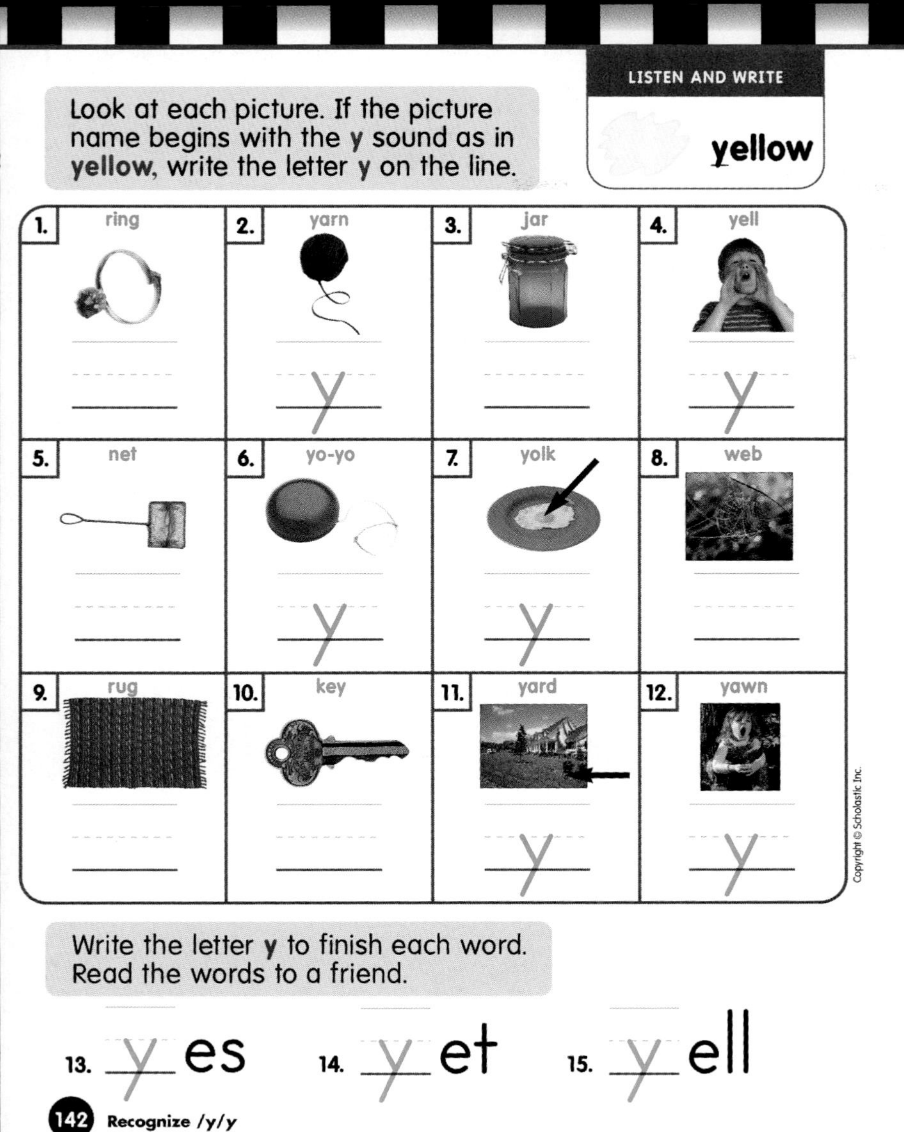

LISTEN AND WRITE

yellow

Look at each picture. If the picture name begins with the y sound as in **yellow**, write the letter **y** on the line.

1. ring	2. yarn	3. jar	4. yell
	y		y
5. net	**6.** yo-yo	**7.** yolk	**8.** web
	y	y	
9. rug	**10.** key	**11.** yard	**12.** yawn
		y	y

Copyright © Scholastic Inc.

Write the letter **y** to finish each word. Read the words to a friend.

13. y es 14. y et 15. y ell

142 Recognize /y/y

Display pictures of objects whose names begin with **/y/**, such as **yarn, yellow, yo-yo, yam,** and **yolk**. Prompt children acquiring English to take turns identifying each object, emphasizing **/y/** as they say each name. Then give each child a self-sticking note with the letter **y** written on it. Have children trace the letter and attach the note to one of the objects as they repeat its name.

CHALLENGE

Have children play a word-building game. To begin, have one child use the magnetic letters from the Phonics and Word Building Kit to make a word with **/y/y**. Then the second child changes one letter to make another word. Model how to begin by changing the ***s*** in ***yes*** to ***t*** to make ***yet***. Have children write a list of the words with **/y/y** that they make.

EXTRA HELP

Use one of the puppets in the Phonemic Awareness Kit to reinforce how to write the letter **y**, to review that **/y/** is written as **y**, and to model blending the word ***yet***.

Integrated Curriculum

SPELLING CONNECTION

WILD ABOUT *Yy* Have children dictate sentences with **/y/** words. Write the sentences on the chalkboard or on chart paper. Then have children take turns identifying the words with **/y/**. Have volunteers underline each **y**.

ART CONNECTION

***Yy* IS FOR YELLOW** Have children make a poster of things that are yellow or have names that have **/y/y**. If possible, make the poster on yellow paper. Children can draw pictures of things such as **yarn, bananas, yams,** and so on. Then children can label or dictate labels for each picture.

Phonics Connection

Literacy Place: *Team Spirit*
Teacher's SourceBook, pp. T210–211;
Literacy-at-Work Book, pp. 46–47

My Book: *Are We There Yet?*

Big Book of Rhymes and Rhythms, 1A: "Yankee Doodle," p. 35.

Chapter Book: *Fun With Zip and Zap,* Chapter 3

Lesson 86

pages 143-144

Blend and Build Words With /y/y

QUICKCHECK ✔

Can children:

✔ **orally blend word parts?**

✔ **identify the sound the letter *y* stands for?**

✔ **blend words with /y/y?**

✔ **build words?**

If YES go to Read and Write.

TEACH

Develop Phonemic Awareness

Oral Blending State aloud the following word parts, **/y/...et, /y/...es, /y/...am, /y/...ell.** Ask children to orally blend them. Provide corrective feedback and modeling when necessary.

Connect Sound-Symbol Review with children that the letter *y* stands for /y/ as in ***yes***. Write the word ***yes*** on the chalkboard, and have a volunteer circle the letter *y*. Then model for children how to blend the word.

I can put the letters *y, e,* and *s* together to make the word *yes.* Let's say the word slowly as I move my finger under the letters.

READ AND WRITE

Blend Words List the following on a chart. Have volunteers read each aloud.

- yes yet yip yap
- They can't get it yet.

Build Words Distribute these letter cards to children: ***a, e, m, s, t,*** and ***y***. Ask children to build as many words as possible and to record them. Have children build more words with the following sets of letter cards: ***a, e, g, n, t, y*** and ***b, i, d, o, p, t***.

Complete Activity Pages Read aloud the directions on pages 143–144. Review each picture name with children.

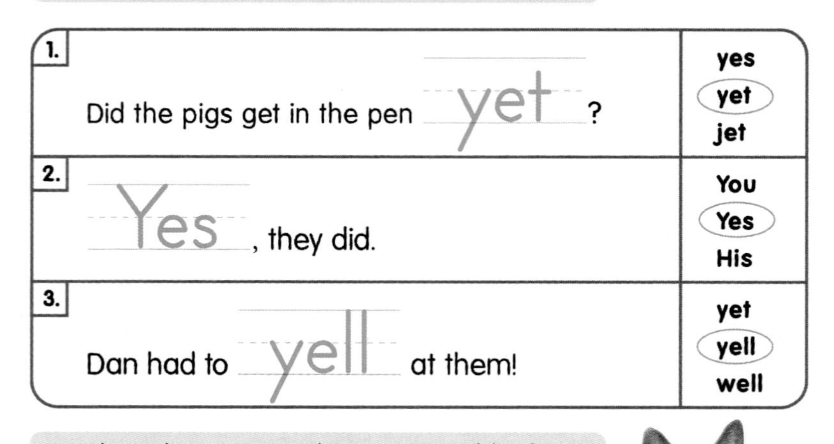

Circle the word that best finishes each sentence. Then write the word on the line.

1.	Did the pigs get in the pen yet?	yes / (yet) / jet
2.	Yes, they did.	You / (Yes) / His
3.	Dan had to yell at them!	yet / (yell) / well

Look at the picture. Then unscramble the words to make a sentence that tells about the picture. Write the sentence on the line.

fed yet? he Did get

4. Did he get fed yet?

Supporting All Learners

KINESTHETIC LEARNERS

WORD SORT Give children the following picture cards from the Phonemic Awareness Kit: **yarn, yellow, wig,** and **window**. Have children sort the cards into two piles based on the sound they hear at the beginning of each picture name.

AUDITORY LEARNERS

RHYME TIME Use the tune and lyrics of "Old MacDonald" to create a silly rhyme song. Have children choose farm animals and sounds to rename with **/y/**. For example, they may call a ***cow*** a ***yow,*** or say ***yoo*** instead of ***moo.*** Then sing the song, ending each verse with a special **/y/** refrain: ***Ye-yi-ye-yi-yo***.

VISUAL LEARNERS

Y HUNT Have children look through books, magazines, and newspapers to find words with **/y/y**. List the words on the chalkboard as children share them. Then have children read the words together.

Look at each picture. Use the letter tiles to write each picture name.

s b n u g

1. b u g
2. s u n

h t a m y

3. h a t
4. h a m
5. y a m

Integrated Curriculum

SPELLING CONNECTION

Write the word ***yet*** on the chalkboard, and model blending. Then tell children that you want them to replace **/y/,** the first sound in ***yet,*** with **/m/** to make a new word. Ask children what letter stands for **/m/**. Then replace the letter ***y*** in ***yet*** with the letter ***m,*** and blend the new word formed. Continue by changing one letter to build new words. You may wish to use this sequence of words:

met
men
pen
pan
ran

SCIENCE CONNECTION

SAND WRITING Have children work in small groups or as partners to take turns writing words with **/y/y** in sand. Have children keep a list of the words they make.

CHALLENGE

Ask children to build as many words with **/y/** as they can. Children can list these words on the chalkboard or on chart paper. Then have children alphabetize the list.

Second-language learners might be confused by the word ***why*** and the letter ***y***. Explain that in English, only the words ***I*** and ***a*** are spelled and pronounced exactly the same way as the letter; all other words that seem to be the same as letters are spelled differently, such as ***be, gee, jay, oh,*** and ***you***.

- **Yellow Ball**
  by Molly Bang
  A yellow ball, left behind at the beach, floats away to another child. **(/y/y)**

Lesson 87

pages 145-146

Recognize and Write /v/v

QUICKCHECK ✔

Can children:

- ✓ **substitute sounds in words?**
- ✓ **write capital and small *Vv*?**
- ✓ **recognize /v/?**
- ✓ **identify the letter that stands for /v/?**

If YES go to Read and Write.

TEACH

Develop Phonemic Awareness

Phonemic Manipulation Explain to children that you will read a list of words. They are to replace the first sound in each word with /v/. For example, if you say ***can,*** children are to say ***van***. Use the following words:

- nine
- nest
- berry
- bet
- case
- few
- coat
- towel
- tan

Write the Letter Explain to children that the letter *v* stands for /v/, the sound they hear in ***van***. Write *Vv* on the chalkboard. Point out the capital and small forms of the letter. Then model for children how to write the letter.

Have children write both forms of the letter in the air with their fingers. They may also practice writing the letter on the chalkboard.

READ AND WRITE

Connect Sound-Symbol Write the word ***van*** on the chalkboard, and have a volunteer circle the letter *v*. Remind children that the letter *v* stands for /v/.

Write this word part on the chalkboard: _*et*. Have a volunteer add the letter *v* to it. Model how to blend the word. Then have children write the word on a sheet of paper.

Complete Activity Pages Read aloud the directions on pages 145–146. Review each picture name with children.

HANDWRITING

Name

Vase begins with the v sound.

V v

V V V

v v v

The vase is pretty.

Write the Letter Vv 145

Supporting All Learners

VISUAL/AUDITORY LEARNERS

FIND *Vv* WORDS Read aloud "Down to the Valley" on page 38 in the *Big Book of Rhymes and Rhythms, 1A*. Have children find all the words with **/v/** in the rhyme. Write these words on the chalkboard. Invite volunteers to add other **/v/** words to the list.

KINESTHETIC/VISUAL LEARNERS

PICTURE RHYMES Have children fold a sheet of paper in half. On each half, children can draw and label pictures of objects, animals, and people whose names rhyme, such as ***dog*** and ***log, pet*** and ***vet, can*** and ***van, sock*** and ***rock,*** and ***bags*** and ***rags***. Collect the pictures, and bind them into a rhyme book for the classroom library.

AUDITORY LEARNERS

LISTENING TIME Have children listen to "Roses Are Red" and "Go Round and Round the Village" on the Sounds of Phonics Audiocassette (Managing Information). Invite children to identify all the words they hear that contain **/v/** in each song. Write the words that children name, and then read the list emphasizing **/v/** in each word.

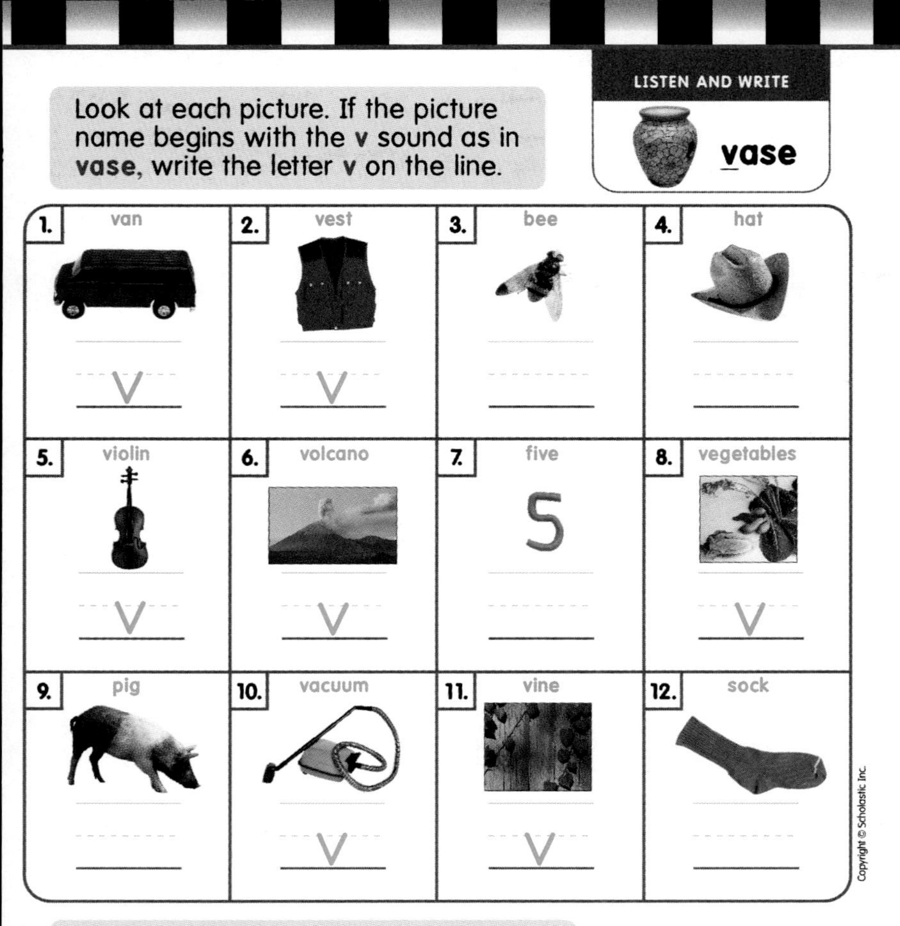
LISTEN AND WRITE

vase

Look at each picture. If the picture name begins with the v sound as in **vase**, write the letter **v** on the line.

1. van — v	2. vest — v	3. bee	4. hat
5. violin — v	6. volcano — v	7. five	8. vegetables — v
9. pig	10. vacuum — v	11. vine — v	12. sock

Write the letter **v** to finish each word. Circle the word that names the picture.

13. v an

14. v et

146 **Recognize /v/v**

Integrated Curriculum

SPELLING CONNECTION

Write the following words on the chalkboard: ***man, best, mine, net,*** and ***towel.*** Have children write the words on paper. Then have children replace one letter in each word with **v** to make a new word.

TECHNOLOGY CONNECTION

For practice with **/z/s, /y/y,** and **/v/v,** have children read the following My Books: *Making Friends, Are We There Yet?* and *Be Very, Very Quiet*. For information on using the My Books on the computer, see the WiggleWorks Plus My Books Teaching Plan.

You may also wish to have children use words from I Can Read! on page 150 to write sentences or stories on the computer using the WiggleWorks writing area. See the WiggleWorks Plus Teaching Plan for ideas.

EXTRA HELP

Help children work together to dictate sentences using words with **/v/v**. Write the sentences on the chalkboard. Have children underline each word with **/v/v**.

CHALLENGE

Suggest that children draw a long vine on a large sheet of paper or on poster board. Have children label the vine. Then children can draw and cut out leaves for the vine. Children should write a word with **/v/v** on each leaf before gluing or taping it to the vine. Challenge children to see how many words they can fit on the vine.

Phonics Connection

Literacy Place: *Team Spirit*
Teacher's SourceBook, pp. T274–275; Literacy-at-Work Book, pp. 58–59

My Book: *Be Very, Very Quiet*

Big Book of Rhymes and Rhythms, 1A: "Down to the Valley," p. 38

Chapter Book: *Fun With Zip and Zap,* Chapter 3

Lesson 88

page 147

Blend Words With /v/v

QUICKCHECK ✔

Can children:

- ✓ **orally blend word parts?**
- ✓ **identify /v/?**
- ✓ **blend words with /v/v?**

If YES go to Read and Write.

TEACH

Develop Phonemic Awareness

Oral Blending State aloud the following word parts, and ask children to orally blend them. Provide corrective feedback and modeling when necessary.

/v/...an	**/v/...et**	**/v/...oice**
/v/...ideo	**/v/...alentine**	**/v/...acuum**

Connect Sound-Symbol Review with children that the letter *v* stands for /v/ as in ***van***. Write the word ***van*** on the chalkboard, and have a volunteer circle the letter *v*. Then model for children how to blend the word.

THINK ALOUD

I can put the letters *v, a,* and *n* together to make the word *van.* Let's say the word slowly as I move my finger under the letters. Listen to how I string together the sound that each letter stands for to make the word: *vvvvaaaannnn, vvaann, van.*

Help children blend the word ***vet.***

READ AND WRITE

BUILT-IN REVIEW

Blend Words List the following words and sentence on a chart. Have volunteers read each aloud. Model blending when necessary.

- **van** **vet**
- **The vet got in her van.**

Complete Activity Page Read aloud the directions on page 147. Review the art with children.

Name

Write the word that names each picture.

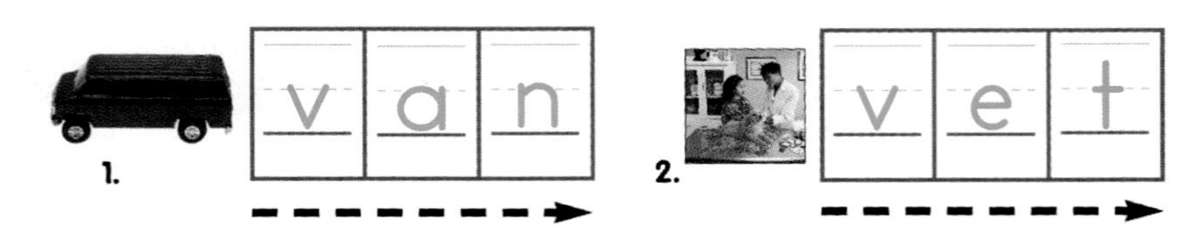

Look at the picture. Then unscramble the words to make a sentence that tells about the picture. Write the sentence on the line.

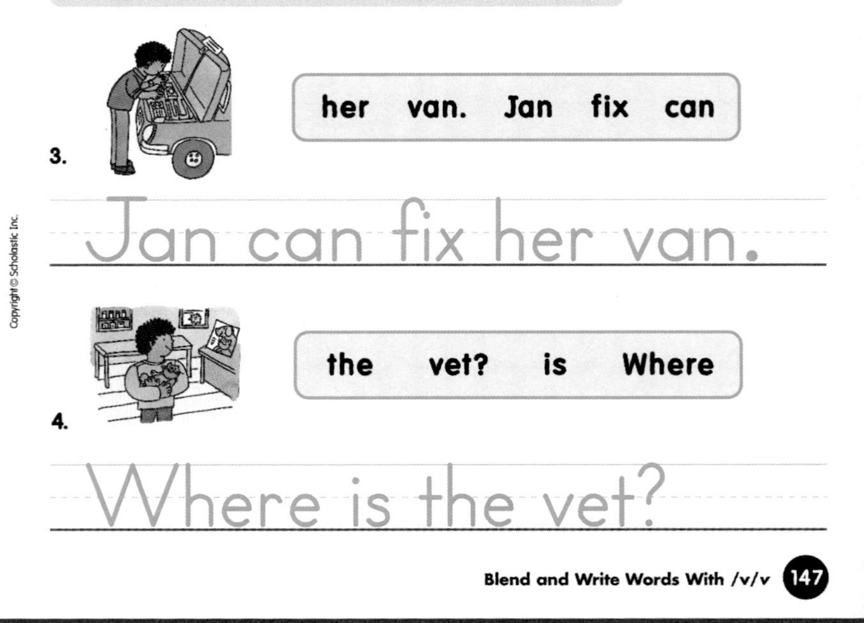

Supporting All Learners

KINESTHETIC/VISUAL LEARNERS

WORD INDEX Invite children to go on a scavenger hunt to collect things or pictures of things whose names contain **/v/**, such as **vest, volleyball, volcano, valentine, vine, vacuum, violet, violin, video,** and perhaps even **vowel**. Display the collected items or pictures. Have children dictate the name of each thing or picture. Write each name on an index card, and display it near the correct picture.

★ The Book Shop ★

- ***Vegetable Soup***
  by Anne Morris
  Children read a page-by-page recipe for making vegetable soup. **(/v/v)**
- ***Arthur's Prize Reader***
  by Lillian Hoban
  Who is a better reader — Arthur or his sister Violet? **(/v/v)**

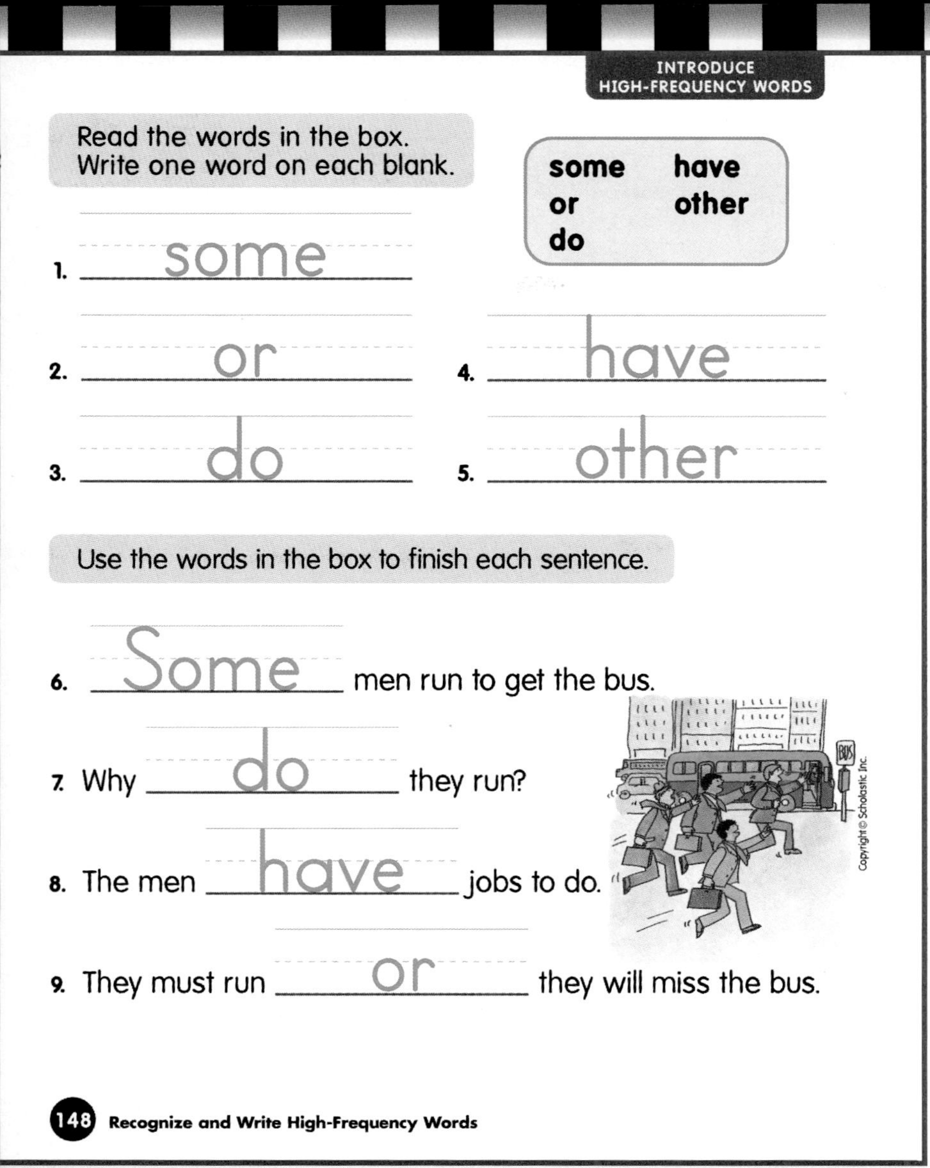
INTRODUCE HIGH-FREQUENCY WORDS

Read the words in the box.
Write one word on each blank.

some	have
or	other
do	

1. some
2. or
3. do
4. have
5. other

Use the words in the box to finish each sentence.

6. Some men run to get the bus.
7. Why do they run?
8. The men have jobs to do.
9. They must run or they will miss the bus.

Copyright © Scholastic Inc.

148 Recognize and Write High-Frequency Words

Supporting All Learners

KINESTHETIC LEARNERS

LETTER TIME Have children write letters to each other using the high-frequency words. Have children fold the letters to look like envelopes and write who the letter should go to on the "envelope." You may also wish to have children write creative addresses on the envelopes, such as: Jo Chen, 4 Desk Street, School City, NY 10001. Children can deliver their letters to each other.

EXTRA HELP

For children who need additional support reading the high-frequency words ***some, do,*** and ***or,*** use Phonics Readers #31, *Two Dogs* and #33, *A Trip in the Van*.

Relate the new word ***or*** to common English sayings children may have already heard or will hear; for example, ***yes or no, now or never,*** and ***sooner or later***. Use each phrase in context, and, if possible, act out a situation or show children an illustration of a situation in which the phrase might be said. For example, someone might say ***now or never*** to a reluctant diver perched at the end of a diving board.

Recognize High-Frequency Words

QUICKCHECK ✔

Can children:

✔ **recognize and write the high-frequency words *some, or, do, have, other*?**

✔ **complete sentences using the high-frequency words?**

If YES go to Read and Write.

TEACH

Introduce the High-Frequency Words

Write the high-frequency words ***some, or, do, have,*** and ***other*** in sentences on the chalkboard. Read the sentences aloud. Underline the high-frequency words, and ask children if they recognize them. You may wish to use the following sentences:

1. **Do the men have cats or dogs?**
2. **Some men have cats.**
3. **Other men have dogs.**

Then ask volunteers to dictate sentences using the high-frequency words. You may wish to begin with the following sentence starters: ***We have some*** ___ or ***Do they like*** ___?

READ AND WRITE

Practice Write each high-frequency word on a note card. Read each word aloud as you display the cards. Then do the following:

- **Mix the cards.**
- **Display one card at a time, and ask children to state each word aloud.**
- **Have children spell each word aloud, clapping on each letter.**
- **Ask children to write each word in the air as they state aloud each letter. Then have them write each word on a sheet of paper.**

Complete Activity Page Read aloud the directions on page 148. Have children complete the page independently.

Lesson 90

page 149

Write and Read Words With /z/s, /y/y, /v/v

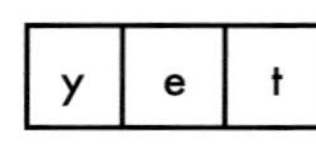

QUICKCHECK ✔

Can children:

✔ spell words with /z/s, /y/y, /v/v?

✔ read a poem?

If YES go to Read and Write.

TEACH

Link to Spelling Review with children the following sound-spelling relationships: /**z**/*s*, /**y**/*y*, and /**v**/*v*. State aloud one of these sounds. Have a volunteer write on the chalkboard the spelling that stands for the sound. Continue with all the sounds. You may also wish to review /**k**/*k*, *ck*; /**u**/*u*; and /**th**/*th*.

Phonemic Awareness Oral Segmentation
State aloud the word *yet*. Have children orally segment the word. *(/y/ /e/ /t/)* Ask them how many sounds the word contains. *(3)* Draw three connected boxes on the chalkboard, and have a volunteer write the spelling that stands for each sound in the word *yet* in the appropriate box. Continue with the words *van* and *has*.

y	e	t

READ AND WRITE

Dictate Dictate the following words and sentence. Have children write them on paper. When they are finished, write the words and sentence on the chalkboard. Have children make any necessary corrections on their papers.

- van vet yet
- some do them
- Some vets do get cats.

Complete Activity Page Read aloud the directions on page 149. Children can read the poem independently or with partners.

POETRY

Name

Read the poem. Write a sentence telling who is on the van.

Who Is on the Van?

Some dogs hop on the van.
They yip and yap and run.
Other dogs hop on,
And they will have some fun!

Some pigs hop on the van.
The van dips and rocks!
Lots of ducks hop on,
And with them is a fox!

Yip, yap! Yip, yap! Yip, yip!
They do have lots of fun!
Will the van dip and tip?
Yes! But it will run.

Answers will vary.

Supporting All Learners

AUDITORY/KINESTHETIC LEARNERS

DRAMATIC REREADING Have children practice and prepare a dramatic rereading of the poem. Children may split up the poem by verses, or read it all together. They can add excitement to the poem by using their voices in loud and soft ways and by adding hand and body movements to show the animal movements and the dipping and tipping of the van.

VISUAL LEARNERS

PORTFOLIO

WHAT IS IN THE VAN? Ask children to draw pictures of vans, one picture per child. Then have children write words with **/z/s, /y/y,** and **/v/v** in their vans. You may wish to display the vans "driving" about the classroom. Call on volunteers to identify the words in the vans with **/z/s, /y/y,** and **/v/v.**

CHALLENGE

Have children generate a list of words that end with **/z/s**. Record these words on the chalkboard. Then have children create a story using as many **/z/s** words on the chalkboard as they can. Begin by generating possible story titles. Write the dictated story on chart paper for group and individual reading. Return to the story, rereading it in subsequent lessons.

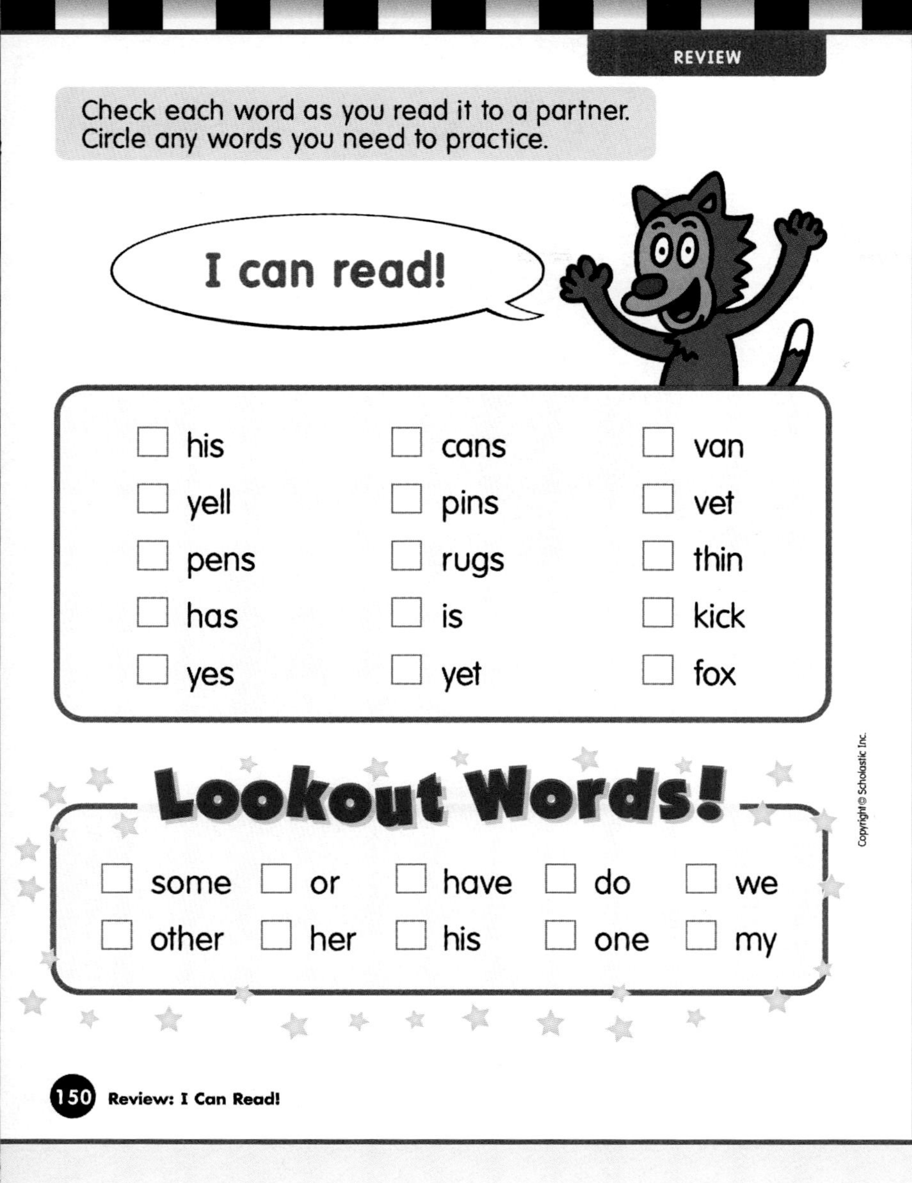

Supporting All Learners

TACTILE LEARNERS

LETTER ART Invite children to write words with ***/z/s, /y/y, /v/v*** in large print on colored construction paper. Then children can decorate the letters ***s, y,*** and ***v***. You may wish to suggest that children draw zig-zags on and around the ***s***'s, use yarn to decorate the ***y***'s, and draw vines around the ***v***'s.

VISUAL LEARNERS

TIC-TAC-TOE Draw three large tic-tac-toe grids on the chalkboard or on chart paper. Write ***s, y,*** and ***v*** over the grids, one letter per grid. Then have children play tic-tac-toe with a twist. After each child writes an ***x*** or an ***o*** in the grid, he or she must suggest a word with ***/z/s, /y/y,*** or ***/v/v,*** depending on the sound and letter on the grid.

EXTRA HELP

Have children use the pocket ABC cards and the pocket chart from the Phonics and Word Building Kit to build words with ***/z/s, /y/y,*** and ***/v/v*** as well as high-frequency words. Once a word is formed, mix the letters, and have a volunteer rebuild it.

Lesson 91

page 150

Read and Review Words With /z/s, /y/y, /v/v

QUICKCHECK ✔

Can children:

✔ read words with /z/s, /y/y, /v/v?

✔ recognize high-frequency words?

If YES go to Read and Write.

TEACH

Review Sound-Spellings Review the sound-spelling relationships from the past few lessons. These include /z/*s*, /y/*y*, and /v/*v*. State aloud one of these sounds. Have a volunteer write the spelling that stands for the sound on the chalkboard. Continue with all the sounds. Then display objects, or pictures of objects, whose names begin with /y/ or /v/ or end with /z/*s*. Have children write the letter that each picture's name begins or ends with as the picture is displayed.

Review High-Frequency Words Review the high-frequency words ***some, do, or, have,*** and ***other.*** Write on the chalkboard sentences containing each high-frequency word. State aloud one word, and have volunteers circle the word in the sentences.

READ AND WRITE

Build Words Distribute the following letter cards to children: ***d, g, i, o, s,*** and ***w***. Provide time for children to build as many words as possible. Suggest that they record their words on a sheet of paper. Have them continue with the letter cards: ***a, e, n, t, v, y*** and ***e, s, t, v, y***.

Build Sentences Write the following words on note cards: ***dogs, have, or, cats, they, some, we, do, pens,*** and ***bags***. Display the cards. Have children make sentences using the words. To get them started, suggest: ***We have dogs***.

Complete Activity Page Read aloud the directions on page 150. Have children complete the page with partners.

Lesson 92

pages 151-152

Read Words in Context

TEACH

Assemble the Story Ask children to remove pages 151–152, and then fold them in half to form the Take-Home Book.

Preview the Story Preview *Hen's Yams,* a story in which two barnyard animals solve a problem. Invite children to browse through the first two pages of the story and to comment on anything they notice. Suggest that they point out any unfamiliar words. Read these words aloud as children repeat them. Then have children predict what they think the story might be about.

READ AND WRITE

Read the Story Read the story aloud, or have volunteers take turns reading aloud a page at a time. Discuss with children anything of interest on each page. Encourage them to help each other with any blending difficulties. The following prompts may help children who need extra support while reading:

- **What letter sounds or word parts do you know in the word?**
- **What do the pictures tell you about the story?**

Reflect and Respond Have children share their reactions to the story. What did they like best about the story? What surprised them in the story? Encourage children to write a different ending to the story.

Develop Fluency You may wish to reread the story as a choral reading, or have partners reread the story independently. Provide time for children to reread the story on subsequent days to develop fluency and increase reading rate.

Ask children how they figure out unfamiliar words when reading. Encourage them to model how they blend words. Continue to review challenging sound-spelling relationships for children needing additional support.

Supporting All Learners

VISUAL LEARNERS

WORD COUNT Have children count the number of **/v/v** words in "Down to the Valley" on page 38 in the *Big Book of Rhymes and Rhythms, 1A.* Then have children count the number of **/v/v** words in the My Book *Be Very, Very Quiet.* Children can also count the number of **/y/y** words in "Yankee Doodle" on page 35 in the *Big Book of Rhymes and Rhythms, 1A.*

AUDITORY LEARNERS

NAME GAME Have children say names with **/z/s, /y/y,** and **/v/v.** Write the names on the chalkboard, and read them aloud with children. You may wish to challenge children to use the names in sentences with other words that have **/z/s, /y/y,** and **/v/v.**

Reflect on Reading

ASSESS COMPREHENSION

To assess their understanding of the story, ask children questions such as:

- **What couldn't Hen find?** ***(her yams)***
- **What did Pig have?** ***(hot buns)***
- **How does the story end?** ***(Pig and Hen find the yams. They eat yams and buns together.)***

I CAN READ! OPTIONS

The I Can Read! page can be used for one or all of the following:

- **paired reading**
- **individual assessment**
- **choral reading**
- **homework practice**
- **program placement**

HOME-SCHOOL CONNECTION

Send home *Hen's Yams*. Encourage children to read the story to a family member.

WRITING CONNECTION

DEAR HEN OR PIG Have children think of something they would like to say to Hen or Pig about what happened in the story. First, children can talk about what they want to say. Then, have children take turns writing or dictating sentences to Hen or Pig. Display the sentences to use when writing the letter. When completed, read the letter aloud with children. Help children to revise the letter if they wish. Then have children identify words with ***/z/s, /y/y***, and ***/v/v***.

EXTRA HELP

Read aloud one or more of the Phonics Readers with children. Then have children read the Phonics Readers together and to each other. Have children identify the words with ***/z/s, /y/y***, and ***/v/v***.

CHALLENGE

Suggest that children make a sounds-and-letters book for ***/z/s, /y/y***, and ***/v/v***. To begin, have children write each letter at the top of a page. Then they can cut out pictures whose names contain each sound and letter. After children paste the pictures on the appropriate pages, they can write or dictate each picture name beneath it. Help children gather the pages into a sounds-and-letters book.

Phonics Connection

Phonics Readers:
#31, *Two Dogs*
#32, *Yip! Yap!*
#33, *A Trip in the Van*

Lesson 93

pages 153-154

Recognize and Write /kw/*qu*

QUICKCHECK ✔

Can children:

✔ **write capital and small *Qq*?**

✔ **recognize /kw/?**

✔ **identify the letters that stand for /kw/?**

If YES go to Read and Write.

TEACH

Develop Phonemic Awareness

Phonemic Manipulation Explain to children that you will say a word and they are to think of a word that rhymes with your word and begins with /**kw**/. For example, ask what word rhymes with ***hit*** and begins with /**kw**/? After children say ***quit,*** continue with the following riddles:

- **It rhymes with *is* and starts with /kw/.**
- **It rhymes with *pick* and starts with /kw/.**
- **It rhymes with *back* and starts with /kw/.**
- **It rhymes with *diet* and starts with /kw/.**
- **It rhymes with *green* and starts with /kw/.**

Write the Letter Explain that the letters ***qu*** stand for /**kw**/, the sounds at the beginning of ***quack***. Write *Qq* on the chalkboard. Point out the capital and small forms of the letter. Then model for children how to write the letter.

Have children write both forms of the letter in the air with their fingers. Volunteers can practice writing the letter on the chalkboard.

READ AND WRITE

Connect Sound-Symbol Write the word ***quick*** on the chalkboard. Have a volunteer circle the letters ***qu***.

Then write the following word parts on the chalkboard: ***_it, _iz, _ack***. Have volunteers add the letters ***qu*** to each one. Model how to blend each word.

Complete Activity Pages Read aloud the directions on pages 153–154. Review each picture name with children.

Name

Q Q Q

q q q

Queen Anne is quiet.

Supporting All Learners

AUDITORY/VISUAL LEARNERS

WORD HUNT Read aloud "I Quickly Do My Homework" on page 39 in the *Big Book of Rhymes and Rhythms, 1A*. Have children find all the words with **/kw/** in the rhyme. Write these words on the chalkboard. Invite volunteers to add other **/kw/** words to the list. Have volunteers circle the letters ***qu*** in each one.

VISUAL/KINESTHETIC LEARNERS

GO TO SCHOOL To review initial consonants, children can once again play "Go to School" on pages 9–12 in *Quick-and-Easy Learning Games: Phonics*.

VISUAL LEARNERS

WORD WALL On a large note card, write the letters ***qu*** and the key word ***queen***. Display the card on the Word Wall. Have children suggest words with ***qu***, write them on small note cards, and add them to the wall. Remind children to look for words that begin with ***qu*** during their reading. Add these words to the wall.

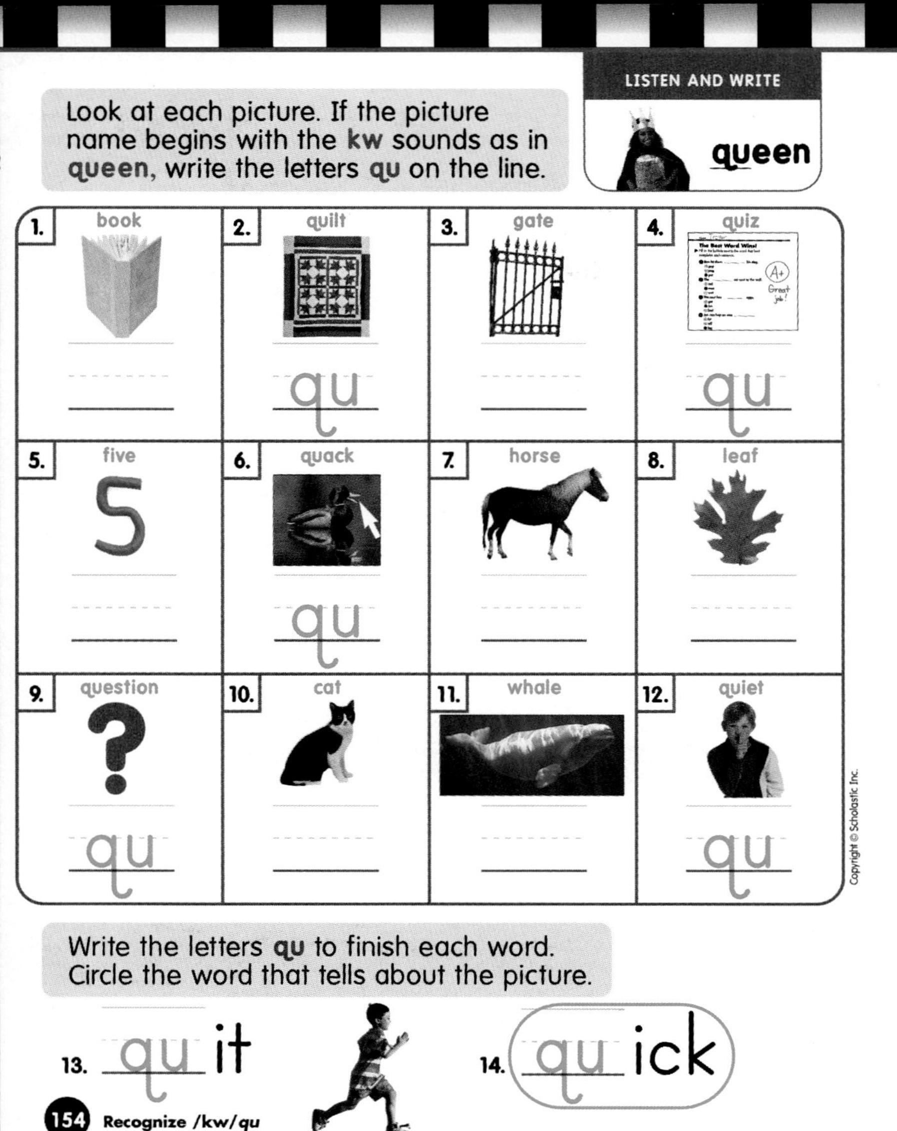
LISTEN AND WRITE

queen

Look at each picture. If the picture name begins with the kw sounds as in queen, write the letters qu on the line.

1. book	2. quilt qu	3. gate	4. quiz qu
5. five	6. quack qu	7. horse	8. leaf
9. question qu	10. cat	11. whale	12. quiet qu

Write the letters qu to finish each word. Circle the word that tells about the picture.

13. qu it 14. qu ick

154 Recognize /kw/qu

Copyright © Scholastic Inc.

Integrated Curriculum

WRITING CONNECTION

SILLY SENTENCES Have children suggest words that begin with ***/kw/qu***. List these words on the chalkboard or on chart paper. Then have children use as many of these words as they can in sentences. Have children dictate these sentences. You may wish to suggest a model sentence such as ***The queen quit sewing the quilt***. Write the sentences on the chalkboard. Have volunteers underline the letters ***qu*** in each word.

MATH CONNECTION

QU* FOR QUART** Write the word ***quart on the chalkboard. Read ***quart*** aloud with children as you underline ***qu***. If possible, display a quart container and a plastic measuring cup. Explain that one quart equals four cups. Fill the cup with sand or water four times to illustrate that four cups make a quart. Have children take turns filling the cup to make a quart.

Phonics Connection

Literacy Place: *Team Spirit*
Teacher's SourceBook, pp. T276–277;
Literacy-at-Work Book, pp. 60–61

My Book: *Quick! Quick! Quick!*

Big Book of Rhymes and Rhythms, 1A: "I Quickly Do My Homework," p. 39

Chapter Book: *Fun With Zip and Zap*, Chapter 10

EXTRA HELP

Have children use their My ABC Chart in the Phonics and Word Building Kit to review the formation of ***Qq*** and any other letters that present difficulties. Provide opportunities for children to write the letters in sand, with paint, or on the chalkboard.

CHALLENGE

Have children look in books, magazines, and newspapers for words with ***qu***. Have children work with a partner to write these words on a chart with the following headings: ***Words in Books, Words in Magazines, Words in Newspapers***.

Lesson 94

pages 155-156

Blend and Build Words

QUICKCHECK ✔

Can children:

✔ **orally blend word parts?**

✔ **blend words with /kw/*qu* and phonograms *-ill, -ick*?**

✔ **build words with phonograms *-ill, -ick*?**

If YES go to Read and Write.

TEACH

Develop Phonemic Awareness

Oral Blending State aloud the following word parts, and ask children to orally blend them. Provide corrective feedback and modeling.

/kw/...een	**/kw/...ick**	**/kw/...ilt**
/kw/...iz	**/kw/...it**	**/kw/...estion**

Connect Sound-Symbol Review that *qu* stands for /kw/ as in *quit*. Write *quit* on the chalkboard. Have a volunteer circle *qu*. Model for children how to blend the word. Continue with *quiz* and *quack*.

Introduce the Phonograms Write the phonograms *-ill* and *-ick* on the chalkboard. Point out the sounds these phonograms stand for. Add *w* to the beginning of *-ill*, and model how to blend the word formed. Then add *s* to the beginning of *-ick*, and model how to blend the word *sick*.

READ AND WRITE

Blend Words List the following words and sentence on a chart. Have volunteers read each aloud. Model blending as necessary.

- **quick** **quit** **quiz**
- **will** **hill** **sick**
- **Will she get sick?**

Complete Activity Pages Read aloud the directions on pages 155 and 156. Review the art with children.

Name

Look at each picture. Circle the word that best finishes each sentence. Then write the word on the line.

1.	Did they quit?	(quit) quiz kite
2.	This is a quiz.	quick (quiz) sock
3.	He is quick.	quit kick (quick)

Supporting All Learners

AUDITORY/KINESTHETIC LEARNERS

ANIMAL SOUNDS Have children work together to draw pictures of animals. Then children can use their knowledge of sound-spelling relationships to say and write the sounds these animals make, such as ***yip yap, quack, woof, baa,*** and ***ssssss.***

AUDITORY/KINESTHETIC LEARNERS

LISTEN UP! Dictate sentences such as ***The queen had a quilt*** and ***Nick was sick.*** Have children write the sentences on paper. Then write the sentences on the chalkboard, and have children make any necesssary corrections.

KINESTHETIC LEARNERS

LETTER QUILT Have children cut out squares of colored construction paper. On each square, children should write a word with **/kw/*qu*.** If possible, children should also illustrate the word. Then have children use a hole punch and yarn to "sew" the squares together to make a letter quilt.

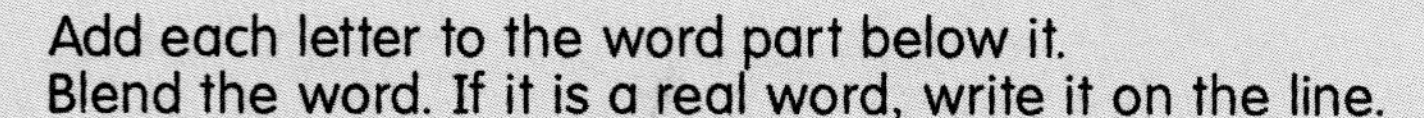
Add each letter to the word part below it.
Blend the word. If it is a real word, write it on the line.

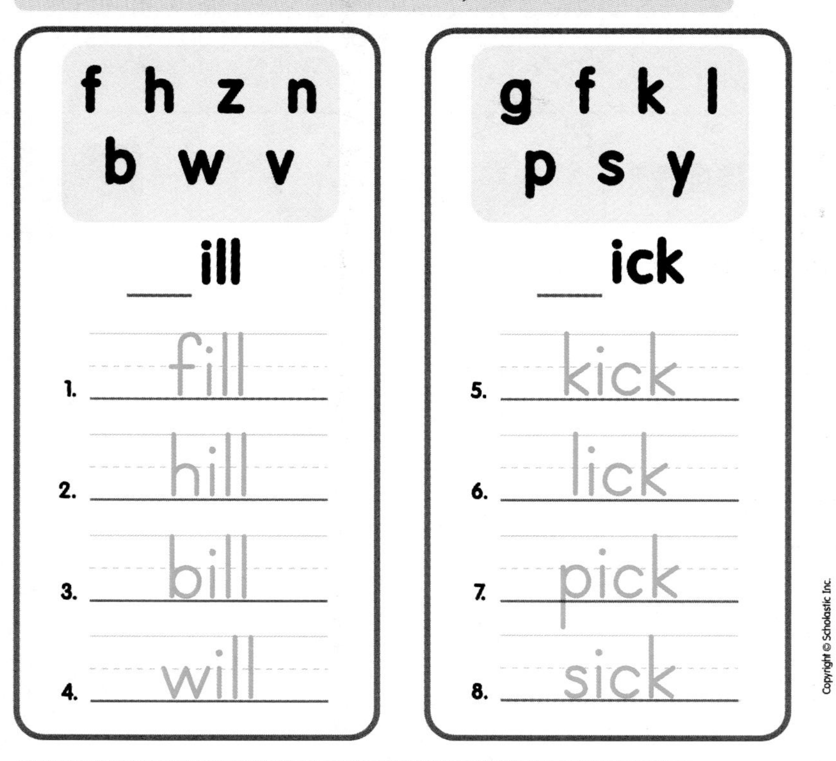

Write a sentence using one of the words you made.

9. Answers will vary.

Integrated Curriculum

SPELLING CONNECTION

RHYME TIME Ask children to suggest words that rhyme with ***will*** and ***sick***. List these words on the chalkboard in separate columns. Have volunteers underline the phonogram ***-ill*** or ***-ick*** in each one. Be sure to include the words ***hill, fill, lick, pick,*** and ***quick***. Blend these words as children repeat them aloud. Encourage children to look for the word parts ***-ill*** and ***-ick*** as they read.

SOCIAL STUDIES CONNECTION

QUIET! Invite children to work together to make a map of the school. Have children write or dictate labels for each room of the school. Then have children write ***quiet*** on sticky notes. Tell children to place the sticky notes on the map, in places or rooms where they need to be quiet. To extend the activity, have children draw maps of their homes and place ***quiet*** sticky notes in places where it is best to be quiet.

CHALLENGE

Children can make up sentences with the phonograms ***-ill*** and ***-ick*** and words with **/kw/**. You may wish to provide a model sentence such as ***The quick queen will pick up the quilt***.

Read aloud one of The Book Shop books. Have children clap whenever they hear a word with **/kw/**. Then have children identify the letters that stand for **/kw/**.

EXTRA HELP

Write the word parts ***-ill*** and ***-ick*** on the chalkboard. (If available, use a pocket chart and letter cards.) Say a word, and have children add a letter to the beginning of each phonogram to form it. Have children replace the initial consonant in the first word to build a second word. Children may build these words: ***sick, kick, thick, quick***.

- **The Queen's Cat**
  by Margaret Mahy
  A queen mistakes a goat for a pet cat.
  (/kw/qu)
- **The Quilt Story**
  by Tony Johnston
  A quilt links two little girls across generations.
  (/kw/qu)

Lesson 95

pages 157-158

Blend Words With /sh/*sh*

QUICKCHECK ✔

Can children:

✔ **replace sounds?**

✔ **identify /sh/?**

✔ **blend words with /sh/*sh*?**

If YES go to Read and Write.

TEACH

Develop Phonemic Awareness

Phonemic Manipulation Explain to children that you are going to read a list of words. Children are to replace the last sound in each word with /**sh**/. Use the following words:

- **wig**
- **dip**
- **crab**
- **fit**
- **dad**
- **rub**
- **hug**
- **rat**
- **trap**

Connect Sound-Symbol Write the words ***sip*** and ***ship*** on the chalkboard as you read each aloud. Ask children what sounds they hear at the beginning of each word and how the words are different. Point out that /**sh**/ at the beginning of ***ship*** is spelled with two letters together, ***s*** and ***h***.

Then write the following word parts on the chalkboard: ***_op, _ot, fi_, ru_***. Have volunteers add the letters ***sh*** to each one. Model for children how to blend each word. Then have children write each word on a sheet of paper.

READ AND WRITE

Blend Words List the following words and sentences on a chart. Model how to blend the words. Then have children blend them.

- **rash** **wish** **shut**
- **shop** **cash** **dish**
- **We shop for a brush.**
- **They rush to the ship.**

Complete Activity Pages Read aloud the directions on pages 157–158. Review each picture name with children.

LISTEN AND WRITE

Name

Look at each picture. If the picture name begins with the **sh** sound as in **shark**, write the letters **sh** on the first line. If it ends in **sh** as in **wish**, write **sh** on the second line.

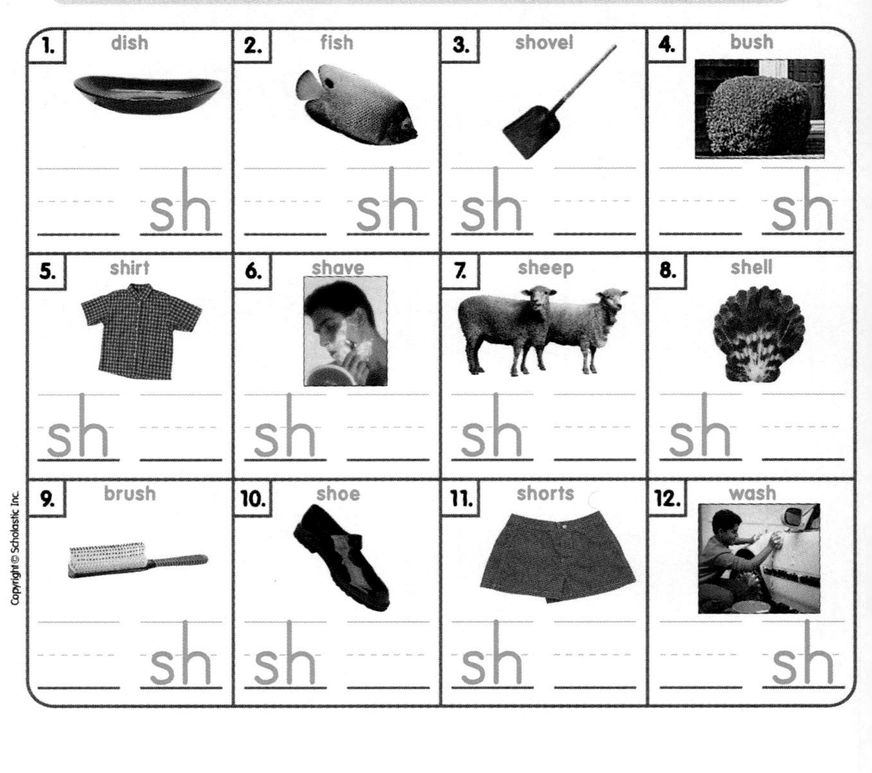

Supporting All Learners

AUDITORY/VISUAL LEARNERS

WORD SEARCH Read aloud "Where Has My Little Dog Gone?" on pages 36–37 in the *Big Book of Rhymes and Rhythms, 1A*. Have children identify all the words with **/sh/** in the rhyme. Write these words on the chalkboard. Have volunteers circle the letters ***sh*** in each one. Invite volunteers to add other **/sh/** words to the list.

VISUAL/KINESTHETIC LEARNERS

WORD CARDS Have children write the words from the Connect Sound-Symbol exercise on note cards. They can use the note cards as flash cards to practice reading the words.

TACTILE LEARNERS

SANDPAPER WORDS Print words with **/sh/** on sandpaper using a marker. Invite children to blend the word and say each letter as they trace it with their fingers.

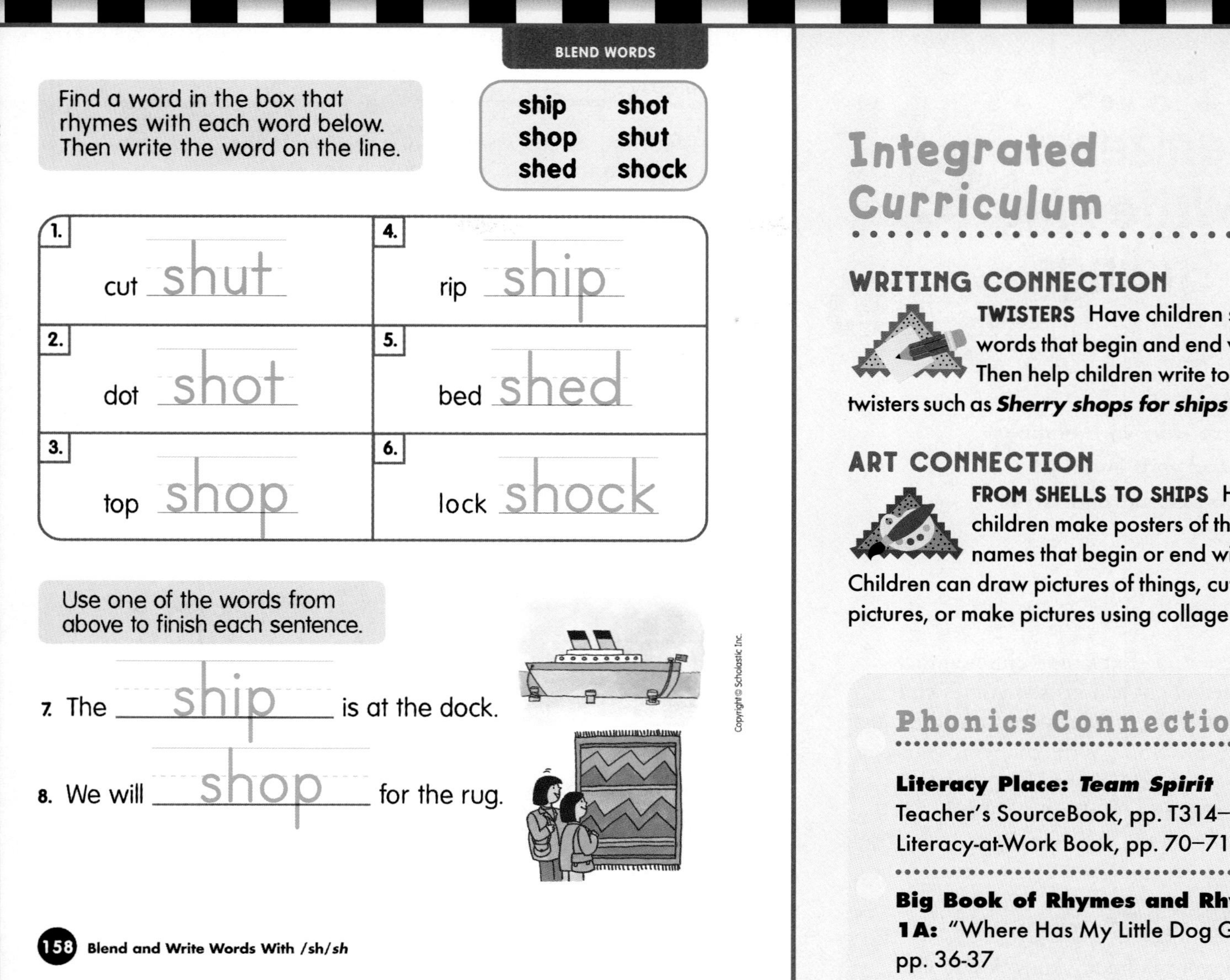
BLEND WORDS

Find a word in the box that rhymes with each word below. Then write the word on the line.

ship	shot
shop	shut
shed	shock

1. cut shut	4. rip ship
2. dot shot	5. bed shed
3. top shop	6. lock shock

Use one of the words from above to finish each sentence.

7. The ship is at the dock.

8. We will shop for the rug.

Copyright © Scholastic Inc.

158 Blend and Write Words With /sh/*sh*

Integrated Curriculum

WRITING CONNECTION

TWISTERS Have children suggest words that begin and end with **/sh/*sh***. Then help children write tongue twisters such as ***Sherry shops for ships and shells.***

ART CONNECTION

FROM SHELLS TO SHIPS Have children make posters of things with names that begin or end with **/sh/*sh***. Children can draw pictures of things, cut out pictures, or make pictures using collage materials.

Phonics Connection

Literacy Place: *Team Spirit*
Teacher's SourceBook, pp. T314–315; Literacy-at-Work Book, pp. 70–71

Big Book of Rhymes and Rhythms, 1A: "Where Has My Little Dog Gone?" pp. 36-37

Chapter Book: *Fun With Zip and Zap*, Chapter 11

ESL

Play a circle game with children to build their oral vocabularies. Go around the circle, and have children name things they like to buy when shopping. Then have children name types of boats, including ships. Finally, have children name farm animals such as sheep. Continue with other topics.

EXTRA HELP

Have children look through books you have in the classroom for pictures of things whose names contain **/sh/**. Then have children work together to find words with **/sh/*sh***. Help children identify the words. Write the words they find on the chalkboard or on chart paper.

- **Sheep on a Ship**
  by Nancy Shaw
  Merry misadventures occur on a pirate ship. **(/sh/*sh*)**

- **Shoes Like Miss Alice's**
  by Angela Johnston
  A girl admires her sister's shoes. **(/sh/*sh*)**

Lesson 96

pages 159-160

Recognize and Write *-ing* Verb Endings

QUICKCHECK ✔

Can children:

✔ **recognize *-ing* verb endings?**

✔ **write *-ing* verb endings?**

If YES go to Read and Write.

TEACH

Introduce *-ing* Explain to children that words that tell about actions sometimes end with *-ing*. Often these words are formed by just adding *-ing* to a word. Write the word ***pack*** on the chalkboard. Use the word in a sentence. Then write ***packing*** on the chalkboard. Ask a volunteer to circle the word ***pack,*** and ask another volunteer to draw a square around ***-ing***.

THINK ALOUD

There is a word I already know inside the word *packing.* It is the word *pack.* I also know there is another word part in the word packing. It is *-ing.* Both words, *pack* and *packing,* name an action. I can *pack* a bag, or I can be *packing* a bag. *Packing* is a form of the word *pack.*

Write the following words on the chalkboard: ***filling, rushing, locking, picking, fixing.*** Have children underline ***-ing*** in each one.

READ AND WRITE

Practice Write the following pairs of sentences on the chalkboard. Ask children to tell what is the same and what is different about the underlined words in each pair.

- **I tell the dog to sit.**
  I am telling the dog to sit.
- **She will fix the van.**
  She is fixing the van.

Complete Activity Pages Read aloud the directions on pages 159 and 160.

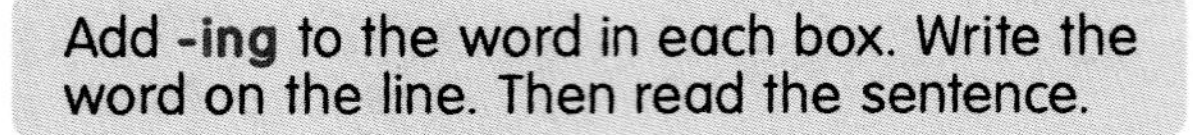

Name

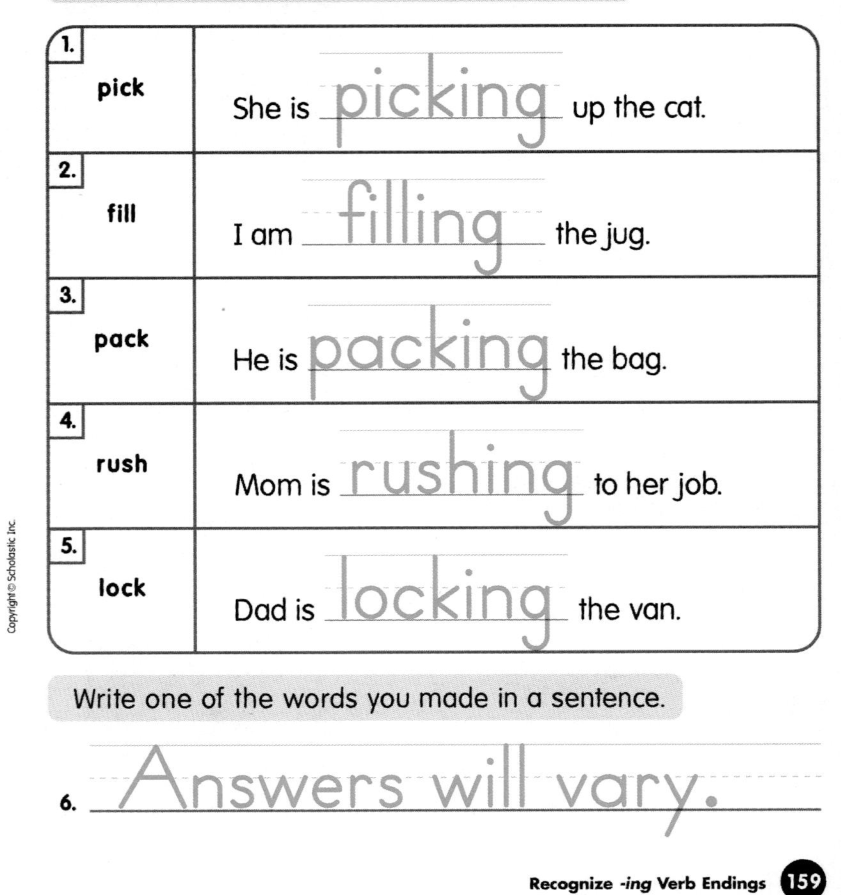

Add **-ing** to the word in each box. Write the word on the line. Then read the sentence.

1. pick	She is picking up the cat.
2. fill	I am filling the jug.
3. pack	He is packing the bag.
4. rush	Mom is rushing to her job.
5. lock	Dad is locking the van.

Write one of the words you made in a sentence.

6. Answers will vary.

Supporting All Learners

KINESTHETIC LEARNERS

CHARADES Write words such as ***packing, licking, fixing, rushing, locking,*** and ***fishing*** on note cards. Invite children to play charades by acting out the words, one at a time. Volunteers can guess each word and spell it.

VISUAL LEARNERS

WORD CHECKERS To review ***-ing*** verb forms and high-frequency words, put a self-sticking note on each square of a checkerboard. On each note, write an ***-ing*** verb form or a high-frequency word. Have children play checkers using the board. Tell them they must read the words on the squares they want to jump over or land on.

Circle the word that best finishes each sentence. Then write the word on the line.

	Sentence	Choices
1.	We can mix up the hats.	(mix) mixing
2.	Jan is fixing the van.	fix (fixing)
3.	The bus is backing up.	back (backing)
4.	Tom is picking up the bat.	pick (picking)
5.	Did he wish for a pet?	(wish) wishing
6.	The dog is licking the cat.	lick (licking)

Integrated Curriculum

WRITING CONNECTION

WRITE A RHYME Have children suggest words with ***-ing.*** Write these words on the chalkboard. Then read aloud the rhyme "The Cat's in the Cupboard" on page 18 in the *Big Book of Rhymes and Rhythms, 1A*. Write the rhyme on the chalkboard. Read it aloud. Have children underline the words with ***-ing***. Then read the rhyme again, but erase the words ***counting*** and ***hiding***. Have children suggest words to complete the sentences ***The cat's in the cupboard*** ____ and ***I'm in the backyard*** ____. You may wish to suggest words such as ***sleeping, eating, hopping, running,*** and ***playing***. Write the words children suggest in the sentences.

MATH CONNECTION

COUNTING Read aloud or have children read aloud books in the classroom. Have children find and count words with ***-ing***. Write the words children find on the chalkboard. Count the number of ***-ing*** words per book. Ask children which book has the most ***-ing*** words and which book has the fewest ***-ing*** words.

CHALLENGE

Have children work together to make rhyming-word puzzles with the verb ending ***-ing.*** You can use the following puzzle as a model: ***fill + ing = filling.***

Make a set of duplicate word cards for ***-ing*** words such as ***packing, picking, filling, telling, mixing, rushing, fishing,*** and ***wishing***. To help children acquiring English become familiar with the words, display each word, and read it aloud chorally. Then pair children with partners whose primary language is English to play Concentration with the cards.

Lesson 97

page 161

Build Words

QUICKCHECK ✔

Can children:

✔ **build words?**

If YES go to Read and Write.

TEACH

Review Sound-Spellings Review with children the sound-spelling relationships from the past few lessons. These include **/sh/*sh*** and **/kw/*qu***. You may also wish to review the phonograms ***-ill*** and ***-ick***.

Then write the word ***pick*** on the chalkboard. Tell children that you want them to replace **/p/**, the first sound in ***pick***, with **/kw/** to make a new word. Ask children what letters stand for **/kw/**. Then replace the letter ***p*** in ***pick*** with the letters ***qu***, and blend the word formed. Continue by changing one letter or letter pair to build new words.

READ AND WRITE

Build Words Distribute the following letter cards to children: ***a, c, h, i, k, o, p, s***. If children have their own set of cards, have them locate the letter set. Provide time for children to build as many words as possible using the letter cards. Suggest that they record their words on a separate sheet of paper.

Then explain to children that you will give them the beginning and ending word of a three- or four-word ladder. They are to change one letter or letter pair in each word to complete the word ladder. Use the following words:

- **will** **(hill)** **hit**
- **ship** **(shop)** **mop**
- **sick** **(pick)** **(pack)** **pad**

Note: The words in parentheses are possible answers.

Complete Activity Page Read aloud the directions on page 161. Review each picture name with children.

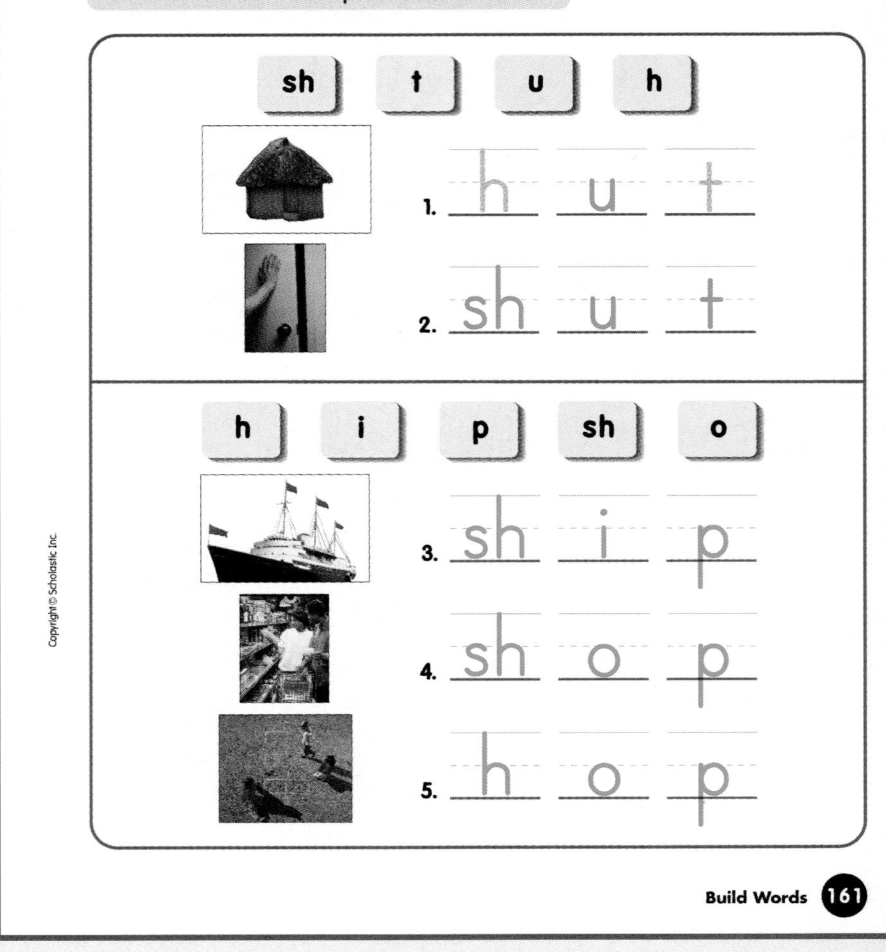
BUILD WORDS

Name

Look at each picture. Use the letter tiles to write each picture name.

sh t u h

1. h u t

2. sh u t

h i p sh o

3. sh i p

4. sh o p

5. h o p

Build Words 161

Supporting All Learners

KINESTHETIC/VISUAL LEARNERS

MAGNETIC LETTERS Using the magnetic letters from the Phonics and Word Building Kit, or other alphabet manipulatives, build the word ***sick***. Model for children how to blend the word. Then ask a volunteer to form the word ***lick*** and to blend it aloud. When completed, mix the letters, and ask another volunteer to reform the word. Continue with other words.

CHALLENGE

To practice building words with phonograms, children might enjoy playing "Build a House" on page 21 in *Quick-and-Easy Learning Games: Phonics*.

POETRY

Read the poem. Write a sentence telling about the boy's wish.

Wishing for a Fish

I am wishing for a fish.
I will not quit.
I am wishing for a fish.
I sit and sit.

I am wishing for a fish.
I will not rush.
I am wishing for a fish.
Shh . . . hush, hush, hush.

I am wishing for a fish.
Will I get one?
I am wishing for a . . . TUG!
Quick! Did I get one?

Answers will vary.

Supporting All Learners

VISUAL LEARNERS

GRAPH LETTERS Have children collect and count words with **/kw/*qu*** and **/sh/*sh***. Children can record the results of their counts in a graph such as the one shown.

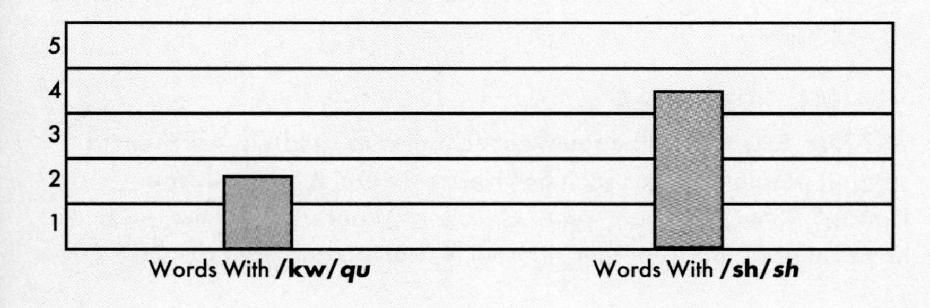

KINESTHETIC/VISUAL LEARNERS

PHONICS AND WORD BUILDING KIT Give children the ***sh*** cards, letter cards, and familiar phonograms from the Phonics and Word Building Kit. Invite them to make as many words as they can that begin or end with ***sh***. Have children share their words and add them to the Word Wall.

Write and Read Words With /kw/*qu* and /sh/*sh*

QUICKCHECK ✔

Can children:

✓ spell words with /kw/*qu* and /sh/*sh*?
✓ read a poem?
If YES go to Read and Write.

TEACH

Link to Spelling Review the following sound-spelling relationships: **/kw/*qu*** and **/sh/*sh***. State aloud one of these sounds. Have a volunteer write the spelling that stands for the sound. For example, the letters ***qu*** stand for **/kw/**. Continue with ***sh***.

Phonemic Awareness Oral Segmentation
State aloud the word ***ship***. Have children orally segment the word. (/sh/ /i/ /p/) Ask them how many sounds the word contains. *(3)* Ask them which sound is shown by two letters. Draw three connected boxes on the chalkboard, and have a volunteer write the spelling that stands for each sound in the word ***ship*** in the appropriate box. Continue with the word ***fish***.

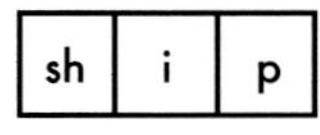

READ AND WRITE

Dictate Dictate the following words and sentence. Have children write them on a sheet of paper. When children are finished, write the words and sentence on the chalkboard, and have children make any necessary corrections on their papers.

- **quick** **fish** **quit**
- **or** **some** **with**
- **Get the big dish.**

Complete Activity Page Read aloud the directions on page 162. Have children read the poem independently or with partners.

Lesson 99

pages 163-164

Read and Review Words With /kw/*qu*, /sh/*sh*

QUICKCHECK ✔

Can children:

✔ read words with /kw/*qu*, /sh/*sh*?

✔ recognize high-frequency words?

If YES go to Read and Write.

TEACH

Review Sound-Spellings Review with children the sound-spelling relationships from the past few lessons. These include **/kw/*qu*** and **/sh/*sh***. State aloud one of these sounds. Have a volunteer write the spelling that stands for the sound on the chalkboard. For example, ***qu*** stands for **/kw/**. Continue with **/sh/**. Then display pictures of objects whose names begin with one of these sounds. Have children write the letters that each picture's name begins with as the picture is displayed.

Review High-Frequency Words Review the high-frequency words ***some, have, do, or,*** and ***other***. Write on the chalkboard sentences containing each high-frequency word. Say one word, and have volunteers circle the word in the sentences.

READ AND WRITE

Build Words Distribute the following letter cards to children: ***f, h, i, p, s, w***. Provide time for children to build as many words as possible. Suggest that they record their words on a sheet of paper. Have them continue with the letter cards ***a, c, d, h, r, s*** and ***h, o, p, s, t, u***.

Build Sentences Write the following words on note cards: ***I, they, quit, wishing, rocking, packing, are, am***. Display the cards. Have children make sentences using the words. To start, suggest the sentence ***I quit***.

Complete Activity Pages Read aloud the directions on pages 163–164. You may wish to review the art with children.

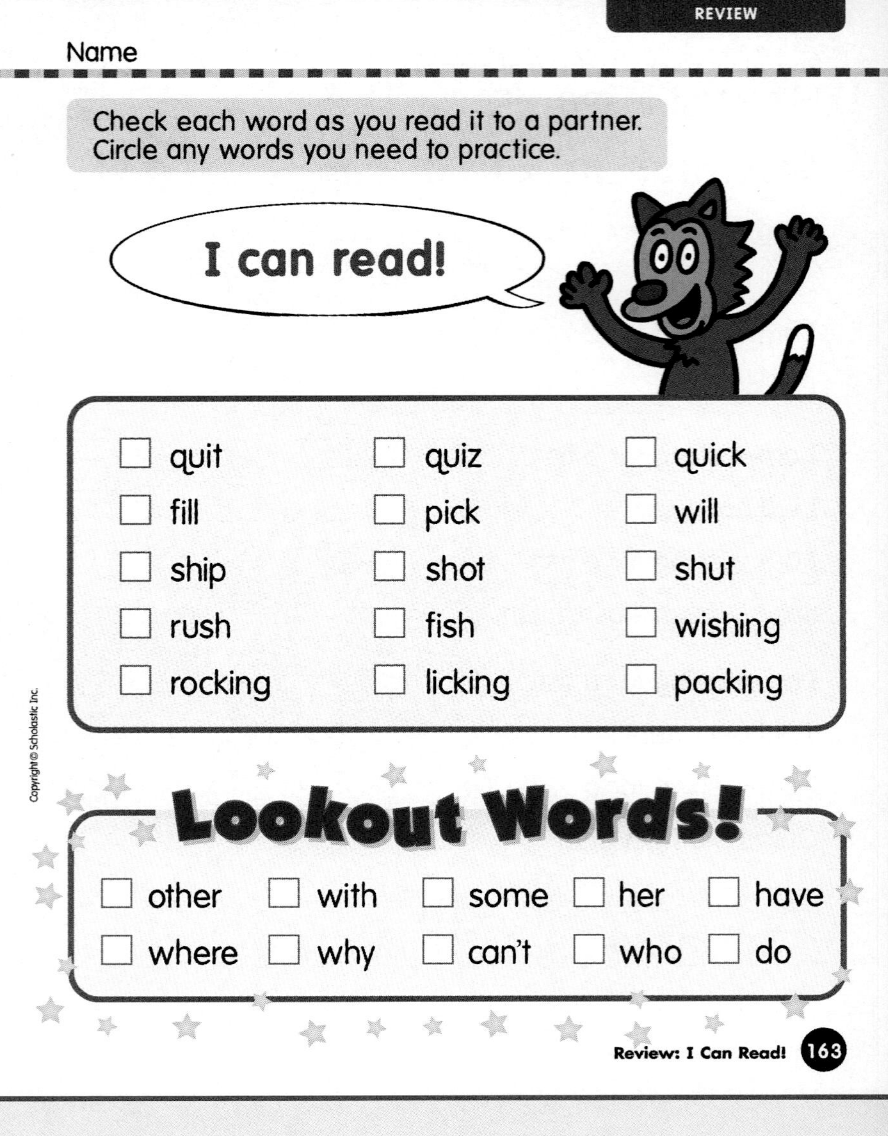

Supporting All Learners

KINESTHETIC/AUDITORY LEARNERS

ADD A RHYME Invite children to sit in a circle. Write the following words on word cards: ***quick, shop, shin, fish, quack,*** and ***quit***. Invite children to read each word and add a rhyming word. Pass the card around. When the card completes the circle, read the words on it aloud, or have a volunteer read the words.

VISUAL LEARNERS

WANTED: SIGNS Walk about your school with children. Ask them to suggest places for signs such as "Please Be Quiet," "Wash Your Hands," "Shut the Door," and "Questions Wanted!" You may wish to have children suggest other signs with **/kw/*qu*** and **/sh/*sh***.

Fill in the bubble next to the sentence that tells about each picture.

1.	● She hugs the pups. ○ The vet looks at the dogs.
2.	○ Bob is picking up the bugs. ● Bob is rushing for the bus.
3.	○ We did not get the quiz yet. ● Quick! Let Meg get in the shed.
4.	○ Max is getting into bed. ● Max is filling up the pots.
5.	● We will lock up the shop. ○ She gets on the ship.
6.	○ The boy gets the pots and pans. ● Yes, that is a big fish in my dish.

Look carefully at the picture before choosing the sentence.

Integrated Curriculum

WRITING CONNECTION

WRITE ABOUT YOU Have children use words with ***/kw/qu*** and ***/sh/sh*** as well as high-frequency words to write or dictate sentences about themselves or their family. Encourage children to use as many words with ***/kw/qu*** and ***/sh/sh*** and high-frequency words as possible. You may wish to have children illustrate their sentences, too. Then children can share their sentences and point out the words with ***/kw/qu, /sh/sh,*** and the high-frequency words.

TECHNOLOGY CONNECTION

MY BOOKS/WIGGLEWORKS PLUS Have children use the Magnet Board to build words with ***qu***. Suggest that they begin with one of the following words: ***quit, quiz, quick***. For additional practice with ***/kw/qu*** and ***/sh/sh,*** have children read the following My Books: *Quick, Quick, Quick* and *Shoes*. For information on using the My Books on the computer, see the WiggleWorks plus My Books Teaching Plan.

I CAN READ! OPTIONS

The I Can Read! page can be used for one or all of the following:

- **paired reading**
- **individual assessment**
- **choral reading**
- **homework practice**
- **program placement**

CHALLENGE

Draw a large ladder on chart paper. Invite children to "climb" the ladder by naming words that begin with ***/kw/qu*** and ***/sh/sh***. Write the words on the ladder, one word per rung, beginning with the bottom rung, and working toward the top. Remind children to look for words with ***/kw/qu*** and ***/sh/sh*** when they read. Make a second word ladder on which to add the words children find. Have children compare their word ladders.

EXTRA HELP

For children needing additional phonemic awareness training, see the Scholastic Phonemic Awareness Kit. The oral segmentation exercises will help children break apart words sound by sound. This is necessary for children to be able to encode, or spell, words while writing.

pages 165-166

Read Words in Context

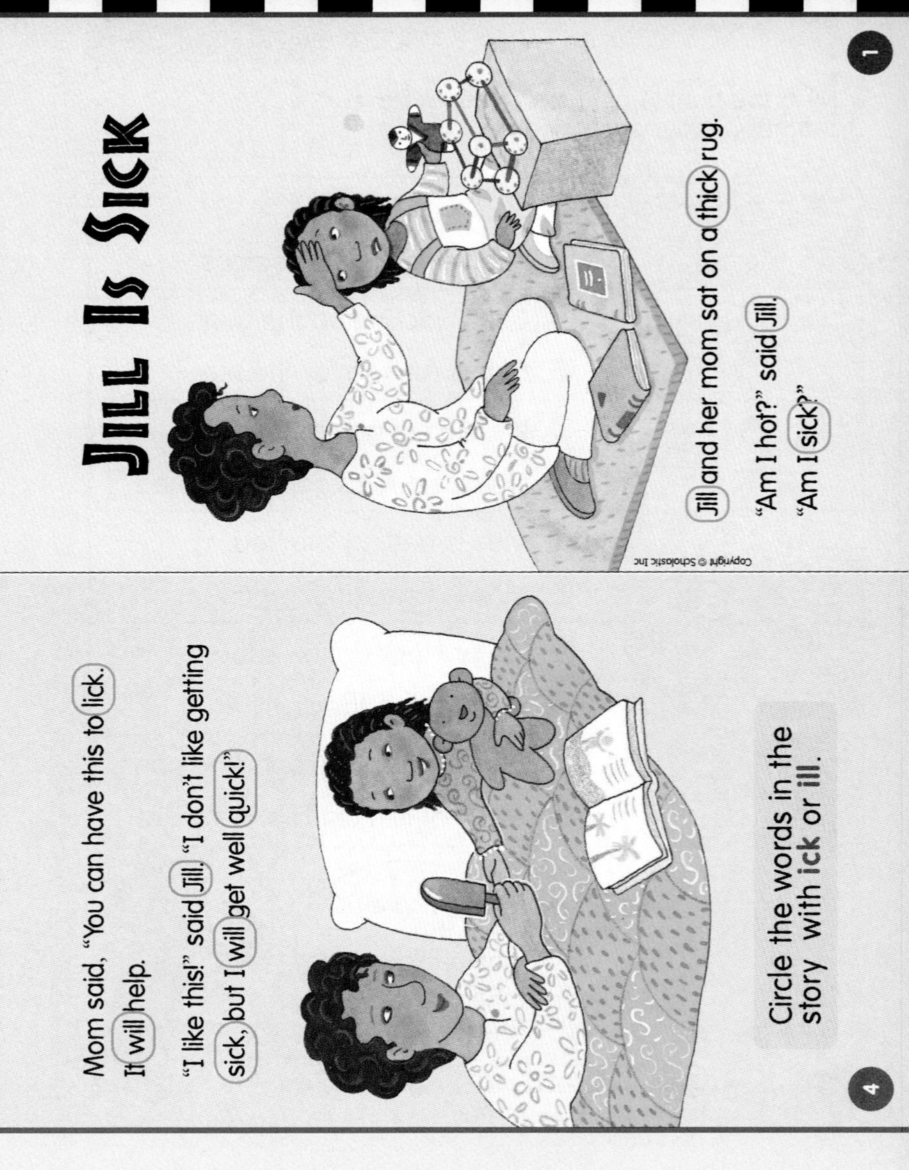

TEACH

Assemble the Story Ask children to remove pages 165–166. Have children fold the pages in half to form the Take-Home Book.

Preview the Story Preview *Jill Is Sick,* a story about a sick girl and the things that will help her get well. Invite children to browse through the first two pages and to comment on anything they notice. Ask children to point out any unfamiliar words. Read these words aloud. Model how to blend them. Then have children predict what the story might be about.

READ AND WRITE

Read the Story Read the story aloud, or have volunteers take turns reading aloud a page at a time. Discuss with children anything of interest on each page. Encourage them to help each other with any blending difficulties. The following prompts may help children who need extra support while reading:

- **What letter sounds do you know in the words?**
- **What word parts do you know?**

Reflect and Respond Have children share their reactions to *Jill Is Sick*. How is it like other stories they have read? What did they like about the story? Encourage children to think of more details about being sick that they would like to add to the story.

Develop Fluency Have children reread the story as a choral reading, or have partners reread the story independently. Provide time for children to reread the story on subsequent days to develop fluency and increase reading rate.

Ask children how they figure out unfamiliar words when reading. Ask them to model how they blend words. Continue to review confusing or challenging sound-spelling relationships for children needing additional support.

Supporting All Learners

VISUAL/AUDITORY LEARNERS

READERS THEATER Encourage children to act out the roles of Jill and her mom. Children can say each speaker's part dramatically, and they may even want to use simple props to suggest a quilt, a bed with pillows, or a frozen-juice pop.

VISUAL LEARNERS

HOW MANY? Invite children to read the Phonics Readers *The Quick Duck* and *Fish Wish*. Ask children to count how many **/kw/qu** words are in *The Quick Duck* and how many **/sh/sh** words are in *Fish Wish*. Have them graph their findings.

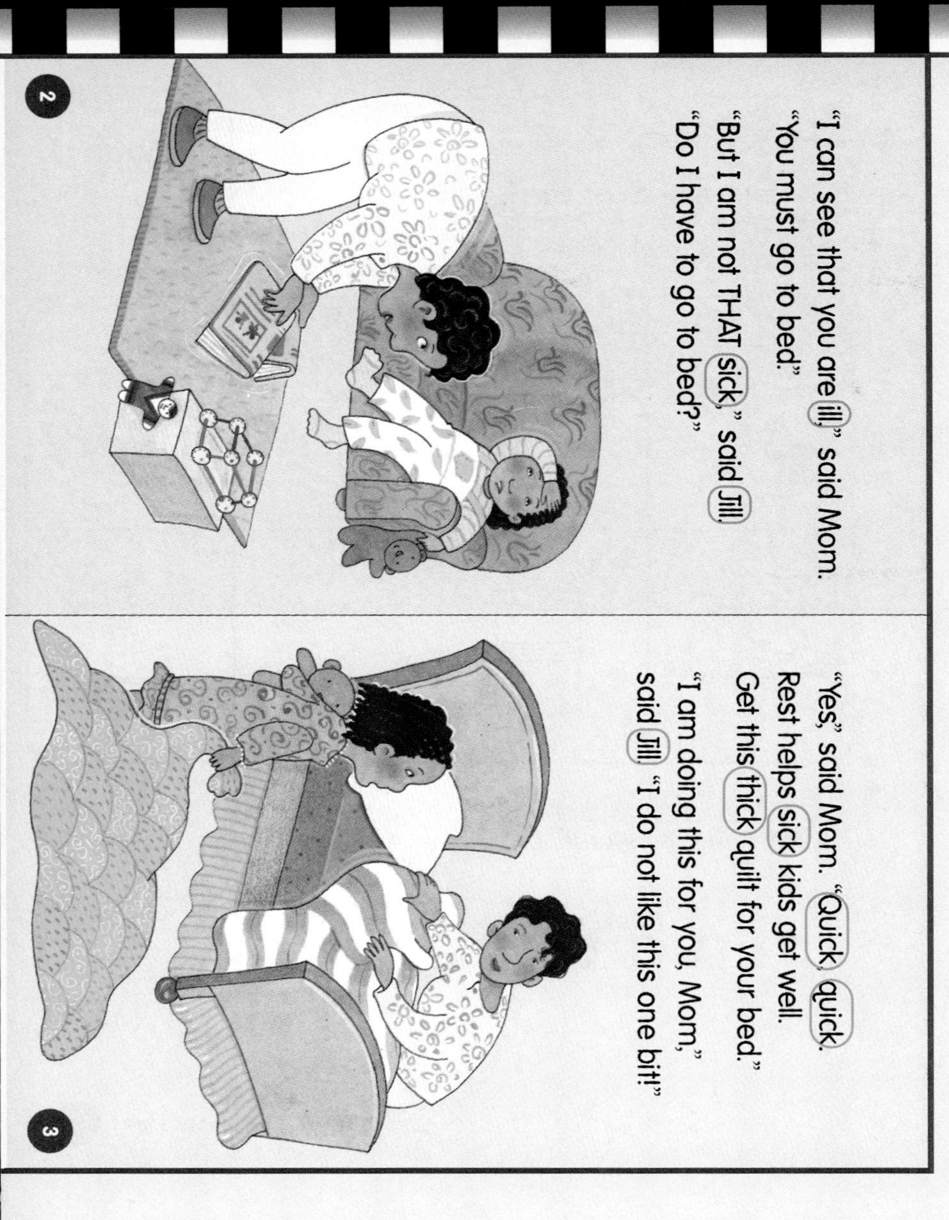

Reflect on Reading

ASSESS COMPREHENSION

To assess their understanding of the story, ask children questions such as:

- **Who are the two people in the story?** ***(Jill and her mom)***
- **What does Jill's mom want her to do?** ***(get bed rest, stay warm under a quilt, and lick a frozen-juice pop)***

HOME-SCHOOL CONNECTION

Send home *Jill Is Sick*. Encourage children to read the story to a family member.

WRITING CONNECTION

WHEN I AM SICK Have children write or dictate stories such as *Jill Is Sick* and use themselves as the main character. Ask children what they say when they are sick and what relatives or friends tell them to do to feel better. Have children illustrate their stories and then take turns sharing them.

CHALLENGE

Have children use words with ***/kw/qu*** and ***/sh/sh*** to add to the story *Jill Is Sick* or write another short, simple story about Jill and her Mom.

EXTRA HELP

Read *Jill Is Sick* aloud with children. Then read the story again with children, but this time, have children change their voices to read the parts of Mom and Jill. Encourage children to ask questions about the story when you are done reading each page.

Phonics Connection

Phonics Readers:
#34, *The Quick Duck*
#35, *Fish Wish*

Lesson 101

pages 167-168

Blend Words With /ā/a-e

QUICKCHECK ✔

Can children:

- ✔ **recognize /ā/?**
- ✔ **identify the sound-spelling pattern *a-e*?**
- ✔ **blend words with /ā/*a-e*?**

If YES go to Read and Write.

TEACH

Develop Phonemic Awareness

Auditory Discrimination Explain to children that they are going to listen for words with /ā/ as in *make*. When they hear /ā/, children are to say /ā/. Use these and other words: ***lake, tape, rain, pan, stay, sad, late, black, name, bake, ran, paint***.

Connect Sound-Symbol Write ***tap*** and ***tape*** on the chalkboard as you read each aloud. Ask children what vowel sound they hear in each word and how the words are different. Point out that the long ***a*** sound in ***tape*** is made by adding ***e***. This is called the ***e***-marker. Write _ ***a*** _ ***e*** on the chalkboard. Have children make words by filling in the blanks with consonants. Use ***make*** to model this procedure.

If I put an *m* in the first blank and a *k* in the second blank, I can make a new word. Listen as I blend the sounds in this word. What word will I form if I use the letters *t* and *k*?

READ AND WRITE

Blend Words List the following on a chart. Have children blend them independently.

- **mad made can cane**
- **take came gate bake**
- **He sat on a rock by the lake.**

Complete Activity Pages Read aloud the directions on pages 167–168. Review each picture name with children.

Name

Look at each picture. If the picture name has the **long a** sound as in **gate**, color the picture.

Write the letters **a** and **e** to finish each word. Read the words to a friend.

9. m a d e   10. l a t e

Supporting All Learners

VISUAL LEARNERS

a-e SEARCH Ask children to suggest other words that contain the long ***a*** sound. List these words on the chalkboard. Have volunteers circle words that follow the ***a-e*** pattern.

KINESTHETIC LEARNERS

VOWEL CHECKERS To practice and review words with the ***e***-marker, children may enjoy playing "Vowel Checkers" on page 26 in *Quick-and-Easy Learning Games: Phonics*.

TACTILE LEARNERS

MAKE WORDS Have children write large **/ā/** words spelled ***a-e***, one word per page. If possible, have children paste dried macaroni on the ***a*** and ***e***. Then children can say the word as they trace the letters with their fingers.

Circle the word that best finishes each sentence. Then write the word on the line.

1. I like to bake cakes.	game (bake) bet
2. Dad and I will bake a cake.	came (cake) can
3. We will make it for my mom.	mat made (make)
4. Then we will take it to her.	date (take) tag
5. She will like the cake we made.	mad (made) mat
6. We will have some cake at the lake.	(lake) late pan

Integrated Curriculum

WRITING CONNECTION

BAKE AN A-E CAKE Have children write or dictate make-believe recipes for silly cakes such as a **game cake** made out of game pieces and flour. Children can illustrate their recipes with drawings of their cakes.

ART CONNECTION

CAKE ART Have children draw, color, and cut out cakes, at least one cake per child. If possible, have children decorate their cakes with collage materials. Then have children write ***a-e*** words on the back of their cakes. You may wish to have a "bake sale," where children display their cakes and share the words they wrote on them.

Phonics Connection

Literacy Place: *Imagine That!*
Teacher's SourceBook, pp. T54–55;
Literacy-at-Work Book, pp. 12–13

My Book: *Make a Face*

EXTRA HELP

For children needing additional phonemic awareness training, see the Phonemic Awareness Kit. The oral blending exercises will help children orally string together sounds to form words. This is necessary for children to be able to decode, or sound out, words while reading.

CHALLENGE

Make or have children assemble flip books with four sections. The top word should be ***made***. Under the ***m,*** have children write ***w, b, f,*** and ***t,*** one letter to a page. Under the ***d,*** write ***t, k, m,*** and ***n,*** one to a page. The ***a*** and the ***e*** remain constant in this book. Children can take turns reading the words in their flip books and making sentences using these words.

- **Jake Baked a Cake**
  by B.G. Hennessy
  The whole town prepares for a wedding. **(/ā/*a-e*)**
- **The Bravest Flute**
  by Ann Grifalconi
  A Mayan boy's flute playing takes on special meaning. **(/ā/*a-e*)**

Lesson 102

pages 169-170

Blend Words With /ī/i-e 9

QUICKCHECK ✔

Can children:

✔ **recognize /ī/?**

✔ **identify the spelling pattern *i-e*?**

✔ **blend words with /ī/*i-e*?**

If YES go to Read and Write.

TEACH

Develop Phonemic Awareness

Oddity Task Show the following picture cards: **kite, dig,** and **five**. Say the picture names. Ask children which two words have the same middle sound. Continue with the following picture cards: **nine, fish, bike; six, mice, dice.**

Connect Sound-Symbol Write *bit* and *bite* on the chalkboard as you read each aloud. Ask children what vowel sound they hear in each word and how the sounds are different. Point out the *e*-marker at the end of *bite*. Write _ *i* _ *e* on the chalkboard. Have children make words by filling in the blanks with consonants. Model this procedure using the word *hide*.

THINK ALOUD

If I put an *h* in the first blank and a *d* in the second blank, I can make a new word. Let's say the sounds in this word as I run my finger under the word. What word will I form if I use the letters *t* and *m*?

READ AND WRITE

BUILT-IN REVIEW

Blend Words List the following words and sentences on a chart. Have volunteers read each aloud. It may be necessary to model how to blend the words.

- **hid** **hide** **kit** **kite**
- **line** **dime** **side** **fine**
- **Tim hid nine times.**
- **That big kite is mine.**

Complete Activity Pages Read aloud the directions on pages 169–170. Review each picture name with children.

Name

9 nine

Look at the picture. If the picture name has the **long i** sound as in **nine**, color the picture.

1. pin	2. line	3. kite	4. mice
5. pig	6. dime	7. five	8. bib
9. six	10. hive	11. fish	12. slide

Write the letters **i** and **e** to finish each word. Circle the word that names the picture.

Supporting All Learners

VISUAL/AUDITORY LEARNERS

FIND /ī/ Read aloud "I Eat My Peas With Honey" on page 5 in the *Big Book of Rhymes and Rhythms, 1B*. Have children find all the words with /ī/ in the rhyme. Write these words on the chalkboard. Invite volunteers to add other /ī/ words to the list.

VISUAL/AUDITORY LEARNERS

i-e SEARCH Ask children to suggest words that contain the long *i* sound. List these words on the chalkboard. Have volunteers circle words with the *i-e* pattern.

VISUAL/KINESTHETIC LEARNERS

MAKE-A-WORD Give two children letter cards for *i* and *e*. Have them stand in front of the room. Give other pairs of children one of these letter card sets: *t* and *m*, *p* and *n*, *l* and *k*, *s* and *d*, *f* and *n*, and *r* and *p*. Invite each of these pairs of children to arrange their letters into a word with the two children already standing at the front of the room. Have the rest of the class read the words.

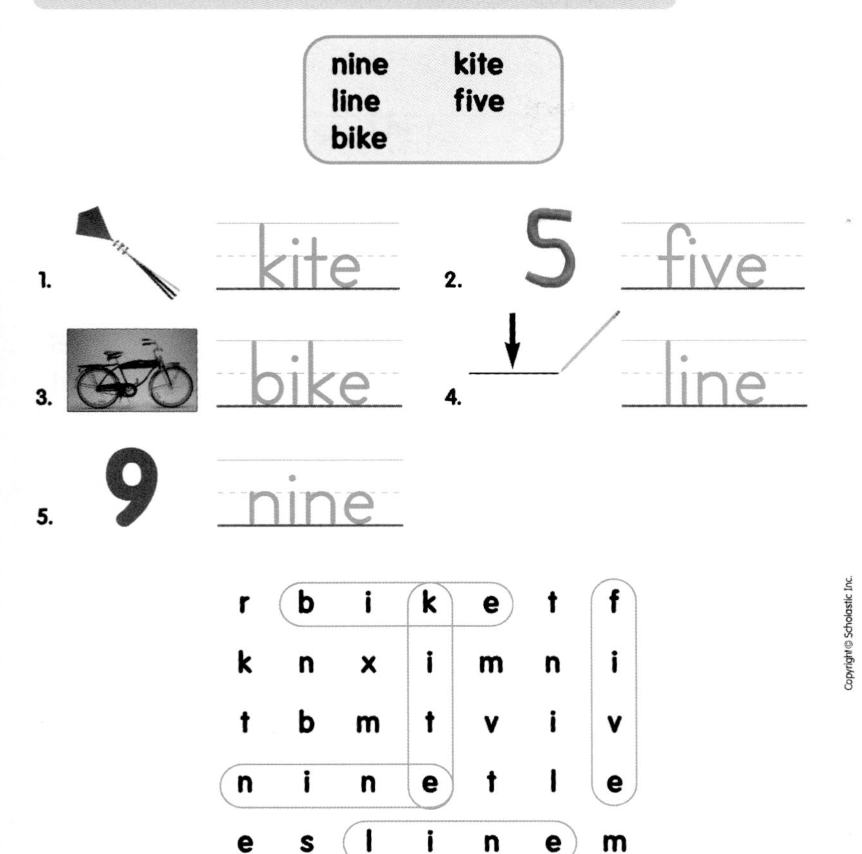
BLEND WORDS

Look at each picture. Write the picture name on the line. Then find the name in the word search.

nine kite
line five
bike

1. kite
2. five
3. bike
4. line
5. nine

r	b	i	k	e	t	f
k	n	x	i	m	n	i
t	b	m	t	v	i	v
n	i	n	e	t	l	e
e	s	l	i	n	e	m

ESL Write ***like, ride,*** and ***bike*** on the chalkboard. Read each word with children, and have them underline the final ***e***. Then sing this song together to the tune of "Row, Row, Row Your Boat" while pantomiming the action: ***Ride, ride, ride a bike/Ride and ride and ride/Ride, ride, ride a bike/How I like to ride!***

EXTRA HELP

Draw a hopscotch grid in the school playground. Instead of numbers, write words with ***i-e*** in each space. Invite children to play hopscotch by tossing a small object and hopping. Each time a child hops on a word, the group should read that word.

Integrated Curriculum

SPELLING CONNECTION

FLY A KITE Children can draw and cut out kite shapes from construction paper. Have children write ___ ***i*** ___ ***e*** on their kites. Then have children write consonants to fill in the blanks and make words. When all the blanks have been filled in, hang the kites and review the words on them.

Phonics Connection

Literacy Place: *Imagine That!*
Teacher's SourceBook, pp. T102–103;
Literacy-at-Work Book, pp. 23–24

My Book: *Mr. and Mrs. Pine*

Chapter Book: *The Puppet Club,* Chapter 1

- **Chicken Soup With Rice**
  by Maurice Sendak
  Celebrate a year of eating soup. **(/ī/*i-e*)**
- **Osa's Pride**
  by Ann Grifaconi
  Osa learns about pride. **(/ī/*i-e*)**

Lesson 103

pages 171-172

Blend Words With /ō/o-e

QUICKCHECK ✔

Can children:

✔ **recognize /ō/**

✔ **identify the sound-spelling pattern *o-e*?**

✔ **blend words with /ō/*o-e*?**

If YES go to Read and Write.

TEACH

Develop Phonemic Awareness

Oddity Task Show pictures of three words, and say their names. Children are to decide which two words have the same vowel sound. Make picture cards or use the following picture cards from the Phonemic Awareness Kit: **dot, bone, rope; nose, home, hot; cone, cot, note; mop, hope, rode.**

Connect Sound-Symbol Write ***hop*** and ***hope*** on the chalkboard as you read each aloud. Ask children what vowel sound they hear in each word and how the sounds are different. Point out the ***e***-marker in ***hope***. Write the long ***o*** pattern ***_o _e*** on the chalkboard. Have children make words by filling in the blanks with consonants. Model this procedure.

THINK ALOUD

If I put a *p* in the first blank and an *l* in the second blank, I can make a word. Let's say the word as I run my finger under it. What word will I form if I use the letters *c* and *d*?

READ AND WRITE

Blend Words List the following words and sentence on a chart. Have children blend them.

- **not** **note** **hop** **hope**
- **rode** **hole** **nose** **vote**
- **We rode home in the bus.**

Complete Activity Pages Read aloud the directions on pages 171–172. Review each picture name with children.

LISTEN AND WRITE

bone

Name

Look at the picture. If the picture name has the **long o** sound as in **bone**, color the picture.

Write the letters **o** and **e** to finish each word. Read the words to a friend.

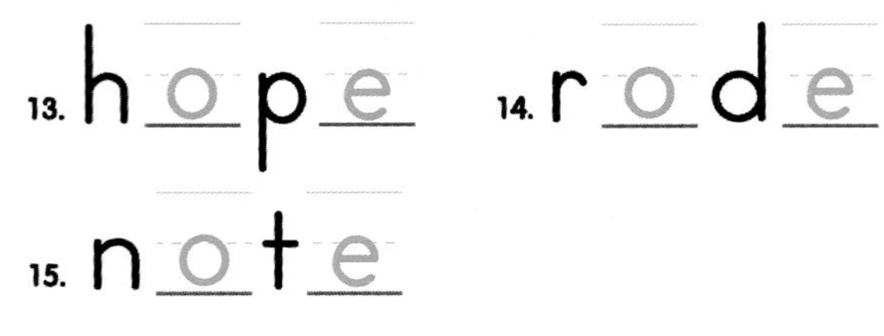

Supporting All Learners

VISUAL LEARNERS

EXPLORE MEANING One by one, call out words to explore, such as ***bone, rope, nose,*** and ***home***. Provide sentence starters such as ***A bone is ____, A rope is like a ____,*** or ***Noses can be ____***. Invite children to dictate serious or silly answers. Children can then copy one of their favorite sentences and illustrate it.

VISUAL LEARNERS

o-e **SEARCH** Ask children to suggest other words that contain the long ***o*** sound. List these words on the chalkboard. Have volunteers circle the words with ***o-e***.

KINESTHETIC/VISUAL LEARNERS

WORD SORT Write the following words on index cards: ***bike, bake, bone, cake, cone, fine, home, take, game, note, line, late, home,*** and ***side***. Have children sort them according to the long vowel sound.

Circle the word that names each picture.
Then write the word on the line.

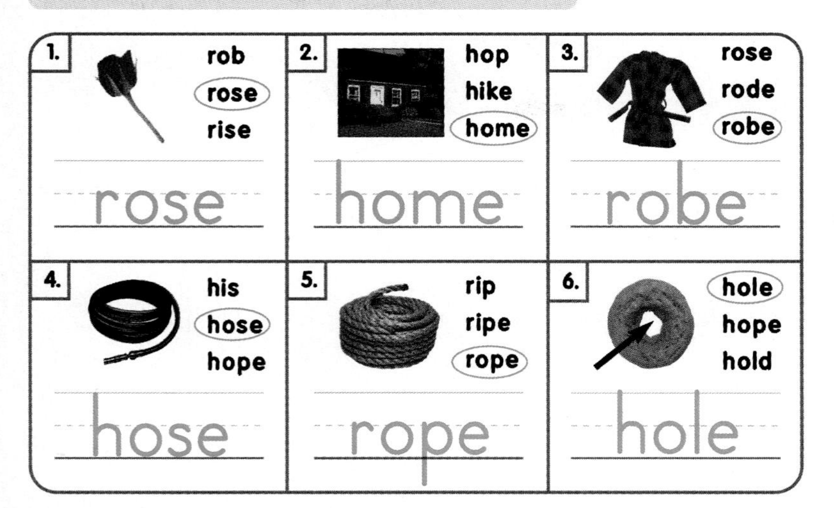

Write the word that names each picture.

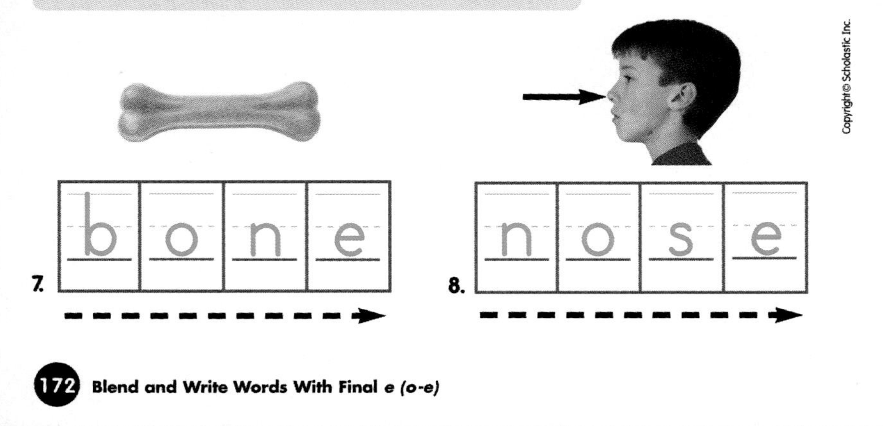

Integrated Curriculum

SPELLING CONNECTION

O-BONES Write the following words on the chalkboard: ***bone, bat, nose, cone, cot, note, shop, home, ride, on, rope***. Read each word aloud, then read it aloud again with children. Have children draw six pictures of bones and print ***o*__*e*** on each bone. Tell children to listen and watch as you say each word and point to it. If the word has the spelling pattern ***o-e,*** children should write the word on one of their bones. After you go through the list of words, have children share what words they wrote on their bones.

SCIENCE CONNECTION

DINOSAUR BONES Write ***bone*** on the chalkboard. Display a picture of a dinosaur skeleton, and point out that it is made up of bones. Have children draw large outlines of dinosaurs, then glue pipe cleaners or toothpicks inside the dinosaur to make its bones. Then have children use the word ***bone*** or ***bones*** to write or dictate sentences about their dinosaur pictures.

CHALLENGE

Children can make up their own sentence starters for words with the spelling pattern ***o-e***. Have children challenge partners to complete the sentence starters.

EXTRA HELP

Have children write the words from the Connect Sound-Symbol exercise on note cards. Suggest that they write other ***o-e*** words, too. They can use the note cards as flash cards to practice reading the words.

Phonics Connection

Literacy Place: *Imagine That!*
Teacher's SourceBook, pp. T102–103;
Literacy-at-Work Book, pp. 23–24

My Book: *Alone*

Chapter Book: *The Puppet Club*, Chapter 1

Lesson 104

page 173

Write and Read Words With Final *e*

QUICKCHECK ✔

Can children:

✔ spell words with final *e*?

✔ read a poem?

If YES go to Read and Write.

TEACH

Link to Spelling Review with children the final *e* spelling patterns. State aloud one of the long vowel sounds. Have a volunteer write on the chalkboard the spelling that stands for the sound. For example, the letters ***i-e*** stand for /ī/. Continue with all the sounds. You may also wish to review /z/*s*, /y/*y*, /v/*v*, and /th/*th*.

Phonemic Awareness Oral Segmentation
Say the word ***pine***. Have children orally segment the word. (/p/ /ī/ /n/) Ask them how many sounds the word contains. *(3)* Have a volunteer write the word ***pine*** on the chalkboard. Continue with the words ***rode*** and ***bake***.

READ AND WRITE

Dictate Dictate the following words and sentence. The words in the first set are decodable based on the sounds previously taught. The words in the second set are high-frequency words from this and previous units. Have children write the words and sentence on a sheet of paper. When children are finished, write the words and sentence on the chalkboard, and have children make any necessary corrections on their papers.

- **quick** **hope** **make**
- **have** **they** **can't**
- **They made a hole.**

Complete Activity Page Read aloud the directions on page 173. Children can read the poem independently or with partners.

POETRY

Name

Read the poem. Finish each sentence using words from the poem.

Bugs

The bugs ride a bike.
The bugs bake a cake.
The bugs take a hike.
They run to the lake.

The bugs see the time.
The bugs have a date.
The bugs run home to dine.
They hate to be late.

Copyright © Scholastic Inc.

1. The bugs run to the lake.
2. They also run home.
3. The bugs hate to be late.

Link to Spelling 173

Supporting All Learners

KINESTHETIC LEARNERS

DRAW A PICTURE AND WRITE A CAPTION Have children draw a picture showing one scene in the poem. Have them label their drawing with the appropriate line from the poem.

VISUAL/KINESTHETIC LEARNERS

BUGS! BUGS! BUGS! Children can write or dictate sentences about other things that bugs can do. Have children draw pictures to go with their sentences. Then give children an opportunity to share their sentences and pictures.

Make a set of duplicate word cards for final **e** words with these spelling patterns: ***a-e, i-e, o-e***. As you display each word, help children read it. Provide visual support when appropriate. Then encourage them to use the cards for a variety of games, such as mixing up the cards and matching the word pairs, sorting the words by spelling pattern, or playing Concentration.

CHALLENGE

Have children generate a list of words with final **e**. Record these words on the chalkboard. Then have children create a story using as many final **e** words on the chalkboard as they can.

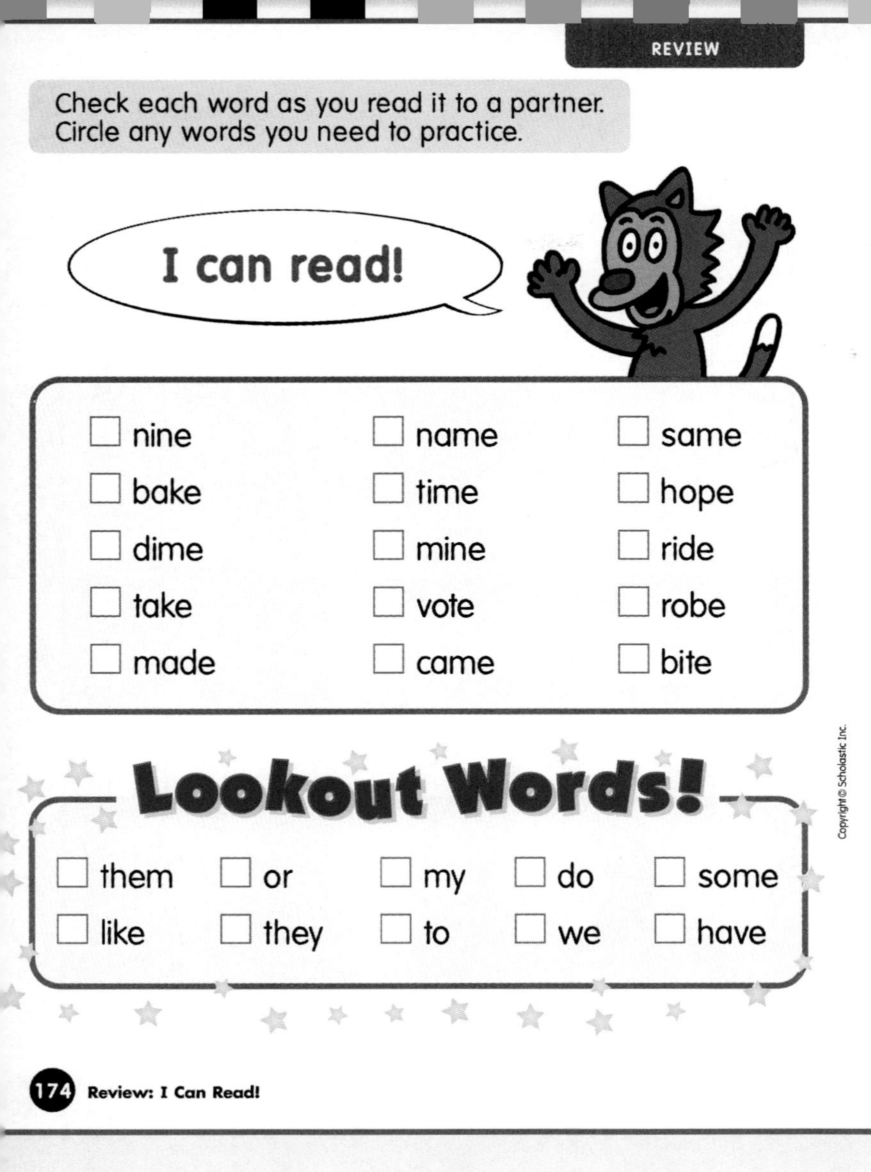

REVIEW

Check each word as you read it to a partner.
Circle any words you need to practice.

I can read!

☐ nine	☐ name	☐ same
☐ bake	☐ time	☐ hope
☐ dime	☐ mine	☐ ride
☐ take	☐ vote	☐ robe
☐ made	☐ came	☐ bite

Lookout Words!

☐ them	☐ or	☐ my	☐ do	☐ some
☐ like	☐ they	☐ to	☐ we	☐ have

Lesson 105

page 174

Read and Review Words With Final *e*

QUICKCHECK ✔

Can children:

✔ read words with final *e*?

✔ recognize high-frequency words?

If YES go to Read and Write.

TEACH

Review Sound-Spellings Review the sound-spelling relationships for words with final *e*. Say one of these sounds. Have a volunteer write the spelling that stands for the sound on the chalkboard. For example, the letters ***a-e*** stand for /ā/. Continue with the other sounds. Then display pictures of objects whose names contain these sounds. Have children write the spelling pattern that each picture name contains as the picture is displayed.

Review High-Frequency Words Review the high-frequency words ***some, or, do, have,*** and ***other***. Write sentences on the chalkboard containing each word. Say one word, and have volunteers circle it in the sentences.

READ AND WRITE

Build Words Distribute the following letter cards or have children use their own set: ***a, e, k, m, r,*** and ***t***. Encourage children to build as many words as possible. Suggest that they record their words. Have partners compare their lists. Continue with additional sets of cards: ***d, e, f, i, l, n*** and ***e, h, l, m, o, p***.

Build Sentences Write these words on note cards: ***I, home, a, cake, like, make, the, hike, take, game***. Have children use the words to make sentences. You might suggest this sentence: ***I take the cake home.***

Complete Activity Pages Read aloud the directions on page 174. Children can complete the page with partners.

Supporting All Learners

VISUAL/KINESTHETIC LEARNERS

SCHOOL OR NEIGHBORHOOD TOUR Walk around your school or neighborhood with children. Ask them to look for and list words with final *e*.

CHALLENGE

Invite children to suggest phrases containing rhyming words with final *e* that they can illustrate. You may wish to suggest some models such as ***a mole hole, a pine with a vine, mice with dice, bake a cake, a rose on a nose,*** or ***a bike I like.***

I CAN READ! OPTIONS

The I Can Read! page can be used for one or all of the following:

- **paired reading**
- **individual assessment**
- **choral reading**
- **homework practice**
- **program placement**

Lesson 106

pages 175-176

Read Words in Context

TEACH

Assemble the Story Ask children to remove pages 175–176. Have children fold the pages in half to form the Take-Home Book.

Preview the Story Preview *At Home,* a nonfiction selection about the places where some animals live. Invite children to browse through the first two pages of the book and to comment on anything they notice. Suggest that they point out any unfamiliar words. Read these words aloud for children to repeat. Then have children predict what the selection might be about.

READ AND WRITE

Read the Story Read the book aloud, or have volunteers take turns reading aloud a page at a time. Discuss with children anything of interest on each page. Encourage them to help each other with any blending difficulties. The following prompts may help children who need extra support while reading:

- **What letter sounds do you know in the word?**
- **Which word parts do you know?**

Reflect and Respond Have children share their reactions to the book. What did they learn about animal homes? Which animal home do they find most interesting? Encourage children to think of one more animal, and its home, that they could add to the story.

Develop Fluency You may wish to reread the book as a choral reading, or have partners reread the book independently. Provide time for children to reread the book on subsequent days to develop fluency and increase reading rate.

Have children share how they figure out unfamiliar words. Ask children to model how they blend words. Encourage children to look for words with final *e* in other stories and to apply what they have learned about these sound-spelling relationships to decode the words.

Supporting All Learners

AUDITORY LEARNERS

RHYME TIME Explain to children that you will play a rhyming game with them. Say the sentence ***I can bake a cake***. Ask children which words rhyme. Repeat this activity with the following sentences: ***I have time to buy a lime; The mole is in the hole; I will rake by the lake.*** Have children try making up their own rhymes using words with the spelling patterns ***o-e, a-e,*** and ***i-e***.

VISUAL/KINESTHETIC LEARNERS

BUILD IT Display the pocket ABC cards and pocket chart from the Phonics and Word Building Kit. Place the cards for the spelling patterns ***o-e, a-e,*** and ***i-e*** in the pocket chart. Have children add consonant letter cards to make words. List the words children make on the chalkboard.

TACTILE LEARNERS

INVISIBLE WORDS Dictate words with the spelling patterns ***o-e, a-e,*** and ***i-e***. Invite children to use their finger to "write" each word on their opposite hand as they say each letter. Write the word on the chalkboard for children to "check" their work.

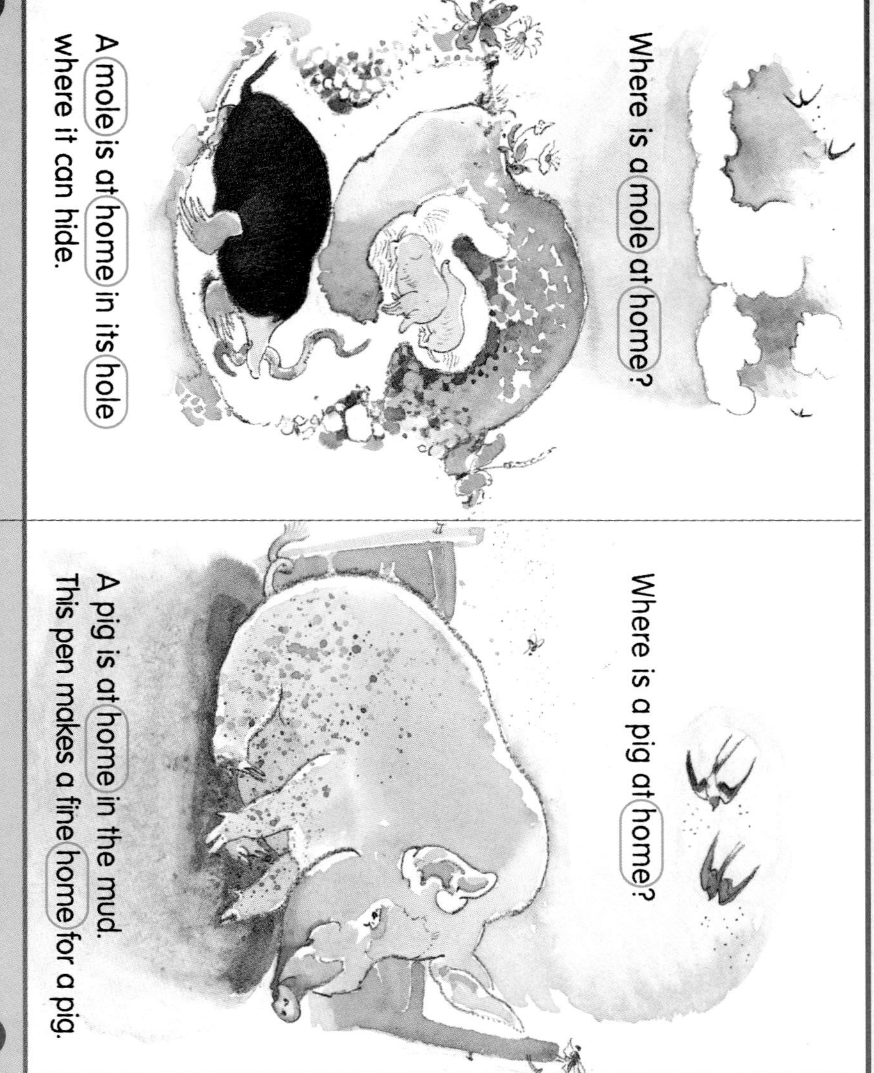

Reflect on Reading

ASSESS COMPREHENSION

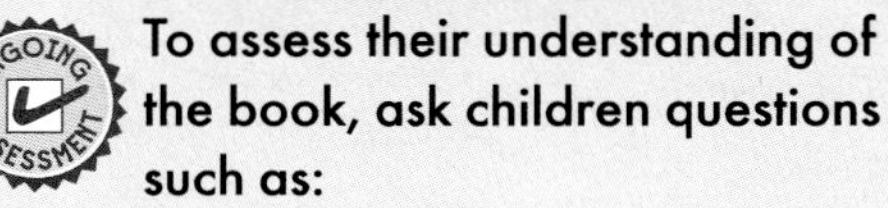

To assess their understanding of the book, ask children questions such as:

- **What kind of home does a mole need?** ***(a hole underground where it can hide)***
- **What kinds of homes do pigs live in?** ***(Pigs like the mud; they sometimes live in pens.)***
- **How is a home for people different from a home for animals?** ***(People live in houses. They often live with other people.)***

HOME-SCHOOL CONNECTION

Send home *At Home*. Encourage children to read the book to a family member.

WRITING CONNECTION

MORE HOMES Ask children to suggest other animals that they know about. Write the names of these animals on the chalkboard. You may wish to suggest animals such as cats, dogs, birds, and horses. Then have children complete the question ***Where is a _____ at home?*** by filling in the name of an animal. Children can write the question, then write or dictate an answer to it.

EXTRA HELP

For additional practice with final ***e,*** have children read the following My Books together: *Make a Face, Alone,* and *Mr. and Mrs. Pine*. For information on using the My Books on the computer, see the WiggleWorks Plus My Books Teaching Plan.

CHALLENGE

Challenge children to think of animals who make their homes in and around the following places: holes, lakes, caves, pines, hives, and vines. Make lists on the chalkboard. Then have children select one animal, research it, and make a mural of the animal and its home.

Phonics Connection

Phonics Reader:
#39, Lime Ice Is Nice

Lesson 107

pages 177-178

Blend Words With /yōō/u-e

QUICKCHECK ✔

Can children:

✔ **recognize /yōō/**

✔ **identify the sound-spelling pattern *u-e*?**

✔ **blend words with /yōō/*u-e*?**

If YES go to Read and Write.

TEACH

Develop Phonemic Awareness

Oddity Task Explain to children that you are going to read three words. Children are to decide which word does not have the same vowel sound. Use these words:

- rug — tub — vacuum
- duck — hug — cube
- unicorn — bugle — cut

Connect Sound-Symbol Write *cut* and *cute* on the chalkboard as you read each aloud. Ask children what vowel sound they hear in the middle of each word. Point out that the long *u* sound in *cute* is made by adding an *e* at the end of *cut*. Write _ *u*_ *e* on the chalkboard. Have children make words by filling in the blanks with consonants. Model this procedure using the word *mule*.

If I put an *m* in the first blank and an *l* in the second blank, I can make a new word. Listen as I blend the sounds in this word. What word will I form if I use the letters *h* and *g*?

READ AND WRITE

BUILT-IN REVIEW

Blend Words List the following on a chart. Model how to blend the words.

- cut — cute — hug — huge
- mule — use — cube — fuse
- She made a huge cut in the cake.

Complete Activity Pages Read aloud the directions on pages 177–178. You may wish to review each picture name with children.

LISTEN AND WRITE

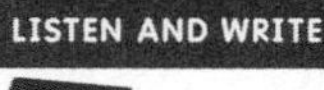

Name

Look at each picture. If the picture name has the **long u** sound as in **use**, color the picture.

1. mule
2. rug
3. cup
4. nut
5. jug
6. cube
7. vacuum
8. duck

Write the letters **u** and **e** to finish each word. Read the words to a friend.

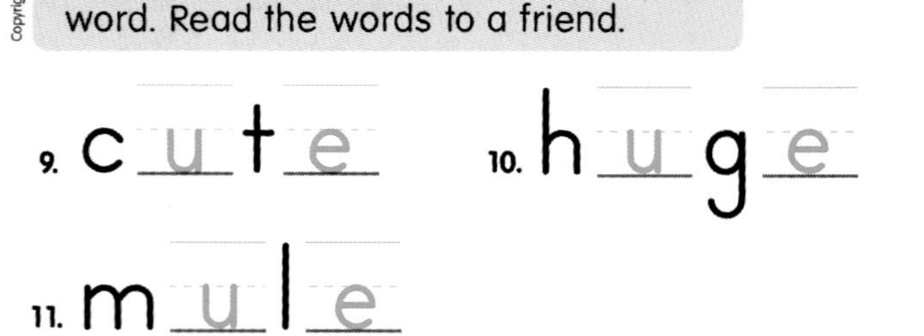

Recognize Final *e* (*u-e*) 177

Supporting All Learners

VISUAL/AUDITORY/KINESTHETIC LEARNERS

WE ARE FAMILY Write the following words on note cards: ***cake, came, take, gate; hike, vine, pine, mile; home, hope, robe, hole; use, mule, cute,*** and ***huge***. Give each child a card. Then invite children to come to the front of the room and group themselves according to the long vowel sound they hear in the middle of each word.

VISUAL/KINESTHETIC LEARNERS

CUTE MULES Have children draw a picture of a cute mule. Have children label their picture "A Cute Mule" by writing on the picture or placing a self-sticking note on it. Challenge children to find and label other "cute" pictures, such as pictures of babies or small animals.

BLEND WORDS

Circle the word that best finishes each sentence. Then write the word on the line.

	Sentence	Choices
1.	She has a cute hat.	cut / (cute) / came
2.	A box is like a cube.	(cube) / cub / cute
3.	I rode the mule home.	mug / mill / (mule)
4.	They made a huge cake.	hug / (huge) / have
5.	He will use his backpack on the hike.	(use) / huge / yet

Integrated Curriculum

SPELLING CONNECTION

Ask children to dictate words that contain the long ***u*** sound. List these words on the chalkboard. Have volunteers circle the words with the ***u-e*** pattern.

ART CONNECTION

CUBE ART If possible, bring in square tissue boxes and wrapping paper. Mention that the boxes are cubes. Have children decorate them with the wrapping paper and collage materials. Then have children write on sticky notes words with the ***u-e*** spelling pattern. Children can stick the words to the sides of the cube they decorated.

Phonics Connection

Literacy Place: *Imagine That!*
Teacher's SourceBook, pp. T102–103

My Book: *Alone*

CHALLENGE

Provide the following letter cards or have children use their own sets: ***b, c, e, m, l, t,*** and ***u***. Invite children to build words with the long ***u*** sound, including ***cube, cute,*** and ***mule***. Then have children use these words to write sentences.

EXTRA HELP

Read aloud the My Book *Alone* with children. Help them to find the word with the spelling pattern ***u-e***. Then have partners read the book together and to each other.

The Book Shop

- **Frog and Toad Are Friends**
  by Arnold Lobel
  Two true friends! **(/yōō/*u-e*)**
- **Going Home**
  by Margaret Wild
  Hugo is in the hospital and missing his home.
  (/yōō/*u-e*)

Lesson 108

page 179

Build Words

QUICKCHECK ✔

Can children:

✓ build words?

✓ recognize sounds previously taught?

If YES go to Read and Write.

TEACH

Review Sound-Spellings Review with children the sound-spelling patterns from the past few lessons. These include ***a-e, i-e, o-e,*** and ***u-e***. Then write ***ride*** on the chalkboard. Tell children that you want them to replace /**d**/, the last sound in ***ride,*** with /**p**/ to make a new word. Ask children what letter stands for /**p**/. Then replace the letter ***d*** in ***ride*** with the letter ***p,*** and blend the new word formed. Continue by changing one letter to build new words. You may wish to use this sequence of words: ***ripe, rope, hope, hole, pole***.

READ AND WRITE

Build Words Distribute the following letter cards to children: ***b, e, h, i, k, l***. Provide time for children to build as many words as possible using the letter cards. Suggest that they record their words on a sheet of paper. Have them continue with the letter cards: ***a, d, e, h, l, t*** and ***e, h, n, o, r, s***.

Then explain to children that you will give them the beginning and ending words of a three- or four-word ladder. They are to change one letter or letter pair in each word to complete the word ladder. Use the following words:

• late	(gate)	game	
• take	(bake)	(bike)	like
• lime	(like)	(lake)	cake

Note: The words in parentheses are possible answers.

Complete Activity Page Read aloud the directions on page 179. You may wish to review each picture name with children.

Supporting All Learners

KINESTHETIC/VISUAL LEARNERS

WORD CARDS Make word cards with these words: ***gate, late, lake, like, bike, hike, hive.*** Mix up the cards. Challenge children to arrange them in a word ladder.

EXTRA HELP

Children who need extra help can use the pocket ABC cards and pocket chart from the Phonics and Word Building Kit to build words with the ***a-e, i-e, o-e,*** and ***u-e*** spelling patterns. Start children out by building the words ***bake, kite, rope,*** and ***cute.*** Have children work together to change the consonants to make new words. Write on the chalkboard the words children make.

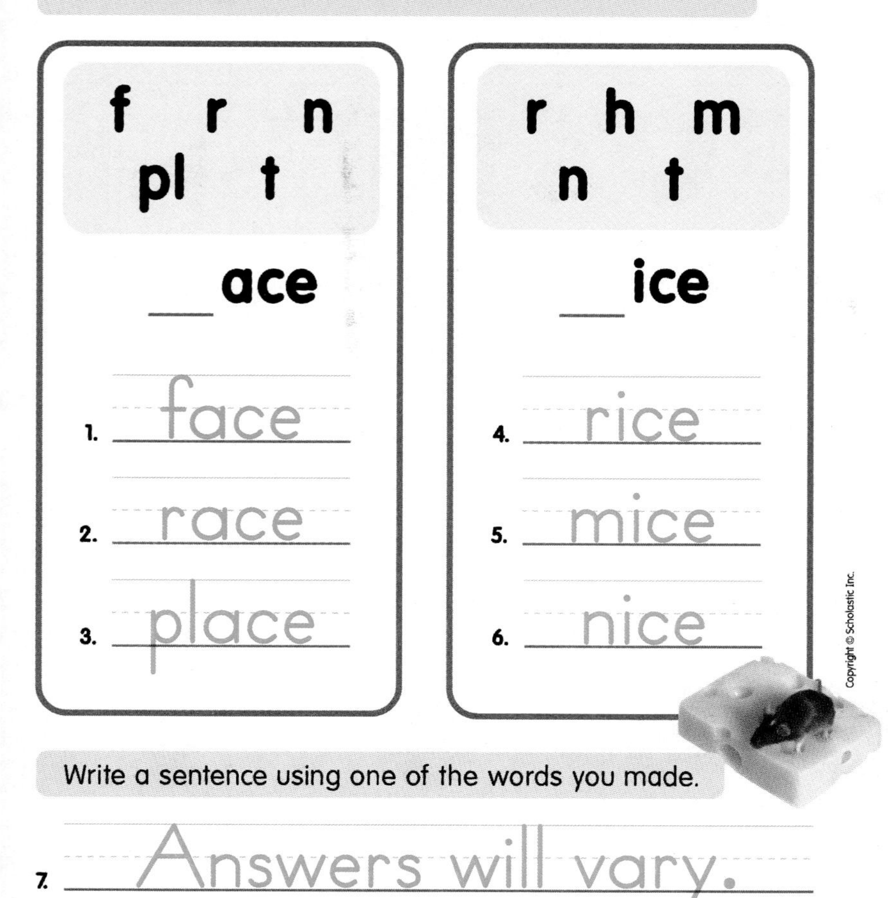

Add each letter or letters to the word part below it. Blend the word. If it is a real word, write it on the line.

f r n
pl t

__ace

1. face
2. race
3. place

r h m
n t

__ice

4. rice
5. mice
6. nice

Write a sentence using one of the words you made.

7. Answers will vary.

page 180

Blend and Build Words With Phonograms

QUICKCHECK ✔

Can children:

✔ **blend words with phonograms *-ace, -ice*?**

✔ **build words with *-ace, -ice*?**

If YES go to Read and Write.

TEACH

Introduce the Phonograms Write the phonograms *-ace* and *-ice* on the chalkboard. Point out the sounds these phonograms stand for. Add the letter *f* to the beginning of *-ace*. Model for children how to blend the word formed. Have children repeat the word *face* aloud as you blend the word again. Then add ***n*** to the beginning of *-ice*, and model for children how to blend the word ***nice***.

Ask children to suggest words that rhyme with *face* and *nice*. List these words on the chalkboard in separate columns. Have volunteers underline the phonogram *-ace* or *-ice* in each word. Be sure to include the words ***lace, race, place, rice,*** and ***mice***. Blend these words for children as they repeat them aloud. Encourage children to look for the word parts *-ace* and *-ice* as they read.

READ AND WRITE

Build Words Write the word parts *-ace* and *-ice* on the chalkboard. (If available, use a pocket chart and letter cards.) Have children add a letter to the beginning of each phonogram to make a new word. Continue by having children replace the initial consonant in the first word to build a second word. For example, children may build the following sequence of words: ***nice, rice, dice, mice***.

Complete Activity Page Read aloud the directions on page 180. Children can complete the page independently.

Supporting All Learners

VISUAL LEARNERS

WORD COUNT Have children read aloud Phonics Reader #39, *Lime Ice Is Nice* and #37, *The Big Race*. Children can count the number of words with ***-ice*** in *Lime Ice Is Nice* and the number of words with ***-ace*** in *The Big Race*.

CHALLENGE

Invite children to make flip books featuring the phonograms ***-ace*** and ***-ice***. The top word may be ***race*** for ***-ace*** and ***rice*** for ***-ice***. Under the ***r*** in ***race***, children can put ***f, l,*** and ***pl*** one letter or letter combination per page. Under the ***r*** in ***rice***, children can put ***d, n,*** and ***m*** one letter per page. Children may take turns reading the words in their flip books and then make up sentences using the words.

ASSESSMENT

ONGOING ASSESSMENT

Continue to keep anecdotal records. Focus on each child's ability to blend, or construct, words, as well as to segment, or take apart, words. Note whether children recognize and use familiar letter combinations, such as digraphs and phonograms, in each process.

Lesson 110

pages 181-182

Blend Words With /ē/

QUICKCHECK ✔

Can children:

✓ **substitute sounds in words?**

✓ **recognize /ē/?**

✓ **blend words with /ē/*ea, ee*?**

If YES go to Read and Write.

TEACH

Develop Phonemic Awareness

Phonemic Manipulation Explain to children that they are going to listen to words with /e/ as in ***met***. Children are to replace the short *e* sound with a long *e* sound. For example, if you say ***met***, children should say ***meet*** (or ***meat***). Use these and other words:

- bet
- sell
- bed
- fed
- led
- red
- wed
- net

Connect Sound-Symbol Write the words ***ten*** and ***teen*** on the chalkboard as you read each aloud. Ask children what sound they hear in the middle of each word and how the words are different. Point out that /e/ in ***ten*** is spelled by *e*, and /ē/ in ***teen*** is spelled by ***ee***. Continue with the words ***net*** and ***neat***.

Ask children to suggest other words that contain /ē/. List these words on the chalkboard. Have volunteers circle the letters *ee* or *ea* in each word that contains one of these spellings.

READ AND WRITE

Blend Words List the following on a chart. Model how to blend the words. Then have children blend them.

- fed feed bet beat
- seat real read leap
- I bet they will beat that team.
- If you feed that seal, you will get wet.

Complete Activity Pages Read aloud the directions on pages 181–182. Review each picture name with children.

LISTEN AND WRITE

leaf

Name

Look at each picture. Write e or ea to finish each picture name.

1. bean — b ea n
2. team — t ea m
3. jet — j e t
4. seal — s ea l
5. meat — m ea t
6. read — r ea d

Look at each picture. Write e or ee to finish each picture name.

7. bed — b e d
8. seed — s ee d
9. queen — qu ee n

Supporting All Learners

AUDITORY/VISUAL LEARNERS

TIME FOR A RHYME Read aloud "Stop! Look! Listen!" on page 8 in the *Big Book of Rhymes and Rhythms, 1B*. Have children find all the words with /ē/ in the rhyme. Write these words on the chalkboard. Invite volunteers to add other /ē/ words to the list.

VISUAL LEARNERS

WORD SEARCH Have children look through books, magazines, and newspapers for words with **/ē/*ee, ea***. Children can add the words they find to the Word Wall.

KINESTHETIC LEARNERS

LEARNING CENTER For practice with a variety of phonics skills, set up Center 7, Activity 1, "Memory Match" from *Quick-and-Easy Learning Centers: Phonics*. This activity focuses on sound-spelling relationships as well as sight-word reading.

Look at each picture. Circle the word that best finishes each sentence. Then write the word on the line.

	Sentence	Choices
1.	He will feed the pigs.	feet / fed / (feed)
2.	The mice peek at Meg.	peel / (peek) / pet
3.	I see a leak in that pot.	let / like / (leak)
4.	She will leap into the lake.	lead / (leap) / leg
5.	The vet can heal the cut.	hill / heat / (heal)
6.	We eat meat, rice, and beans.	meal / met / (meat)

Integrated Curriculum

SPELLING CONNECTION

SENTENCES WITH *ea* AND *ee* Have children dictate sentences using words that contain the long ***e*** sound spelled ***ea*** or ***ee***. Write these sentences on the chalkboard. Continue by having children spell the following words: ***feet, read, seat,*** and ***seed***. Write the words on the chalkboard, and have children self-correct their work.

Phonics Connection

Literacy Place: *Imagine That!* Teacher's SourceBook, pp. T164–165; Literacy-at-Work Book, pp. 34–35

My Book: *What Do You See?*

Chapter Book: *The Puppet Club,* Chapter 2

ESL Help children acquiring English become familiar with the ***ee*** and ***ea*** spelling patterns by completing long ***e*** words. Write ***ee*** and ***ea*** in two different colors as headings, and write incomplete words such as these in each column: ***ee: f__ __t, s__ __d, f__ __l, qu__ __n; ea: t__ __m, r__ __l, n__ __t, b__ __n***. For each column, have children use the matching color to fill in the missing letters and then read the completed words.

EXTRA HELP

Read aloud one or more of the books suggested under The Book Shop. Read each book aloud again, with children. Ask children to point out words with ***/ē/ea, ee***. Then partners or small groups of children can revisit the books to reread them on their own.

CHALLENGE

Have children use the Magnet Board to build words with ***ee*** and ***ea***. Suggest that children begin with one of the following words: ***heal, feet, beat, seen***.

- **Each Peach, Pear, Plum**
  by Janet and Allan Ahlberg
  Children play I Spy.
  (/ē/ea, ee)
- **We Scream for Ice Cream**
  by Bernice Chardiet
  What happens when Brenda picks up Truman's ice cream? **(/ē/ea, ee)**

Lesson 111

page 183

Recognize High-Frequency Words

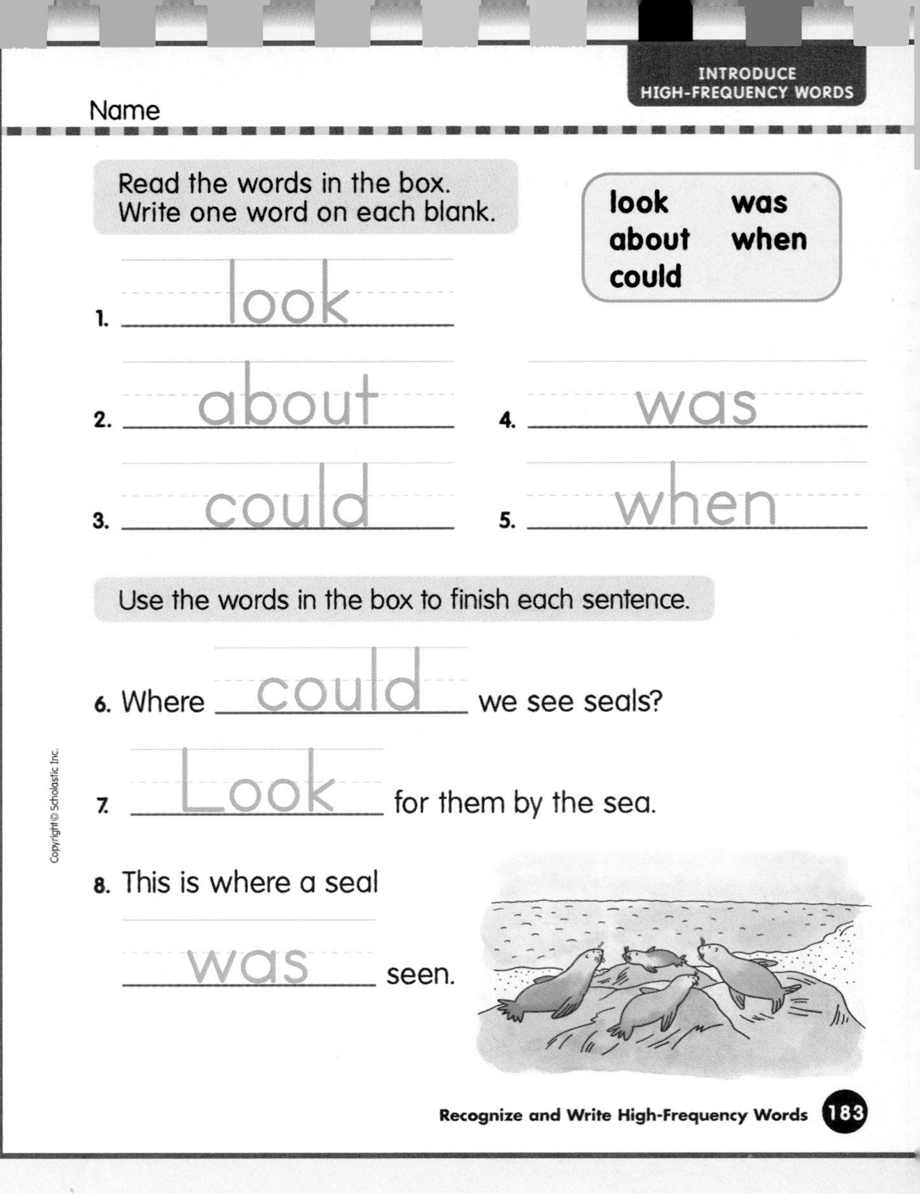

INTRODUCE HIGH-FREQUENCY WORDS

Name

Read the words in the box.
Write one word on each blank.

look	was
about	when
could	

1. look
2. about
3. could
4. was
5. when

Use the words in the box to finish each sentence.

6. Where could we see seals?
7. Look for them by the sea.
8. This is where a seal was seen.

Copyright © Scholastic Inc.

Recognize and Write High-Frequency Words 183

QUICKCHECK ✔

Can children:

✓ **recognize and write the high-frequency words *look, about, could, was, when*?**

✓ **complete sentences using the high-frequency words?**

If YES go to Read and Write.

TEACH

Introduce the High-Frequency Words

Write the high-frequency words ***look, about, could, was,*** and ***when*** in sentences on the chalkboard. Read aloud the sentences. Underline the high-frequency words, and ask children if they recognize them. You may wish to use the following sentences:

1. **Look at that dog.**
2. **Could it be Dot?**
3. **When did we last see her?**
4. **It was about six years past!**

Ask volunteers to dictate additional sentences using the high-frequency words.

READ AND WRITE

Practice Write each high-frequency word on a note card. Read each word aloud as you display the cards. Then do the following:

- **Mix the cards.**
- **Display one card at a time, and ask children to state each word aloud.**
- **Have children spell each word aloud, clapping on each letter.**
- **Ask children to write each word in the air as they state aloud each letter. Then have them write each word on a sheet of paper.**

Complete Activity Page Read aloud the directions on page 183. Have children complete the page independently.

Supporting All Learners

KINESTHETIC/VISUAL LEARNERS

CONCENTRATION Write each high-frequency word from this lesson and others you wish to review on two word cards. Shuffle the cards, and lay them facedown. Invite children to take turns picking up two cards at a time, trying to match two words. When children find two words that match, they must read them aloud.

VISUAL LEARNERS

ANIMAL MOBILE Ask children to draw or cut out pictures of animals they know something about. Have children use the high-frequency words they have learned to write about the animals on the back of the pictures. Use thread or yarn to hang each animal picture from a hanger to make an animal mobile.

EXTRA HELP

Make word cards for the high-frequency words you have taught so far. Give each child one word card. Invite that child to stand when you call out the word. Have children spell aloud the word and then use it in a sentence.

POETRY

Read the poem. Write a sentence telling about Bat's jeep.

Beep, Beep!

Bat had a jeep.
Seal said, "I could use a ride!"
"Beep, beep!" said the jeep.
And Seal got inside.

Bat had a jeep.
Pig said, "I could use a ride!"
"Beep, beep!" said the jeep.
And Pig got inside.

Bat had a jeep.
Mule said, "I could use a ride!"
"Beep, beep!" said the jeep.
And Mule got inside.

"Quick, quick!" they said.
"To the bat cave."
"Beep, beep!" said the jeep.
See them wave!

Answers will vary.

Supporting All Learners

VISUAL/AUDITORY LEARNERS

WRITING EXTENSION Have children generate a list of words that have **/ē/ea, ee** in the middle. Record these words on the chalkboard. Then have children create a story, using as many /ē/ words on the chalkboard as they can. Write the dictated story on chart paper for group and individual reading. Return to the story, rereading it in subsequent lessons.

AUDITORY/VISUAL/KINESTHETIC LEARNERS

SONG BINGO To practice and review words with **/ē/ee, ea,** children may enjoy playing "Sound Bingo" on pages 28–29 in *Quick-and-Easy Learning Games: Phonics*. Adapt Game 1 by substituting these words for the long *a* words: ***weed, need, lead, seen, teen, mean, seem, deep, keep, leap, heat,*** and ***real***.

CHALLENGE

Have children list as many words with ***u-e*** and ***/ē/ee, ea*** as they can. Then have children write word problems such as ***meet – ee + a = mat***. Have children cover the answers to their word problems with sticky notes. Then children can challenge others to figure out the answers.

Write and Read Words With *u-e* and /ē/*ee, ea*

QUICKCHECK ✔

Can children:

✔ **spell words with *u-e* and /ē/*ee, ea*?**

✔ **read a poem?**

If YES go to Read and Write.

TEACH

Link to Spelling Review with children the following sound-spelling relationships: ***u-e*** and /ē/***ee, ea***. Say the long *e* sound aloud. Have a volunteer write on the chalkboard the two spellings that stand for the sound. Then review the ***u-e*** spelling pattern. You may also wish to review the spelling patterns ***a-e, i-e,*** and ***o-e***.

Phonemic Awareness Oral Segmentation
Say aloud the word ***cute***. Have children orally segment the word. (**/k/ /yōō/ /t/**) Ask a volunteer to write the word ***cute*** on the chalkboard. Point out that the final ***e*** is silent. Continue with the words ***cube*** and ***huge***.

READ AND WRITE

Dictate Dictate the following words and sentences. The words in the first two sets are decodable based on the sounds previously taught. The words in the third set are high-frequency words from this and previous lessons. Have children write the words and sentences on a sheet of paper. When children are finished, write the words and sentences on the chalkboard, and have children make any necessary corrections on their papers.

- meet deal feel real
- weak seem meat peep
- was when could some
- I could feel that the seal was sick.
- Some dogs seem mean.

Complete Activity Page Read aloud the directions on page 184. Children may read the poem independently or with partners.

Lesson 113

pages 185-186

Read and Review Words With *u-e* and /ē/*ee, ea*

QUICKCHECK ✔

Can children:

✔ **read words with u-e and /ē/*ee, ea*?**

✔ **recognize high-frequency words?**

If YES go to Read and Write.

TEACH

Review Sound-Spellings Review with children the sound-spelling relationships from the past few lessons. These include the *u-e* spelling pattern and /ē/*ee, ea*. Say aloud one of these sounds. Have a volunteer write the spelling(s) that stands for the sound on the chalkboard. For example, the *u-e* spelling pattern stands for /yōō/. Continue with all the sounds. Then display pictures of objects whose names contain one of these sounds. Have children write the spelling that each picture name contains as the picture is displayed.

Review High-Frequency Words Review the high-frequency words *look, about, could, was,* and *when*. Write on the chalkboard sentences containing each high-frequency word. Say aloud one of the words, and have volunteers circle the word in the sentences.

READ AND WRITE

Build Words Distribute the following letter cards to children: *b, c, e, s, t, u*. Provide time for children to build as many words as possible. Suggest that they record their words on a sheet of paper. Have children continue with the letters: *a, e, l, m, n, t, s* and *e, e, m, n, s, t*.

Build Sentences Write the following words on note cards: *the, look, need, at, feed, we, dog, they, to, cats, seals, could, mule*. Display the cards, and have children make sentences using the words.

Complete Activity Pages Read aloud the directions on pages 185–186. You may wish to review the art with children.

REVIEW

Name

Check each word as you read it to a partner.
Circle any words you need to practice.

I can read!

☐ use	☐ mule	☐ cute
☐ huge	☐ need	☐ heat
☐ meet	☐ week	☐ team
☐ deep	☐ neat	☐ seat
☐ feet	☐ read	☐ feed

Lookout Words!

☐ look	☐ about	☐ could	☐ was	☐ do
☐ have	☐ when	☐ with	☐ one	☐ my

Supporting All Learners

KINESTHETIC/VISUAL LEARNERS

A TREE AND LEAVES Draw or cut out the shape of a large tree with a big trunk and many bare branches for leaves. Display the tree on a wall or bulletin board. Have a volunteer label the tree. Invite children to write words with ***ee*** on sticky notes, one word per note, and attach them to the tree. Then have children draw and cut out leaves. Children can write ***ea*** words on the leaves, one word per leaf, and attach the leaves to the tree.

AUDITORY/KINESTHETIC LEARNERS

LETTER PLAY Write the following list of words on the chalkboard: ***meat, set, seat, seal, sell, fell, feel, feet, feed, need, nut, neat,*** and ***net***. Then use the pocket ABC cards and pocket chart from the Phonics and Word Building Kit to make the word ***met***. Read the first word from the list. Ask children if ***meat*** has the long ***e*** sound spelled ***ea*** or ***ee***. If it does, have children use the pocket ABC cards to change the word ***met*** to the word ***meat***. Continue in this way by saying each word in the list, in order.

Fill in the bubble next to the sentence that tells about each picture.

1.
- ○ He eats the peas.
- ● He reads about peas.

2.
- ○ The teens need jeans.
- ● They like to eat beans.

3.
- ○ Sam will use the mule.
- ● Sam will ride his bike.

4.
- ● That is a big weed.
- ○ Look at the seeds.

5.
- ● Jill made a deep hole.
- ○ Jill is late for the game.

6.
- ○ The dog meets a huge cat.
- ● The dog likes to eat meat.

7.
- ● The mice run and hide.
- ○ The nice man will help us.

EXTRA HELP

For children needing additional phonemic awareness training, see the Scholastic Phonemic Awareness Kit. The oral blending exercises will help children orally string together sounds to form words. This is necessary for children to be able to decode, or sound out, words while reading.

CHALLENGE

Children can write or dictate their own rhyming sentences using words with ***u-e*** and **/ē/*ee, ea***. You may wish to provide a model such as ***Neat Pete eats meat***. Have children illustrate their rhymes, then combine them to form a book of rhymes.

Integrated Curriculum

SPELLING CONNECTION

WHAT'S IN A NAME? Have children write or dictate a list of names with ***u-e*** and **/ē/*ee, ea***. Then have children write sentences using the names and other words with ***u-e*** and **/e/*ee, ea***.

SOCIAL STUDIES CONNECTION

A SCHOOL AND NEIGHBORHOOD TOUR Have children look in school and in the neighborhood for words with ***u-e,*** words with **/ē/*ee, ea,*** and high-frequency words. Children should write a list of the words they find. Then they can create two webs on the chalkboard, one labeled ***school*** and one labeled ***neighborhood***. Children can write the words they found in the school around the school web, and words they found in the neighborhood around the neighborhood web.

I CAN READ! OPTIONS

The I Can Read! page can be used for one or all of the following:

- **paired reading**
- **individual assessment**
- **choral reading**
- **homework practice**
- **program placement**

Lesson 114

pages 187-188

Read Words in Context

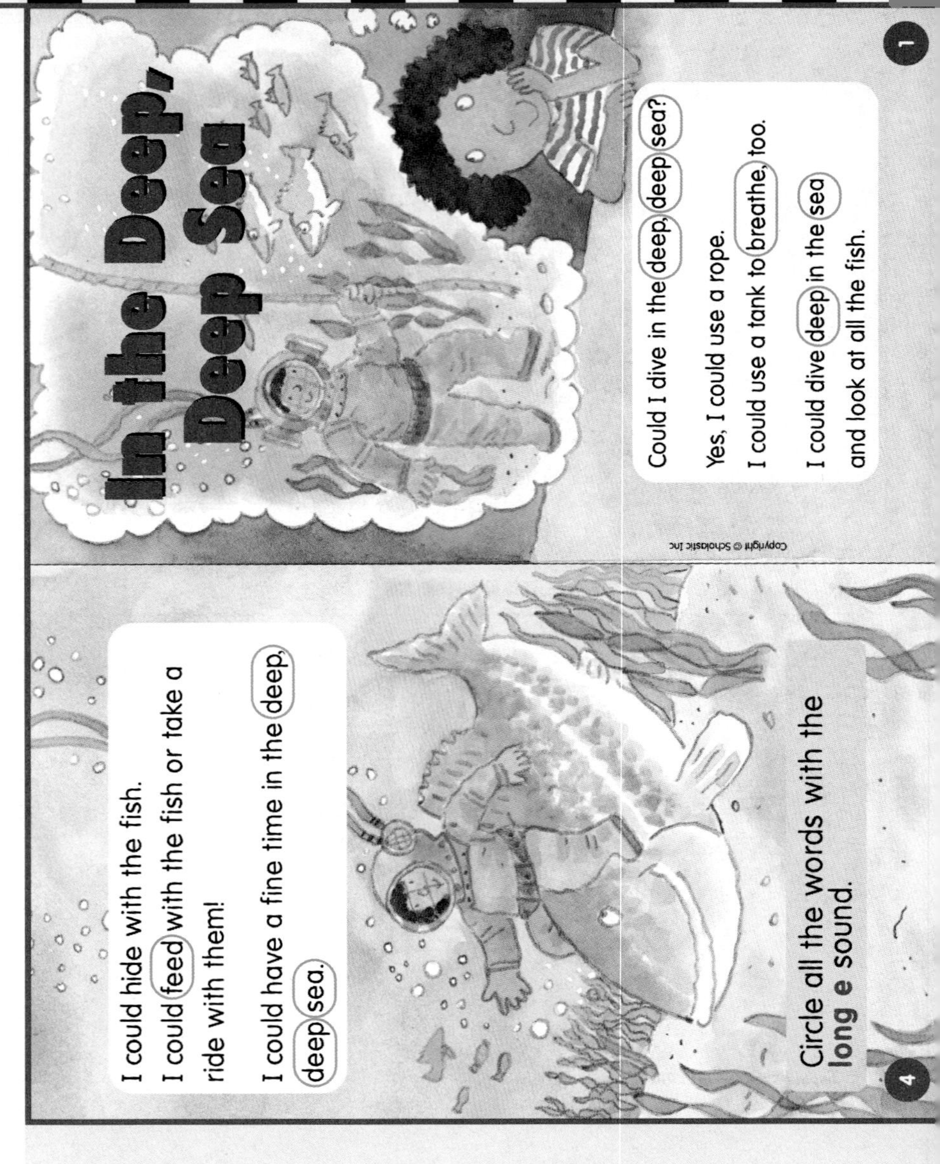

TEACH

Assemble the Story Ask children to remove pages 187–188. Have children fold the pages in half to form the Take-Home Book.

Preview the Story Preview *In the Deep, Deep Sea,* a story about a girl who imagines diving into the sea and swimming with the fish. Invite children to browse through the first two pages of the story and to comment on anything they notice. Suggest that they point out any unfamiliar words. Read these words aloud. Have children blend them. Then have children predict what they think the story might be about.

READ AND WRITE

Read the Story Read the story aloud, or have volunteers take turns reading aloud a page at a time. Discuss with children anything of interest on each page, and encourage them to help each other with any blending difficulties. The following prompts may help children who need extra support while reading:

- **What letter sounds do you know in the words?**
- **What do the pictures tell you about the story?**

Reflect and Respond Have children share their reactions to *In the Deep, Deep Sea.* Does it remind them of other stories they have read? What did they like best about the story? Encourage children to think of more details about the sea that they would like to add to the story.

Develop Fluency Have children reread the story as a choral reading, or have partners reread the story independently. Provide time for children to reread the story on subsequent days to develop fluency and increase reading rate.

Have children model how they blend words when reading other stories. Continue to review confusing or challenging sound-spelling relationships for children needing additional support.

Supporting All Learners

VISUAL/AUDITORY LEARNERS

MY BOOKS/TECHNOLOGY For additional practice with ***/ē/ea, ee,*** have children read the following My Book: *What Do You See?* For information on using the My Books on the computer, see the WiggleWorks Plus My Books Teaching Plan.

AUDITORY LEARNERS

RHYME TIME Invite children to choose a word that contains ***/ē/ee, ea*** from *In the Deep, Deep Sea.* Ask children to write a list of words that rhymes with the word they chose. Then challenge children to write a two-line rhyme using some of the words.

VISUAL LEARNERS

COUNT HOW MANY Have children collect and count words with ***/ē/ee, ea*** in *In the Deep, Deep Sea.* Children can list ***ee*** words in one column and ***ea*** words in another column. They can record the results of their counts using tallies. For example:

deep	𝍸 //	feed	/
feel	//	sea	𝍸 /
weed	/	breathe	/

Reflect on Reading

ASSESS COMPREHENSION

To assess their understanding of the story, ask children questions such as:

- **What does the girl imagine doing in the sea?** ***(She imagines going down on a line, looking around, exploring, hiding, feeding, and even riding the fish.)***
- **What does the girl think she will see in the sea?** ***(the sea bed, sea weed, and an eel)***
- **How does the girl feel about the sea?** ***(It is a fun and interesting place.)***

HOME-SCHOOL CONNECTION

Send home *In the Deep, Deep Sea.* Encourage children to read the story to a family member.

WRITING CONNECTION

OUR DEEP SEA Have children write or dictate sentences about what they would do or like to do in a deep, deep sea. Have children write their sentences on large sheets of paper. Children can illustrate their sentences, then take turns sharing them. Combine the pages into a book. Ask children to help you write a title for the cover of the book.

ESL Draw a web on the chalkboard. Write ***The Sea*** in the central circle. In two circles that radiate from the central circle, write the words ***see*** and ***feel***. Invite children to name things they may see and feel if they were in the sea. List each thing under the appropriate heading. Display sea pictures to help children.

CHALLENGE

Have children think of other words with ***/ē/ee, ea*** to add to *In the Deep, Deep Sea*. Encourage children to use the words to write another page for the story.

Phonics Connection

Phonics Readers:
#39, Lime Ice Is Nice
#40, Hen Pen's Joke
#41, The Three Little Pigs

Lesson 115

pages 189-190

Blend Words With *r*-Blends *(br, gr, tr)*

QUICKCHECK ✔

Can children:

- ✓ **orally segment words?**
- ✓ **identify the sounds *r*-blends *(br, gr, tr)* stand for?**
- ✓ **blend words with *br, gr,* and *tr*?**

If YES go to Read and Write.

TEACH

Develop Phonemic Awareness

Oral Segmentation Explain to children that you are going to say words by saying their parts. Children will listen to the parts and then say the whole word. For example, say the sounds /l//ē/ /f/. Guide children to say the whole word *leaf*. Continue with the following word parts:

/br/ /e/ /d/	/br/ /i/ /k/
/gr/ /ā/ /p/	/gr/ /a/ /b/
/tr/ /i/ /p/	/tr/ /ē/ /z/

Connect Sound-Symbol Write *tap* and *trap* on the chalkboard as you read each aloud. Ask children what sounds they hear at the beginning of each word and how the words are different. Point out that the *r*-blend at the beginning of *trip* is spelled by two letters, *t* and *r*.

Repeat this procedure with *brake* and *bake*.

READ AND WRITE

BUILT-IN REVIEW

Blend Words List the following words and sentences on a chart. Have volunteers read each aloud.

•grin	brag	trick
•trip	grab	brake

- **That man likes to grin.**
- **I can trick him.**

Complete Activity Pages Read aloud the directions on pages 189–190. Review each picture name with children.

Name

Look at each picture. Write the letters that stand for the first two sounds in the picture name.

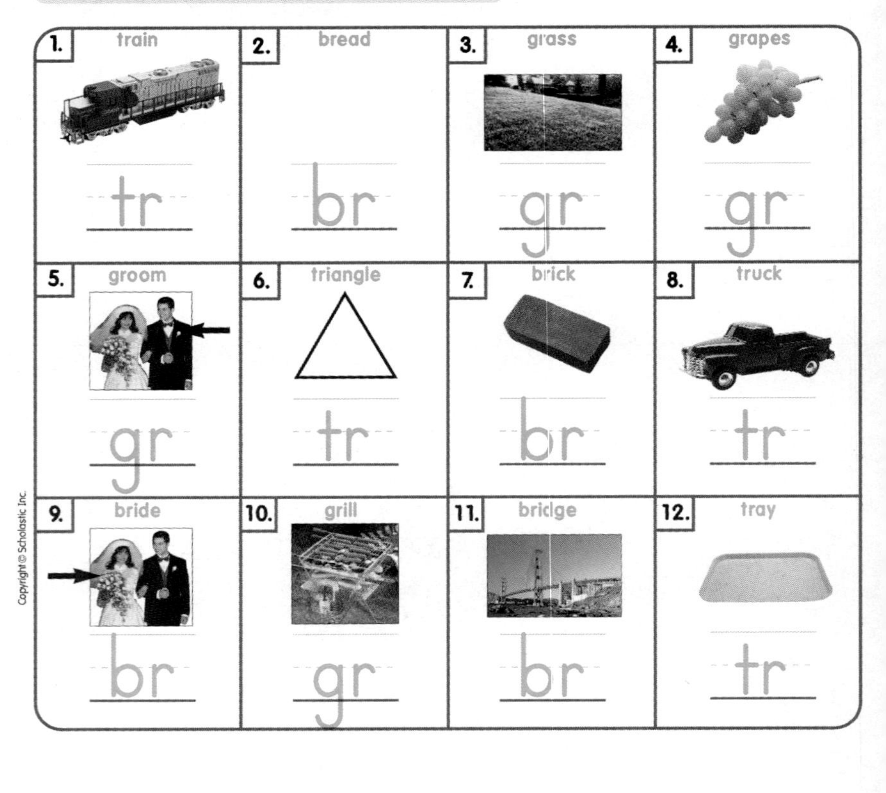

Supporting All Learners

VISUAL LEARNERS

CLASS DICTIONARY As you add words with **br, gr,** and **tr** to the class dictionary, talk about alphabetizing to the second letter. Children may also compare and contrast words entered for **b, g,** and **t** with those they are adding for **br, gr,** and **tr.**

AUDITORY LEARNERS

ADD-A-SOUND GAME Explain to children that you will pronounce the end of a word and that children are to add **br, gr,** or **tr** to the beginning to make one or more words. You may wish to use these and other word parts: ***ick, uck, ade, at, ab, ap, ip, ot, een, eet, eat, ace, ill, ake,*** and ***ide.***

AUDITORY/VISUAL LEARNERS

WORD SEARCH Read aloud "Five Little Peas" on page 9 in the *Big Book of Rhymes and Rhythms, 1B*. Have children find the word with **gr** in the rhyme ***(grew)***. Write the word on the chalkboard. Invite volunteers to add other *r*-blend ***(br, gr, tr)*** words to the list.

Read each sentence. Find the missing word in the box. Write the word on the line and in the puzzle.

brick	grapes	bride	trip	grade

DOWN

1. He likes to eat grapes.

3. She got the best grade on the test.

ACROSS

2. My home is made out of brick.

4. Do not trip on that log!

5. The bride wore a nice, white dress.

						1. g			
					2. b	r	i	c	k
				3. g	■	a			
			4. t	r	i	p			
				a		e			
				d		s			
5. b	r	i	d	e					

Integrated Curriculum

SPELLING CONNECTION

MAKE THE WORD Write the following word parts on the chalkboard: ***_ush, _ick***. Have children add the letters ***br*** to each. Do the same with the word parts ***_ap, _ot*** and ***tr*** and the word parts ***_in, _ab*** and ***gr***. Model for children how to blend each word. Then have children write each word on a sheet of paper. Ask children to dictate other words that begin with ***br, gr,*** and ***tr***. List these words on the chalkboard. Have volunteers circle the letters ***br, gr,*** or ***tr*** in each.

ART CONNECTION

Children can make a collage of words, pictures, and drawings of things whose names begin with ***br, gr,*** and ***tr***. Have children look through magazines and newspapers for words and pictures of things whose names begin with ***br, gr,*** and ***tr***. Children can cut out these words and pictures or write and draw their own words and pictures. Children can glue the words and pictures onto a large sheet of paper. Ask children to use a green crayon to label the words, pictures, and drawings of things whose names begin with ***gr***, a brown crayon to label things whose names begin with ***br***, and any other color crayon to label things whose names begin with ***tr***.

CHALLENGE

Draw a large ladder on chart paper. Invite children to "climb" the ladder by naming words that begin with ***br, gr,*** or ***tr***. Write the words on the ladder, one word per rung, beginning with the bottom rung, and working toward the top. Remind children to look for words that begin with ***br, gr,*** or ***tr*** when they read. Make a second word ladder on which to add the words children find.

EXTRA HELP

Review the class dictionary with children. Ask them to find words with ***br, gr,*** and ***tr*** in the dictionary. Then have children look through other books in the classroom for words with ***br, gr,*** and ***tr***.

Phonics Connection

Literacy Place: *Imagine That!*
Teacher's SourceBook, pp. T206–207;
Literacy-at-Work Book, pp. 44–45

My Book: *Grandpa Gray*

Big Book of Rhymes and Rhythms: 1B: "Five Little Peas," p. 9.

Chapter Book: *The Puppet Club*, Chapter 3

Lesson 116

pages 191-192

Blend Words With *l*-Blends *(bl, cl, pl)*

QUICKCHECK ✔

Can children:

✔ **delete sounds in words?**

✔ **recognize *l*-blends *(bl, cl, pl)*?**

✔ **blend words with *bl, cl, pl*?**

If YES go to Read and Write.

TEACH

Develop Phonemic Awareness

Phonemic Manipulation Explain to children that you are going to say a word. Children should repeat the word without the first sound. For example, if you say ***plane,*** children are to say ***lane***. Continue with the following words:

- clip black place
- plate block blame

Connect Sound-Symbol Write the words ***lap*** and ***clap*** on the chalkboard as you read each aloud. Ask children what sounds they hear at the beginning of each word and then tell how the words are different. Point out that the ***l***-blend at the beginning of ***clap*** is spelled by two letters together, ***c*** and ***l***. Repeat with the words ***block*** and ***plate***.

Write these word parts on the chalkboard: *_ip, _am*. Have volunteers add the letters ***cl*** to each one. Model how to blend the words. Repeat with ***bl*** and the word parts *_ack, _ame,* and ***pl*** and the word parts *_ace, _ug*.

READ AND WRITE

Blend Words List the following on a chart. Have volunteers read each aloud.

- clock blame plan
- We clap for the black dog and his tricks.

Complete Activity Pages Read aloud the directions on pages 191–192. Review each picture name with children.

Name

Look at each picture. Write the letters that stand for the first two sounds in the picture name.

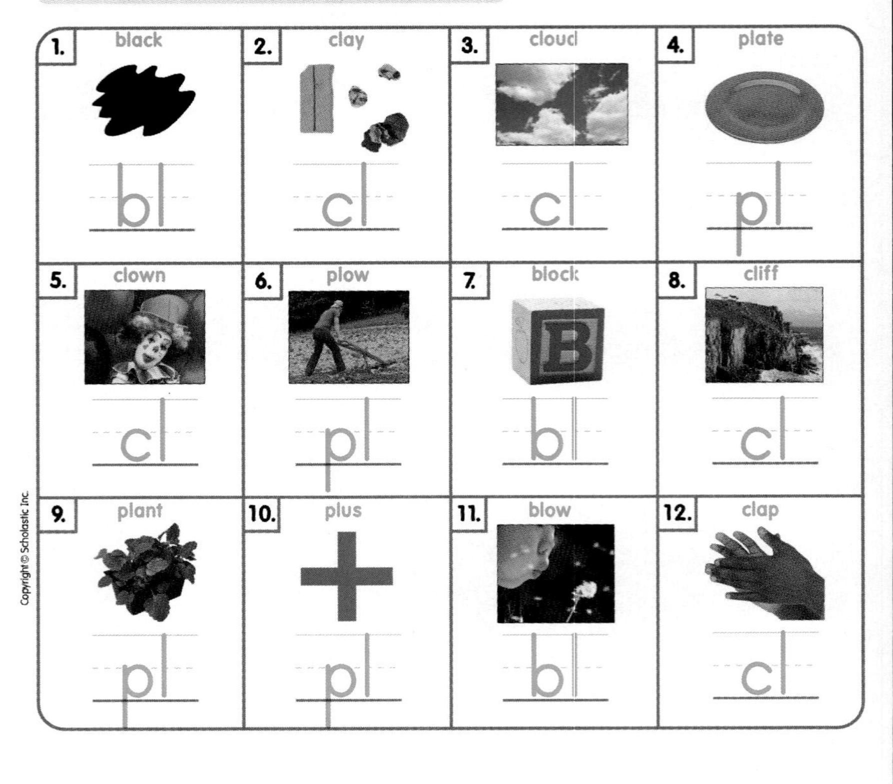

Supporting All Learners

AUDITORY/VISUAL LEARNERS

LOOKING FOR BLENDS Read aloud "If You're Happy and You Know It" on page 10 in the *Big Book of Rhymes and Rhythms, 1B*. Have children find the words with ***cl*** and ***bl*** in the rhyme. Write these words on the chalkboard. Invite volunteers to add other ***l***-blend ***(bl, cl, pl)*** words to the list.

KINESTHETIC LEARNERS

CHARADES Write words with ***l***-blends such as ***clap, clock, block, black, play,*** and ***plane*** on index cards. Invite children to play charades by acting out the words, one at a time. Volunteers can figure out each word and spell it.

TACTILE LEARNERS

SAY IT WITH CLAY Have children form words with the ***l***-blends ***bl, cl,*** and ***pl*** out of clay. Then partners can challenge each other to identify the words by feeling and tracing the letters while blindfolded.

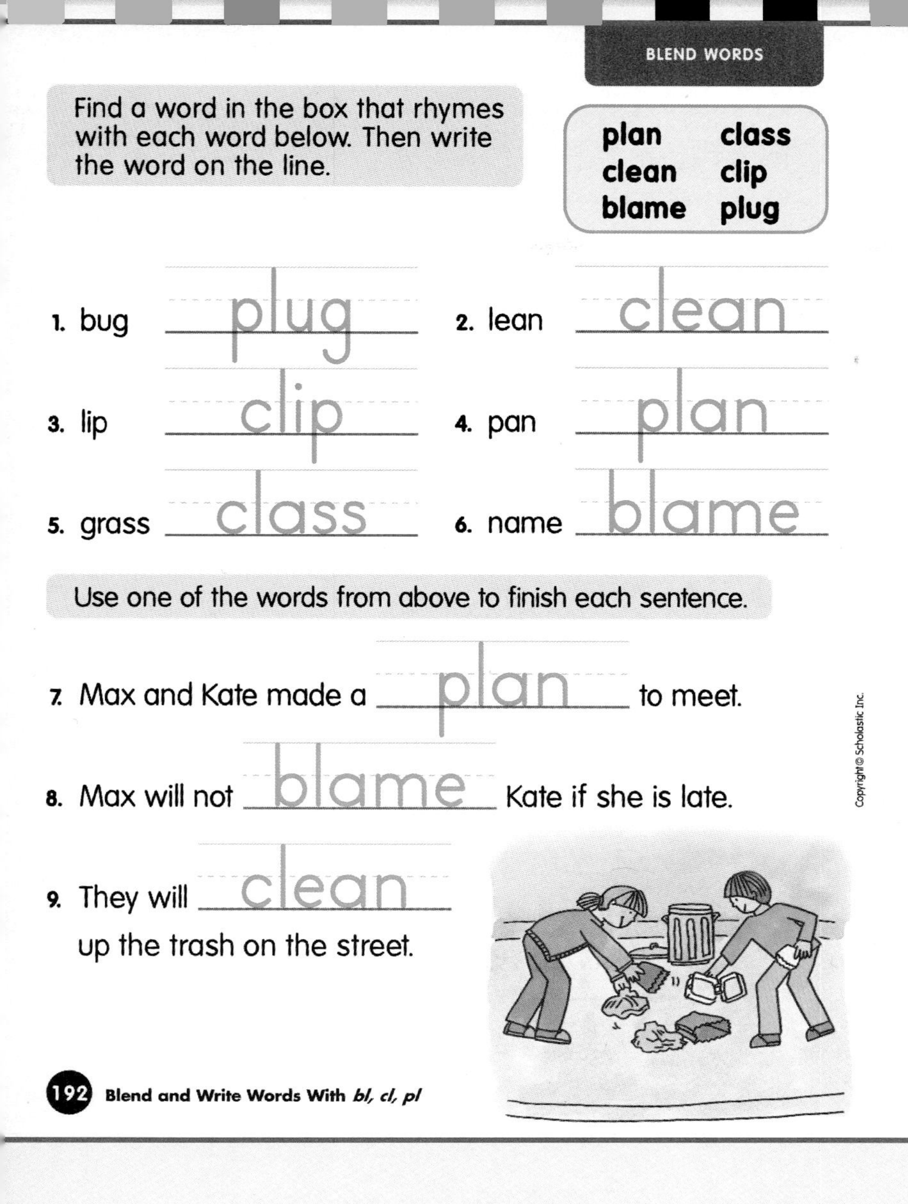

BLEND WORDS

Find a word in the box that rhymes with each word below. Then write the word on the line.

plan	class
clean	clip
blame	plug

1. bug plug
2. lean clean
3. lip clip
4. pan plan
5. grass class
6. name blame

Use one of the words from above to finish each sentence.

7. Max and Kate made a plan to meet.
8. Max will not blame Kate if she is late.
9. They will clean up the trash on the street.

192 Blend and Write Words With *bl, cl, pl*

Integrated Curriculum

SCIENCE CONNECTION

NATURE WORDS Write the following on strips of paper: ***blueberries, bluebirds, blackberries, clouds, clams, claws, planets,*** and ***plants.*** Have children pick a strip, read the item, and talk about it.

Phonics Connection

Literacy Place: *Imagine That!*
Teacher's SourceBook, pp. T208–209;
Literacy-at-Work Book, pp. 46–47

My Book: *Maggie Bloom's Messy Room*

Big Book of Rhymes and Rhythms, 1B: "If You're Happy and You Know It," p. 10

Chapter Book: *The Puppet Club,* Chapter 4

CHALLENGE

Children can write silly sentences using *l*-blends ***(bl, cl, pl)***. You may wish to provide a model sentence such as ***The black clock is on the blue plane***. Have children share their sentences and count how many blends are in each.

As children acquiring English learn to encode and decode new words, relate new words to known words. For example, the word ***clock*** may be clustered with ***time, tick,*** and ***tock,*** and the word ***truck*** may be clustered with ***ride, beep,*** and ***jeep***. Use subclusters and simple labels in each diagram to show relationships.

EXTRA HELP

Have children write ***cl, bl,*** and ***pl*** on index cards, one blend to a card. Ask children to listen for the beginning sounds in each of the following words: ***clown, plant, blue, place, clip, blow***. Children are to hold up the appropriate *l*-blend to show the beginning sounds in each word.

The Book Shop

- **The Day I Had to Play With My Sister**
  by Crosby Bonsall
  A boy discovers that his sister can be fun. **(/pl/pl)**
- **This Is the Place for Me**
  by Joanna Cole
  A bear searches for the perfect place to live. **(/pl/pl)**

Lesson 117

pages 191-192

Build Words

QUICKCHECK ✔

Can children:

✓ build words?

✓ recognize sounds previously taught?

If YES go to Read and Write.

TEACH

Review Sound-Spellings Review with children the sound-spelling relationships from the past few lessons. These include ***u-e*** and /ē/***ee, ea***. Then write the word ***seed*** on the chalkboard. Tell children that you want them to replace /s/, the first sound in ***seed,*** with /f/ to make a new word. Ask children what letter stands for /f/. Then replace the letter ***s*** in ***seed*** with the letter ***f***, and blend the new word formed. Continue by changing one letter or letter pair to build new words. You may wish to use this sequence of words: ***feed, feet, meet, meat, treat***.

READ AND WRITE

Build Words Distribute the following letter cards to children: ***a, c, i, l, m, p***. If children have their own set of cards, have them locate the letter set. Provide time for children to build as many words as possible using the letter cards. Suggest that they record their words on a separate sheet of paper. Continue building words with the following sets of letter cards: ***a, b, c, k, l, o; a, c, e, k, r, t***.

Then explain to children that you will give them the beginning and ending words of a three- or four-word ladder. They are to change one letter or letter pair in each word to complete the word ladder. Use the following words:

- keep (beep) beet
- trap (clap) clip
- truck (track) trap

Note: The words in parentheses are possible answers.

Complete Activity Page Read aloud the directions on page 193. Children can complete the page independently.

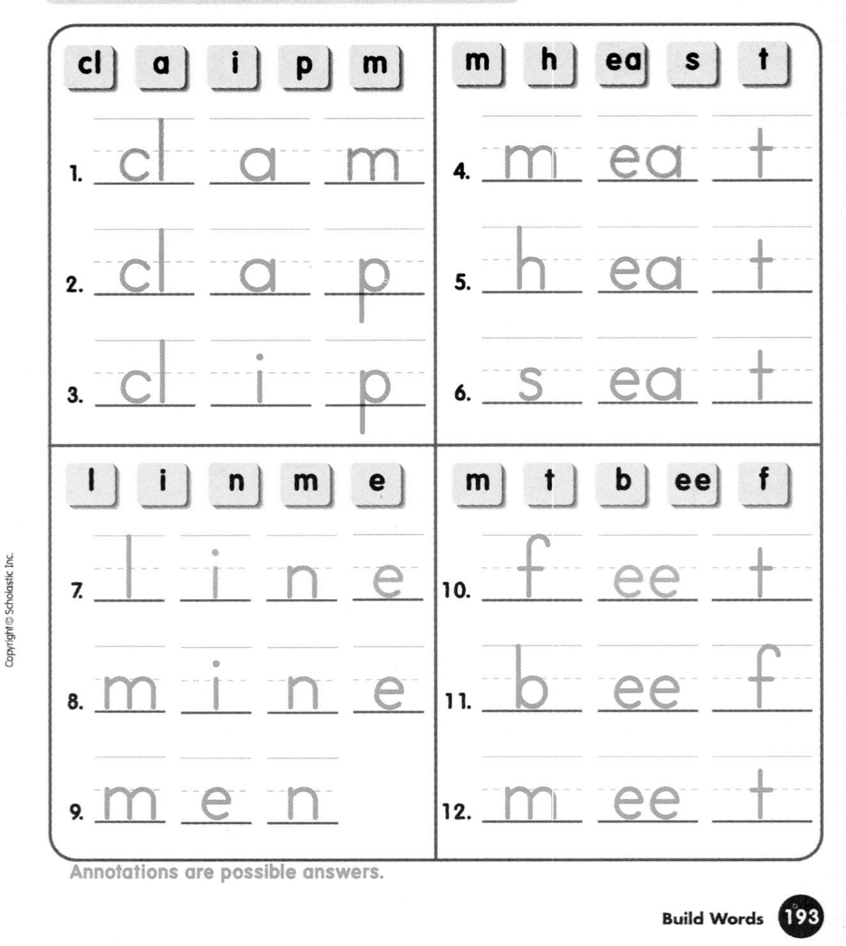
BUILD WORDS

Name

Use the letter tiles to make words.

cl a i p m

1. cl a m
2. cl a p
3. cl i p

m h ea s t

4. m ea t
5. h ea t
6. s ea t

l i n m e

7. l i n e
8. m i n e
9. m e n

m t b ee f

10. f ee t
11. b ee f
12. m ee t

Annotations are possible answers.

Supporting All Learners

KINESTHETIC LEARNERS

BE A WORD LADDER Write the words below on index cards. Give one card to each child, and invite children to arrange themselves in three-word word ladders: ***plate, place, trace; greet, green, seen; brick, click, clock; block, black, track.***

VISUAL/AUDITORY LEARNERS

NAME GAME Have each child write his or her first name on an index card. Put the cards in a bag. Then have children take turns pulling out cards, one at a time. Have children take turns reading each name aloud. Then children can identify any names spelled with ***ee, ea,*** ***r***-blends, and ***l***-blends.

EXTRA HELP

Help children to build words with ***u-e*** and ***/ē/ee, ea*** by using the magnetic letters or the pocket ABC cards and pocket chart from the Phonics and Word Building Kit.

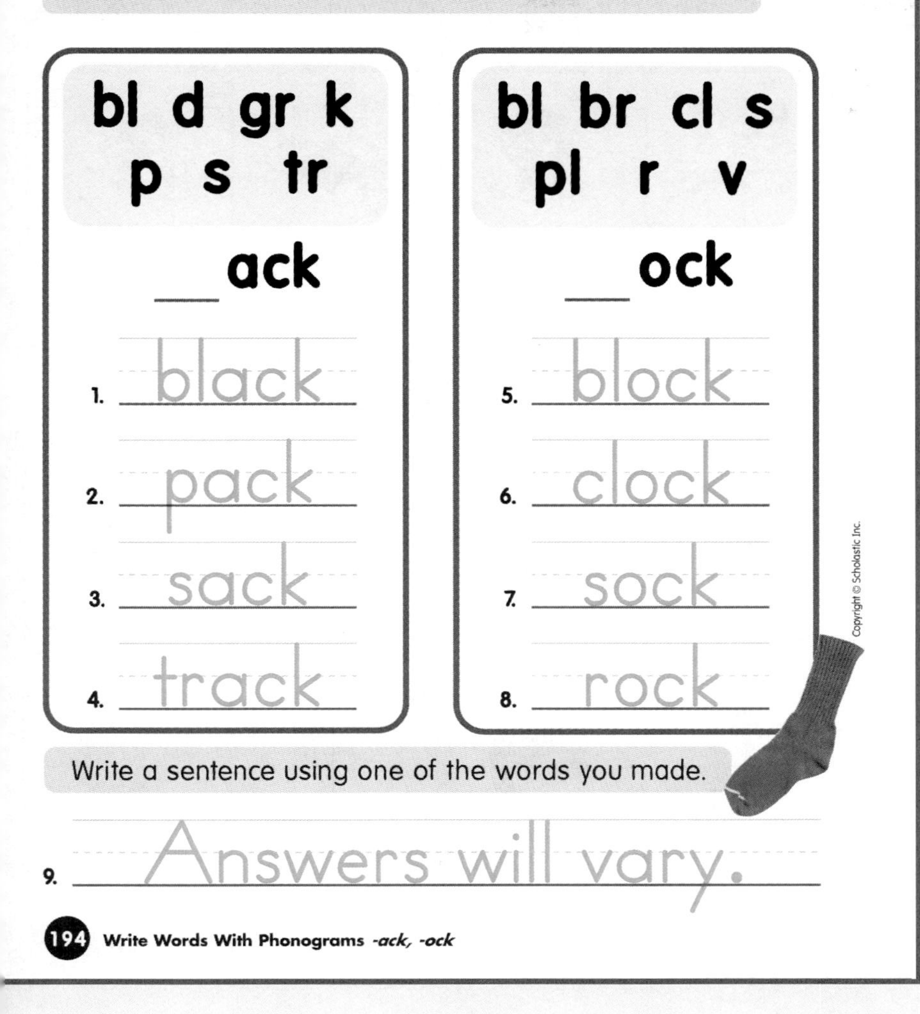
PHONOGRAMS

Add each letter or letters to the word part below it. Blend the word. If it is a real word, write it on the line.

bl d gr k p s tr	bl br cl s pl r v
___ack	___ock
1. black	5. block
2. pack	6. clock
3. sack	7. sock
4. track	8. rock

Write a sentence using one of the words you made.

9. Answers will vary.

194 Write Words With Phonograms *-ack, -ock*

Supporting All Learners

TACTILE LEARNERS

SAND WRITING Print words with ***-ack*** and ***-ock*** on large index cards. Have children select a card and "trace" the word using a glue stick or paste. Immediately have them sprinkle sand over the glue, then shake off the excess. When the glue dries, children can trace the letters with their fingers to practice writing the word.

EXTRA HELP

For additional practice, set up Center 6, Activity 2, "Cloze-ing Song" from *Quick-and-Easy Learning Centers: Phonics*. This activity focuses on using cloze techniques to draw attention to word endings in familiar songs.

CHALLENGE

Assemble the following objects: **clock, block, rock,** and **sock**. Have children say the name of each object and then orally segment the words according to the sounds they hear. Children can write labels for each object.

Lesson 118

page 194

Blend and Build Words With Phonograms

QUICKCHECK ✔

Can children:

- ✔ **blend words with phonograms *-ack, -ock*?**
- ✔ **build words with *-ack, -ock*?**

If YES go to Read and Write.

TEACH

Introduce the Phonograms Write the phonograms *-ack* and *-ock* on the chalkboard. Point out the sounds these phonograms stand for. Add the letter *s* to the beginning of *-ack*, and model for children how to blend the word formed. Have children repeat the word ***sack*** aloud as you blend the word again. Then add ***r*** to the beginning of *-ock*, and model for children how to blend the word ***rock***.

Ask children to suggest words that rhyme with ***sack*** and ***rock***. List these words on the chalkboard in separate columns, and have volunteers underline the phonogram *-ack* or *-ock* in each one. Be sure to include the words ***track, pack, black, lock, clock,*** and ***sock***. Blend these words for children as they repeat them aloud. Encourage children to look for the word parts *-ack* and *-ock* as they read.

READ AND WRITE

Build Words Write the word parts *-ack* and *-ock* on the chalkboard. (If available, use a pocket chart and letter cards.) Have children add a letter or a letter pair to the beginning of each phonogram to make a new word. Continue by having children replace the initial consonant in the first word to build a second word. Children may build the following sequence of words: ***sock, rock, block***.

Complete Activity Page Read aloud the directions on page 194. Have children complete the page independently.

Lesson 119

pages 195–196

Blend Words With *s*-blends *(sp, st)*

QUICKCHECK ✔

Can children:

- ✔ **substitute sounds in words?**
- ✔ **recognize s-blends *(sp, st)*?**
- ✔ **blend words with *sp* and *st*?**

If YES go to Read and Write.

TEACH

Develop Phonemic Awareness

Phonemic Manipulation Explain to children that they are going to play a consonant riddle game. You will say a word. They will think of a word that rhymes with your word and begins with /**sp**/ or /**st**/. For example, ask what word rhymes with ***hot*** and starts with /**sp**/. After children say ***spot***, continue with the following riddles:

- **What word rhymes with *tell* and starts with /sp/?**
- **What word rhymes with *part* and starts with /st/?**
- **What word rhymes with *cone* and starts with /st/?**

Connect Sound-Symbol Write ***pin*** and ***spin*** on the chalkboard as you read each aloud. Ask children what sounds they hear at the beginning of each word. Point out that the *s*-blend *(sp)* at the beginning of ***spin*** is spelled by two letters together, *s* and ***p***.

READ AND WRITE

Blend Words List the following on a chart. Model how to blend the words. Then have children blend them.

• rest	stick	stone
• spin	spill	nest

- **He made a stack of sticks.**

Complete Activity Pages Read aloud the directions on pages 195–196. You may wish to review each picture name with children.

Name

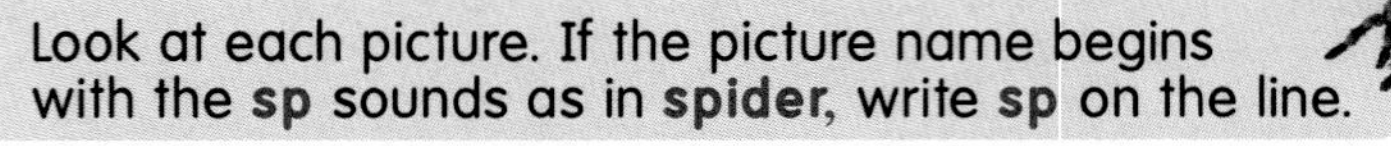

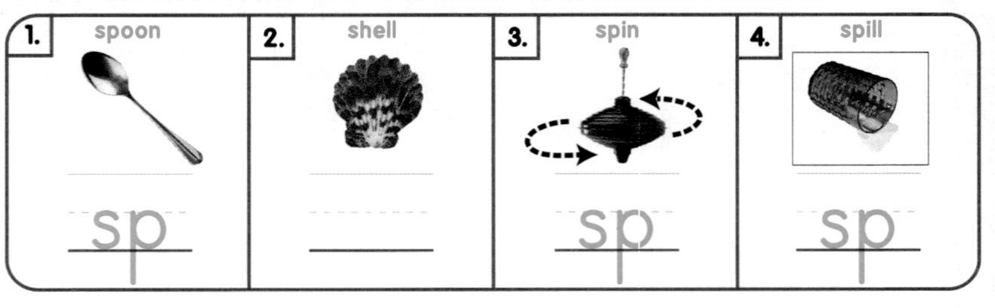

If the picture name begins with the **st** sounds as in **stamp**, write **st** on the first line. If it ends with **st** as in **test**, write **st** on the second line.

5. stump — st ___	6. nest — ___ st	7. star — st ___	8. vest — ___ st
9. stir — st ___	10. toast — ___ st	11. wrist — ___ st	12. stop — st ___

STOP

Supporting All Learners

AUDITORY LEARNERS

WHAT'S THE MISSING WORD? Invite children to name some ***s***-blend words ***(sp, st)***. List them on the chalkboard. Then make up a sentence with one of the words. Say the sentence aloud, omitting the word with the ***s***-blend. For example: ___ *and look before you cross the street.* ***(Stop)*** Invite children to fill in the blank with the correct ***s***-blend word. Repeat with other sentences.

AUDITORY/VISUAL LEARNERS

WORD HUNT Read aloud "Hey, Diddle, Diddle" on pages 12–13 in the *Big Book of Rhymes and Rhythms, IB*. Have children find the words with ***sp*** in the rhyme. Write these words on the chalkboard. Invite volunteers to add other ***s***-blend ***(sp, st)*** words to the list.

KINESTHETIC/VISUAL LEARNERS

WORD SORT Write the following words on note cards: ***spoke, spin, spun, stick, still, stone, spell, spill,*** and ***stack***. Have children sort the words according to whether they begin with ***sp*** or ***st***. Then gather the ***st*** word cards, and add the following words: ***best, list, least, beast, just, must, vest***. Have children sort this new group of words according to whether they begin or end with ***st***.

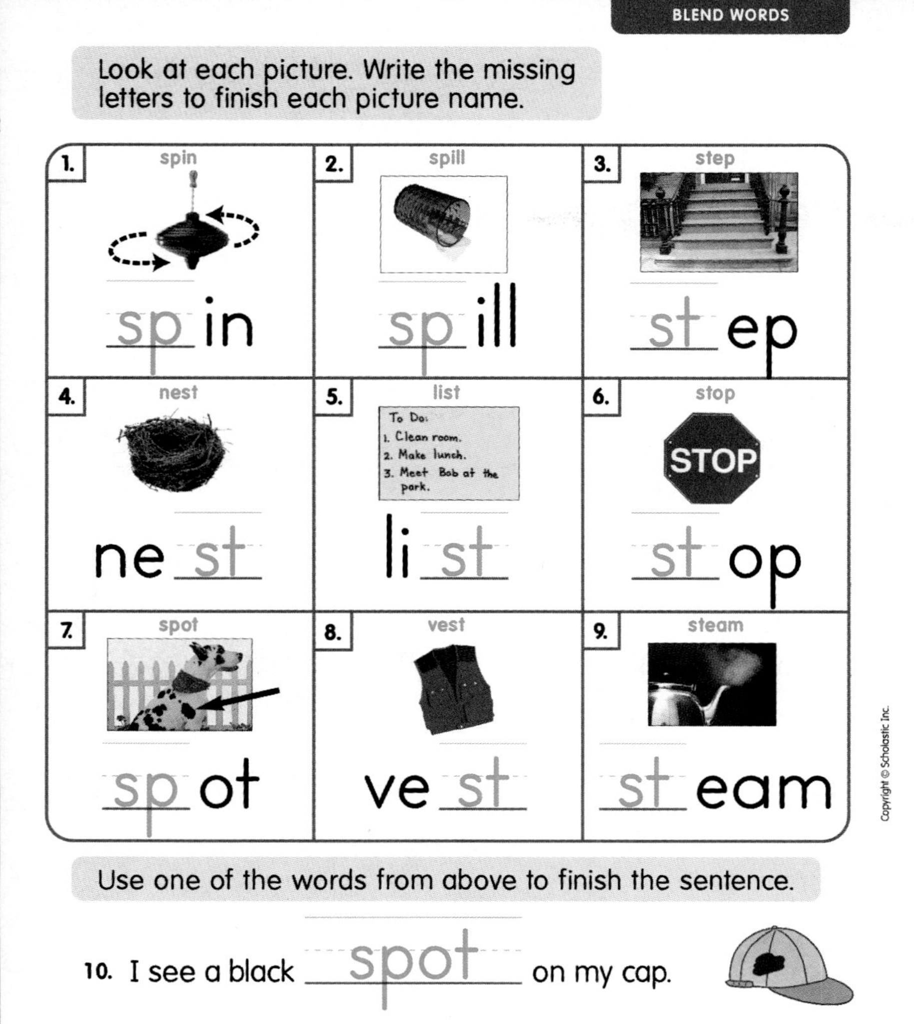
BLEND WORDS

Look at each picture. Write the missing letters to finish each picture name.

1. spin	2. spill	3. step
sp in	sp ill	st ep
4. nest	**5. list**	**6. stop**
ne st	li st	st op
7. spot	**8. vest**	**9. steam**
sp ot	ve st	st eam

Use one of the words from above to finish the sentence.

10. I see a black spot on my cap.

ESL Display objects (or pictures of objects) whose names begin with **/sp/** and **/st/**. You may wish to use objects such as a **spoon, spool, sponge, spelling list, spot, star, stick, stone, stapler,** and **stamps**. Help children acquiring English name each object, emphasizing the beginning sounds. Give children self-sticking notes to write or trace ***sp*** and ***st***. Children can attach their notes to the objects and then sort them according to the ***s***-blend.

EXTRA HELP

Use the magnetic letters from the Phonics and Word Building Kit to build the word ***spill***. Read the word aloud. Have children work together to change the letters in ***spill*** to make other ***s***-blend words. Write the words children make in a list.

Integrated Curriculum

SPELLING CONNECTION

MAKE WORDS Write the following word parts on the chalkboard: ***__ill, __ot, __ oke; __ ill, __ ick, __ ack; ve__, li__, pa__***. Invite children to add ***sp*** to the first set and ***st*** to the last two sets. Help children blend each word formed.

Phonics Connection

Literacy Place: *Imagine That!*
Teacher's SourceBook, pp. T270–271;
Literacy-at-Work Book, pp. 59–60

My Book: *The Slippery, Sloppy Day*

Big Book of Rhymes and Rhythms, 1B: "Hey, Diddle, Diddle," pp. 12-13.

Chapter Book: *The Puppet Club*, Chapter 5

Lesson 120

page 197

Write and Read Words With *r*-, *l*-, and *s*-blends

QUICKCHECK ✔

Can children:

✔ spell words with *r*-, *l*-, and *s*-blends?

✔ read a poem?

If YES go to Read and Write.

TEACH

Link to Spelling Review with children *r*-, *l*-, and *s*-blends. Say aloud one of these sounds. Have a volunteer write on the chalkboard the spelling that stands for the sound. For example, the letters ***bl*** stand for the blend at the beginning of ***black***. Continue with all the sounds. You may also wish to review /ē/*ee, ea* and the ***u-e*** spelling pattern.

Phonemic Awareness Oral Segmentation

Say aloud the word ***stop***. Have children orally segment the word. *(/s/ /t/ /o/ /p/)* Ask them how many sounds the word contains. *(4)* Draw four connected boxes on the chalkboard, and have a volunteer write the spelling that stands for each sound in the word ***stop*** in the appropriate box. Continue with the words ***spin*** and ***spot***.

s	t	o	p

READ AND WRITE

Dictate Dictate the following words and sentence. Have children write the words and sentence on a sheet of paper. When children are finished, write them on the chalkboard, and have children make any necessary corrections on their papers.

- **spot** **still** **best**
- **do** **was** **other**
- **Look at the other test.**

Complete Activity Page Read aloud the directions on page 197. Have children read the poem independently or with partners.

POETRY

Name

Read the poem. Finish each sentence using a word from the poem.

Rocking to the Beat

The duck quacks.
The mule taps his feet.
The clams click.
As we step to the beat.

The mice hip-hop.
The cats clap and grin.
The hens cluck.
As we trot and spin.

1. We like to step to the beat.
2. We like to trot and spin.

Link to Spelling 197

Supporting All Learners

AUDITORY/VISUAL LEARNERS

WRITING EXTENSION Have children generate a list of words that begin and end with **/st/**. Record these words on the chalkboard. Then have children create a story using as many **/st/** words as they can. Write the dictated story on chart paper for group and individual reading. Return to the story, rereading it in subsequent lessons.

EXTRA HELP

Build background and provide an opportunity for children to use new words in context by asking children to write their idea of the "best" in each of the following categories: ***pet, game, treat,*** and ***place***. Children can write each category and then write or dictate their answers.

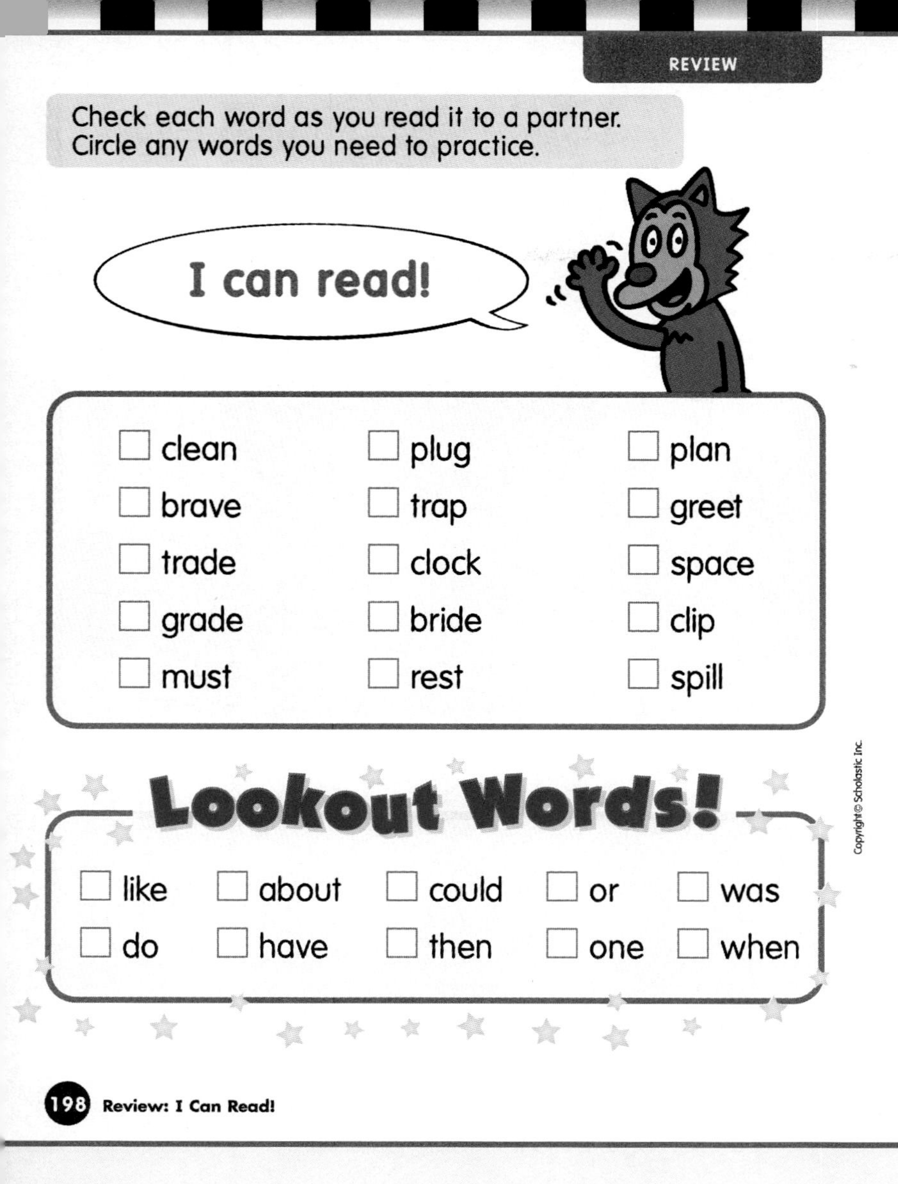

REVIEW

Check each word as you read it to a partner.
Circle any words you need to practice.

I can read!

☐ clean	☐ plug	☐ plan
☐ brave	☐ trap	☐ greet
☐ trade	☐ clock	☐ space
☐ grade	☐ bride	☐ clip
☐ must	☐ rest	☐ spill

Lookout Words!

☐ like	☐ about	☐ could	☐ or	☐ was
☐ do	☐ have	☐ then	☐ one	☐ when

Read and Review Words With *r*-, *l*-, and *s*-blends

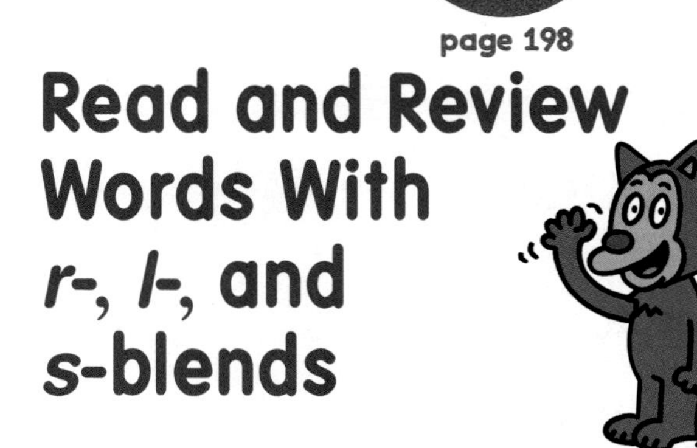

QUICKCHECK ✔

Can children:

✔ **read words with *r*-, *l*-, and *s*-blends?**

✔ **recognize high-frequency words?**

If YES go to Read and Write.

TEACH

Review Sound-Spellings Review with children the sound-spelling relationships from the past few lessons. These include *r*-blends, *l*-blends, and *s*-blends. Say aloud one of these sounds. Have a volunteer write the spelling that stands for the sound on the chalkboard. For example, the letters ***tr*** stand for /**tr**/. Continue with all the sounds. Then display pictures of objects whose names begin or end with one of these spellings. Have children write the spelling that each picture's name begins or ends with as the picture is displayed.

Review High-Frequency Words Review the high-frequency words ***look, about, could, was,*** and ***when***. Write sentences containing each high-frequency word. Say aloud one of the words. Have volunteers circle the word.

READ AND WRITE

Build Words Distribute these letter cards: ***a, i, m, p, r, t***. Ask children to build as many words as possible and to record them. Have children build more words with these letter cards: ***a, c, i, l, m, p*** and ***b, r, e, p, s, t***.

Build Sentences Write these words on note cards: ***the, get, have, look, club, we, clam, at, trap, could, truck***. Have children make sentences using the words. To get them started, suggest this sentence: ***Can we get the truck?***

Complete Activity Page Read aloud the directions on page 198.

Supporting All Learners

VISUAL LEARNERS

FIND BLENDS! Ask children to look on the playground or in their neighborhood for things whose names contain *r*-blends, *l*-blends, and *s*-blends. You may wish to get children started by suggesting that they look for trucks, grass, glass, and steps.

EXTRA HELP

Help children make high-frequency words and words with *r*-blends, *l*-blends, and *s*-blends using the pocket ABC cards and pocket chart. Once a word is formed, mix the letters, and have children rebuild it.

I CAN READ! OPTIONS

The I Can Read! page can be used for one or all of the following:

- **paired reading**
- **individual assessment**
- **choral reading**
- **homework practice**
- **program placement**

Lesson 122

pages 199-200

Read Words in Context

TEACH

Assemble the Story Ask children to remove pages 199–200. Have children fold the pages in half to form the Take-Home Book.

Preview the Story Preview *A Trap,* a nonfiction story about a spider's web. Invite children to browse through the first two pages of the story and to comment on anything they notice. Suggest that they point out any unfamiliar words. Read these words aloud. Have children blend them. Then have children predict what they think the selection might be about.

READ AND WRITE

Read the Story Read the story aloud, or have volunteers take turns reading aloud a page at a time. Discuss with children anything of interest on each page. Encourage them to help each other with any blending difficulties. The following prompts may help children who need extra support while reading:

- **What letter sounds do you know in the words?**
- **Are there any word parts you know?**
- **What do the pictures tell you about the story?**

Reflect and Respond Have children share their reactions to the selection. What did they learn? What would they like to add to the story? Encourage children to write another sentence to add to the story.

Develop Fluency Have children reread the story as a choral reading, or have partners reread the story independently. Provide time for children to reread the story on subsequent days to develop fluency and increase reading rate.

Have children share how they figure out unfamiliar words. Ask children to model how they blend words. Continue to review challenging or confusing sound-spelling relationships for children needing additional support.

Supporting All Learners

VISUAL LEARNERS

MY BOOKS/TECHNOLOGY For additional practice with ***r***-blends, ***l***-blends, and ***s***-blends, have children read the following My Books: *Grandpa Gray, Maggie Bloom's Messy Room,* and *Still Snoring*. For information on using the My Books on the computer, see the WiggleWorks Plus My Books Teaching Plan.

TACTILE LEARNERS

WEAVE A WEB Give children yarn, scissors, glue, and oaktag or construction paper. Have them make a web. You may want to show them how to start by first drawing a pattern of spoke lines and connecting lines. Then, when children are finished, ask them to write a sentence about what their web is like, what it can do, or who or what can get stuck in it.

EXTRA HELP

Read *A Trap* aloud with children. Then have four small groups of children each pick a page to illustrate. When the illustrations are done, children can take turns holding up their pictures and reading aloud the text for that picture.

CHALLENGE

Children can read the story *A Trap* aloud and explain the pictures to groups of children, or even other classes.

Reflect on Reading

ASSESS COMPREHENSION

To assess their understanding of the story, ask children questions such as:

- **How can a web trick a bug?** ***(A bug that steps in a web will get stuck in it.)***
- **Why does a spider make a web?** ***(to trap insects to eat)***
- **What does a web look like?** ***(It looks like a net made out of thin lines.)***

HOME-SCHOOL CONNECTION

Send home *A Trap*. Encourage children to read the story to a family member. Then children can search at home for words with blends, including ***br, gr, tr, bl, cl,*** and ***pl***. Children can look on refrigerator notes, in junk mail, and in other printed matter in their homes. Adults and siblings at home may also join in the search.

WRITING CONNECTION

Have children pick a bug, animal, or other thing in nature that they know about. Children can take turns talking about what they know about their subject. Then have children draw pictures about the bugs, animals, or things in nature and write sentences to go with those pictures. Give children an opportunity to share their pictures and sentences.

Phonics Connection

Phonics Readers:
#43, *Troll Tricks*
#44, *Play the Animal Game!*
#45, *Slip Slide Baseball Jokes*

Lesson 123

pages 201-202

Blend Words With *s*-blends *(sm, sn)*

QUICKCHECK ✔

Can children:

- ✓ **substitute sounds in words?**
- ✓ **recognize *s*-blends *(sm, sn)*?**
- ✓ **blend words with *sm* and *sn*?**

If YES go to Read and Write.

TEACH

Develop Phonemic Awareness

Phonemic Manipulation Explain to children that they are going to play a consonant riddle game. Point out that you will say a word. Children should think of a word that rhymes with your word and begins with /**sm**/ or /**sn**/. For example, ask children what word rhymes with ***bell*** and starts with /**sm**/. After they say ***smell,*** continue with the following riddles:

- What word rhymes with *tail* and starts with *sn*?
- What word rhymes with *bake* and starts with *sn*?
- What word rhymes with *cash* and starts with *sm*?
- What word rhymes with *pile* and starts with *sm*?

Connect Sound-Symbol Write the words ***nail*** and ***snail*** on the chalkboard. Read each word aloud. Ask children what sounds they hear at the beginning of each word and how the words are different. Point out that the *s*-blend /**sn**/ at the beginning of ***snail*** is spelled by putting two letters together: ***s*** and ***n***. Repeat with ***smock*** and ***sock***.

READ AND WRITE

BUILT-IN REVIEW

Blend Words List the following words and sentence on a chart. Have volunteers read each aloud. Model how to blend the words.

- smell smile smoke
- snap sneak snake
- We smile at snack time.

Complete Activity Pages Read aloud the directions on pages 201–202. Review each picture name with children.

LISTEN AND WRITE

Name

Look at each picture. Write the letters that stand for the first two sounds in the picture name.

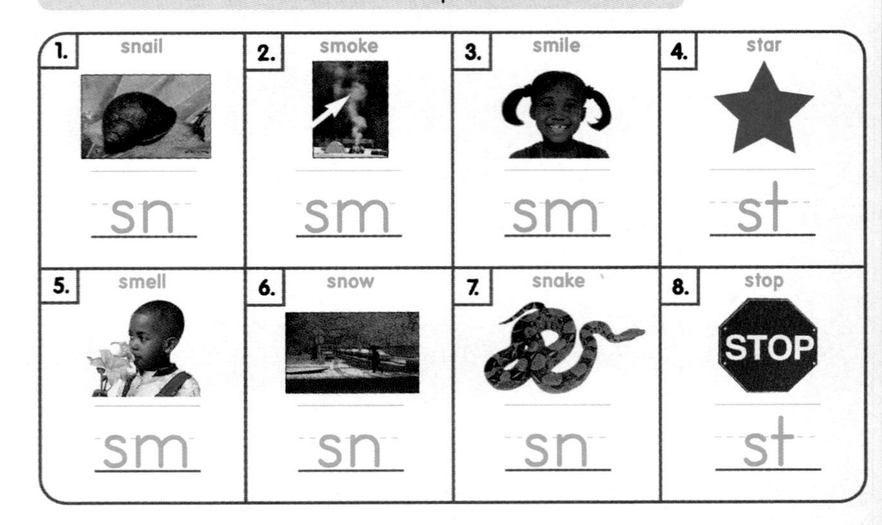

Write the letters **sm** or **sn** to finish each word. Read the words to a friend.

9. sm ash 10. sn ap

11. sn eak

Recognize *s*-Blends 201

Supporting All Learners

AUDITORY/KINESTHETIC LEARNERS

PLAY BINGO To practice and review words with *s*-blends, children may enjoy playing "Sound Bingo" on pages 28–29 in *Quick-and-Easy Learning Games: Phonics*. Use the variation of the game that lists blends, and adapt the word list as needed to reflect sounds and blends already studied.

VISUAL/KINESTHETIC LEARNERS

SMILE! Invite children to draw smiley faces and to label them with simple statements about what makes them smile. If you wish, provide a sentence starter such as ***I smile when*** __ or __ ***makes me smile.***

VISUAL LEARNERS

SMALL SNAKE WORDS Write ***Is the snake small?*** on the chalkboard. Read the sentence aloud. Ask children what sounds they hear at the beginning of ***snake*** and ***small.*** Circle the letters ***sn*** and ***sm***. Ask children to suggest other words that contain **/sn/** and **/sm/**. List these words on the chalkboard. Have volunteers circle the letters ***sn*** or ***sm*** in each word.

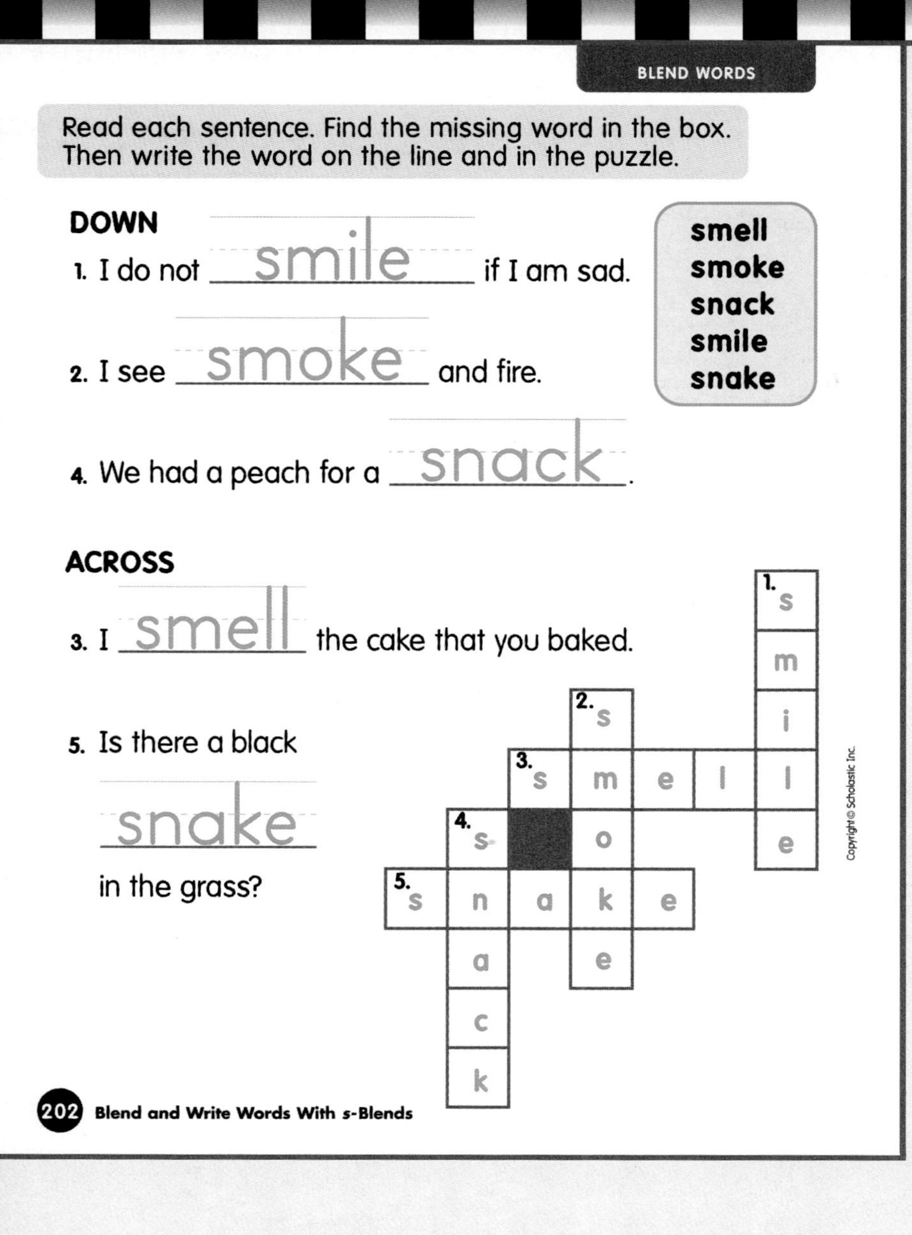
BLEND WORDS

Read each sentence. Find the missing word in the box.
Then write the word on the line and in the puzzle.

smell
smoke
snack
smile
snake

DOWN

1. I do not smile if I am sad.
2. I see smoke and fire.
4. We had a peach for a snack.

ACROSS

3. I smell the cake that you baked.
5. Is there a black snake in the grass?

Copyright © Scholastic Inc.

202 Blend and Write Words With *s*-Blends

Integrated Curriculum

WRITING CONNECTION

WRITING WORDS Write these word parts on the chalkboard: ***_ap, _eak, _ack***. Have volunteers add the letters ***sn*** to each. Then write the word parts ***_ell, _ash,*** and ***_og***. Have volunteers add the letters ***sm*** to each. Model for children how to blend each word.

Phonics Connection

Literacy Place: *Imagine That!*
Teacher's SourceBook, pp. T270–271;
Literacy-at-Work Book, pp. 59–60

My Book: *Still Snoring*

Chapter Book: *The Puppet Club,* Chapter 5

Invite children acquiring English to play "Smack or Snap." Write ***smack*** and ***snap*** on the chalkboard. As you read each word with children, emphasize the blend and demonstrate the action: smack one hand lightly on your knee; snap your fingers. Then, for each word they hear you say that begins with **/sm/,** they can "smack," and for each word beginning with **/sn/,** they can "snap." Use words such as ***smile, snip, snore, smell, snake, smash, snow, snail,*** and ***smart***.

CHALLENGE

Invite children to name words that begin with ***sm*** and ***sn*** to rhyme with the following words: ***tail, cake, low, part, jug, fall, rag, tell, pile, tap, dog,*** and ***back***.

- **Snow on Snow on Snow**
  by Cheryl Chapman
  A boy and dog have fun. **(/sn/sn)**
- **The Jacket I Wear in the Snow**
  by Shirley Neitzel
  An easy-to-read tale. **(/sn/sn)**

Lesson 124

page 203

Recognize High-Frequency Words

QUICKCHECK ✔

Can children:

✔ **recognize and write the high-frequency words *many, which, down, no, out*?**

✔ **complete sentences using the high-frequency words?**

If YES go to Read and Write.

TEACH

Introduce the High-Frequency Words

Write sentences containing *many, which, down, no,* and *out* as you underline the high-frequency words You may wish to use the following sentences:

1. **Which steps lead out to the track?**
2. **Take those steps down.**
3. **Many steps lead to the track.**
4. **No one is on the track yet.**

Ask volunteers to dictate additional sentences using high-frequency words.

READ AND WRITE

Practice Write each high-frequency word on a note card. Read each word aloud as you display the cards. Then do the following:

- Mix the cards.
- Display one card at a time, and ask children to state each word aloud.
- Have children spell each word aloud, clapping on each letter.
- Ask children to write each word in the air as they state aloud each letter. Then have children write each word on paper.

Complete Activity Page Read aloud the directions on page 203. Have children complete the page independently.

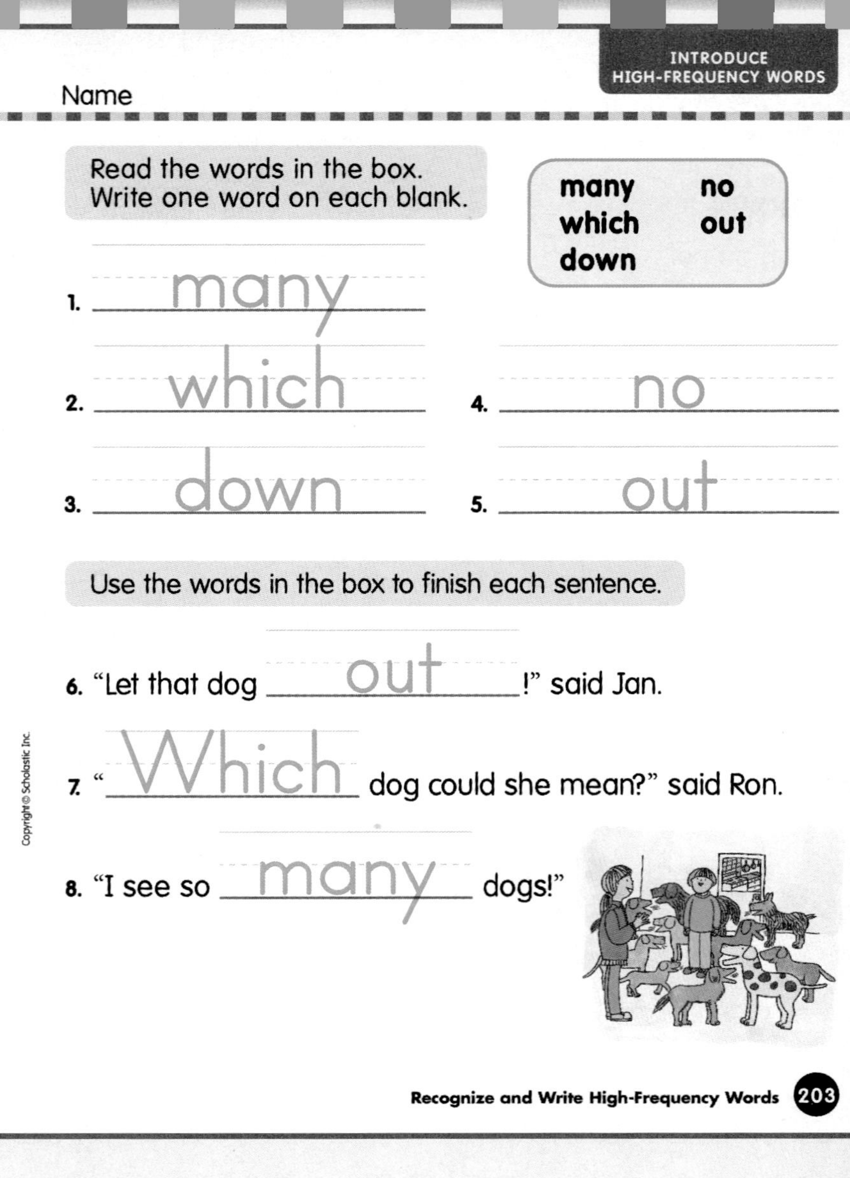
INTRODUCE HIGH-FREQUENCY WORDS

Name

Read the words in the box. Write one word on each blank.

many	no
which	out
down	

1. many
2. which
3. down
4. no
5. out

Use the words in the box to finish each sentence.

6. "Let that dog out!" said Jan.
7. "Which dog could she mean?" said Ron.
8. "I see so many dogs!"

Copyright © Scholastic Inc.

Recognize and Write High-Frequency Words 203

Supporting All Learners

VISUAL/AUDITORY LEARNERS

LEARNING CENTER To help build sight-word awareness, set up Center 6, Activity 1 "Songs in Sequence" from *Quick-and-Easy Learning Centers: Phonics.* This activity focuses on fostering sequencing skills and helping children recognize high-frequency words.

CHALLENGE

Have children make high-frequency word books by writing sentences with the high-frequency words ***many, which, down, no,*** and ***out***. Ask children to write one sentence per page, then illustrate the sentence. Help children bind their pages to create the high-frequency word books. Encourage children to share their books with others and to place the books in a library or book corner.

EXTRA HELP

For children who need additional support reading the high-frequency word ***which,*** read aloud Phonics Reader #45, *Slip Slide Baseball Jokes*. Then have children read the Phonics Reader to each other.

PLURALS

Circle the word that names each picture.
Then write the word on the line.

1. dog / (dogs) — dogs	2. (pig) / pigs — pig	3. nut / (nuts) — nuts
4. can / (cans) — cans	5. top / (tops) — tops	6. (tree) / trees — tree
7. (clock) / clocks — clock	8. brick / (bricks) — bricks	9. bike / (bikes) — bikes

Recognize and Write Plurals

QUICKCHECK ✔

Can children:

✔ recognize plurals?

✔ write plurals?

If YES go to Read and Write.

TEACH

Introduce Plurals Explain to children that many words that name more than one thing are formed by adding *s*. Write the word ***joke*** on the chalkboard. Ask a volunteer to read it. Use the word ***joke*** in a sentence; for example, ***I tell her a joke.*** Then write ***jokes*** on the chalkboard, and ask a volunteer to circle ***joke***. Ask another volunteer to circle the *s* ending. Use the word ***jokes*** in a sentence; for example, ***I tell her jokes.***

THINK ALOUD

I know there is a word I already know inside the word *jokes*. It is the word *joke*. This word tells about one joke. If I want to tell about more than one joke, I add the letter *s* to *joke* to make *jokes*.

Write the following words on the chalkboard: ***caps, pots, bikes, homes,*** and ***steps***. Have children circle the words that mean "one" in each of these words. Then have them underline the *s* ending in each word. Ask children to tell the difference in meaning between the words when they are written without the *s* and with the *s*.

READ AND WRITE

Practice Ask children what is the same and what is different about the underlined words in each pair.

- **The dog runs.**
  The dogs run.
- **He picked up the snake.**
  He picked up the snakes.

Complete Activity Page Read aloud the directions on page 204. Review the picture names with children.

Supporting All Learners

VISUAL/KINESTHETIC LEARNERS

WORD CARDS Write the following words on note cards, and mix them up: ***logs, hogs, log, hog, pigs, wigs, pig, wig, cans, fans, can, fan, lines, vines, line,*** and ***vine***. Have children sort them three ways: First, have them sort the cards by putting them into pairs that name the same object or animal. Then, have children sort the cards into pairs of rhyming words. Finally, have children sort the cards into two piles—words that name one and words that name more than one.

AUDITORY LEARNERS

LISTEN UP! Have children take turns saying sentences with singular and plural words. After each sentence, children listening to the sentences should say whether it was about one thing or more than one thing. You may wish to use sentences such as the following:

- **The cat will sleep on my bed.**
- **My books are on my desk.**
- **I will be home for two days.**
- **The bus is late.**
- **The doors are open.**

Invite children to find pictures in books and other print materials that show one and more than one. Children can use self-sticking notes to label the pictures.

Lesson 126

pages 205-206

Blend Words With /ch/*ch*

QUICKCHECK ✔

Can children:

- ✔ **orally segment words?**
- ✔ **recognize /ch/?**
- ✔ **blend words with /ch/*ch*?**

If YES go to Read and Write.

TEACH

Develop Phonemic Awareness

Oral Segmentation Explain to children that you are going to say words by saying their parts. Children will listen to the parts and then say the whole word. For example, say the sounds /**ch**/ /**i**/ /**n**/. Guide children to say the word ***chin***. Continue with the following word parts.

/ch/ /i/ /p/	/ch/ /o/ /p/
/ē/ /ch/	/t/ /ē /ch/
/m/ /u/ /ch/	/ch/ /a/ /t/

Connect Sound-Symbol Write the words ***eat*** and ***each*** on the chalkboard, and read each word aloud. Ask children what sounds they hear at the end of each word and how the words are different. Point out that /**ch**/ at the end of the word ***each*** is spelled by two letters together, ***c*** and ***h***. Write the words ***cat*** and ***chat*** on the chalkboard as you read each aloud. Repeat the above procedure.

READ AND WRITE

Blend Words List the following words and sentences on a chart. It may be necessary to model how to blend the words. Then have children blend them.

- chop check lunch
- chest reach bunch
- They ate lunch on the bench.
- I reach for the branch.

Complete Activity Pages Read aloud the directions on pages 205–206. Review each picture name with children.

Name

Look at each picture. If the picture name begins with the **ch** sound as in **cheese**, write the letters **ch** on the first line. If it ends in **ch** as in **inch**, write **ch** on the second line.

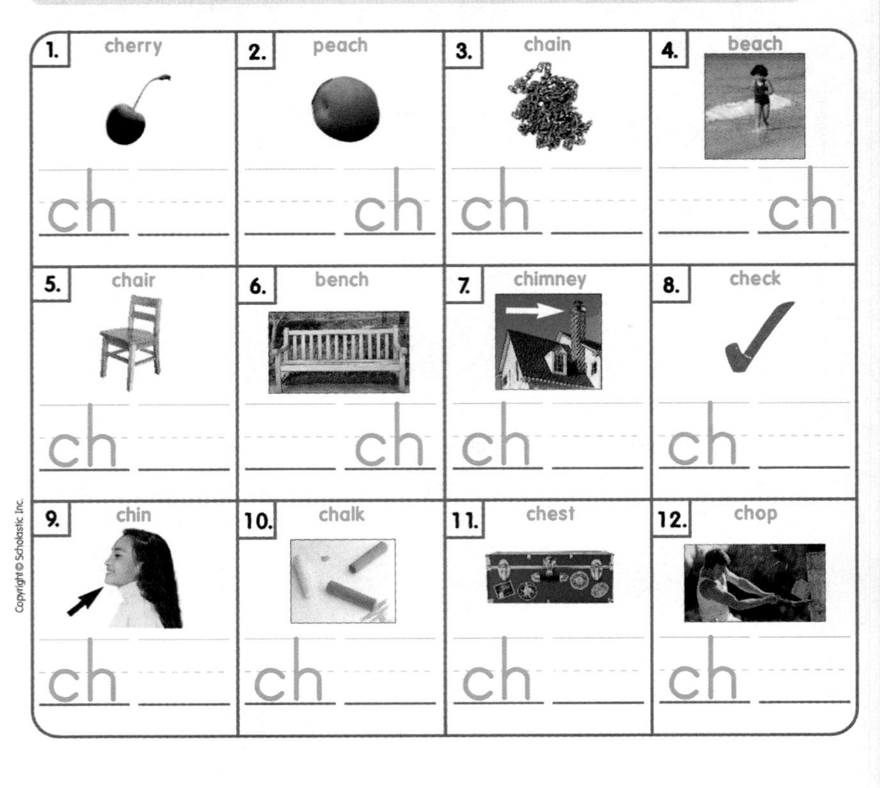

Supporting All Learners

AUDITORY LEARNERS

BEGINNING OR END Have children line up and then sit on the floor in line. Divide the line in half, designating which children are at the beginning and which are at the end. Say words such as the following: ***reach, chin, child, latch, cheap,*** and ***rich***. If the word you say begins with **/ch/,** children at the beginning of the line should stand. If the word ends with **/ch/,** children at the end should stand.

AUDITORY/VISUAL LEARNERS

WORD FIND Read aloud "The Wheels on the Bus" on pages 6–7 in the *Big Book of Rhymes and Rhythms, 1B*. Have children find the word with **/ch/** in the rhyme. Write the word on the chalkboard. Invite volunteers to add other **/ch/** words to the list.

VISUAL/KINESTHETIC LEARNERS

WORD HUNT Invite children to go on a scavenger hunt to collect pictures of things whose names begin or end with **/ch/,** such as a peach, lunch, and a chick. Display the collected pictures. Then have children place a self-sticking note on each picture. Ask children to write or dictate the name of each thing on the self-sticking note.

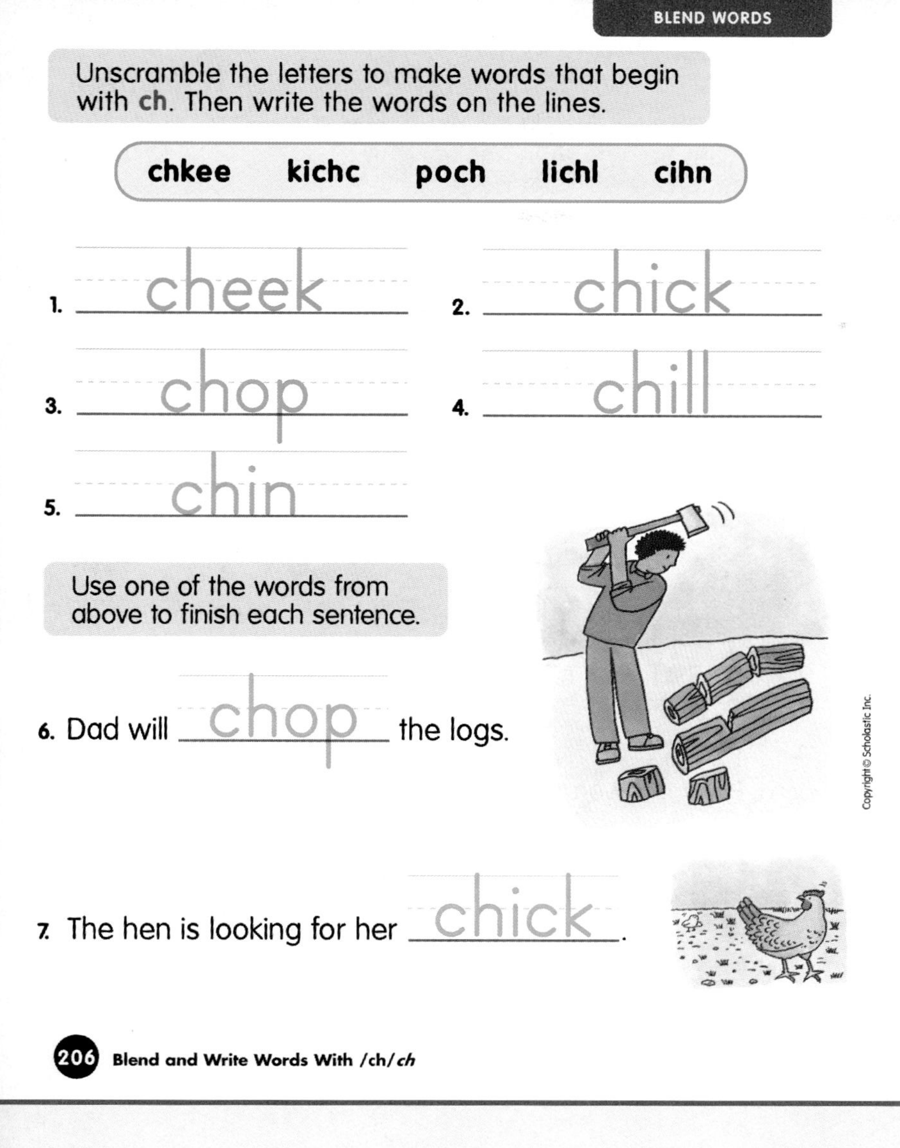
BLEND WORDS

Unscramble the letters to make words that begin with ch. Then write the words on the lines.

chkee kichc poch lichl cihn

1. cheek
2. chick
3. chop
4. chill
5. chin

Use one of the words from above to finish each sentence.

6. Dad will chop the logs.

7. The hen is looking for her chick.

Copyright © Scholastic Inc.

206 Blend and Write Words With /ch/*ch*

Integrated Curriculum

SPELLING CONNECTION

SPELL TIME Write the following word parts on the chalkboard: ***bea__, pea__, __ill, __op***. Have children write each on their own papers. Have children add the letters ***ch*** to each word part. Then ask children to illustrate each word.

MATH CONNECTION

WORD COUNT Have children count the number of words with **/ch/*ch*** in *The Puppet Club* and in each of The Book Shop suggestions. Children can compare the number of **/ch/*ch*** words in each book.

Phonics Connection

Literacy Place: *Imagine That!*
Teacher's SourceBook, pp. T100–101;
Literacy-at-Work Book, pp. 21–22

Chapter Book: *The Puppet Club*, Chapter 6

EXTRA HELP

Have children look for words with **/ch/** in their neighborhood. Children can list the words they find on paper or in a journal. You may wish to add the words found to the Word Wall. Help children blend each word.

CHALLENGE

Invite children to make their own books of favorite, challenging, or useful **/ch/** words. Ask children to list **/ch/** words they want to include in their books. Then have children write the meaning of each word and use it in a sentence. Children can also illustrate each word in ways that are meaningful to them.

- **Shy Charles**
  by Rosemary Wells
  Timid Charles heroically rescues his babysitter. **(/ch/*ch*)**
- **Max's Chocolate Chicken**
  by Rosemary Wells
  Max gets a treat. **(/ch/*ch*)**

Lesson 127

pages 207-208

Blend Words With /hw/*wh*

QUICKCHECK ✔

Can children:

✔ orally segment words?

✔ recognize /hw/?

✔ blend words with *wh*?

If YES go to Read and Write.

TEACH

Develop Phonemic Awareness

Oral Segmentation Explain to children that you are going to say a word. Children will tell you how many sounds they hear. For example, say the word ***chop***, and then say it sound by sound: /**ch**/ /**o**/ /**p**/. Guide children to understand that the word ***chop***, contains three separate sounds. Continue with the following words.

- which
- white
- when
- wheat
- wheel
- whale

Connect Sound-Symbol Write the words ***hen*** and ***when*** on the chalkboard as you read each aloud. Ask children what sounds they hear at the beginning of each word and how the words are different. Point out that /**hw**/ at the beginning of ***when*** is spelled with two letters, ***w*** and ***h***.

Write the following word parts on the chalkboard: ***_eel, _ile, _ite***. Have volunteers add the letters ***wh*** to each. Model how to blend each word.

READ AND WRITE

Blend Words List the following words and sentences on a chart. Have volunteers read each aloud.

- when
- wheel
- whale
- wheat
- white
- which
- We saw a white whale.
- The bike has just one wheel.

Complete Activity Pages Read aloud the directions on pages 207–208. Review each picture name with children.

Name

wheel

Look at each picture. If the picture name begins with the **wh** sound as in **wheel**, write the letters **wh** on the line.

1. cheese	2. whistle	3. plane	4. white
	wh		wh
5. whale	**6. boat**	**7. wheelchair**	**8. shoe**
wh		wh	

Write the letters **wh** to finish each word. Read the words to a friend.

9. wh ich 10. wh en 11. wh ile

12. wh y 13. wh ere

Recognize /hw/ *wh* 207

Supporting All Learners

VISUAL LEARNERS

LISTEN FOR /hw/ Read aloud "The Wheels on the Bus" on pages 6–7 in the *Big Book of Rhymes and Rhythms, 1B*. Have children find all the words with **/hw/** in the rhyme. ***(wheels, When's)*** Write these words on the chalkboard. Invite volunteers to add other **/hw/** words to the list.

AUDITORY/KINESTHETIC LEARNERS

WHALE WORDS Write the word ***whale*** on the chalkboard. Model for children how to blend the word. Invite a volunteer to circle the letters that stand for the beginning sound. Ask children to suggest other words that contain **/hw/**. List these words on the chalkboard, and have volunteers circle the letters ***wh*** in each one. Then have children write each word on a sheet of paper.

KINESTHETIC LEARNERS

BUILD-A-WORD GAME To practice and review words with digraphs, children may enjoy playing "Build-a-Word" on pages 33–35 in *Quick-and-Easy Learning Games: Phonics.*

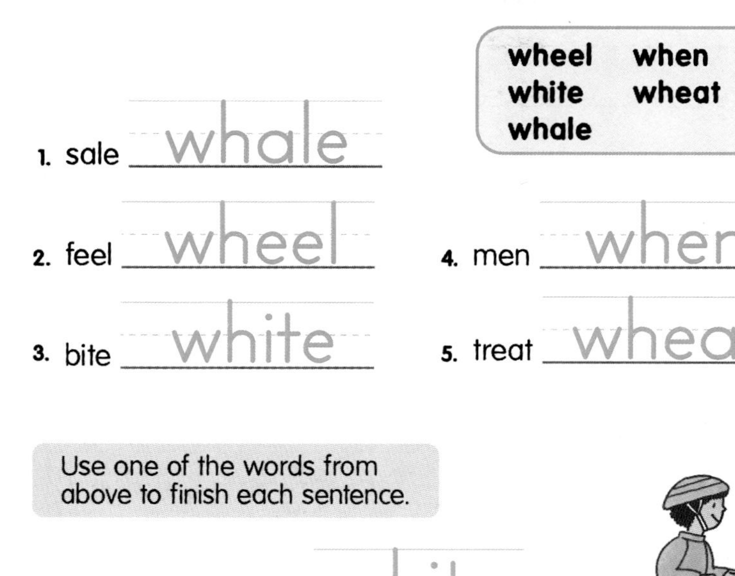

Find a word in the box that rhymes with each word. Then write the word on the line.

wheel	when
white	wheat
whale	

1. sale whale
2. feel wheel
3. bite white
4. men when
5. treat wheat

Use one of the words from above to finish each sentence.

6. Rick has a red and white bike.
7. His bike needs a wheel.
8. Rick will get it when his Mom takes him to the bike shop.

Assign each child to an English-speaking classmate. Partners can ask each other questions using ***what, when,*** and ***where***. Model asking ***wh*** questions, such as ***When is your birthday?*** Encourage children to answer their partner's questions using complete sentences.

CHALLENGE

Challenge children to find words with ***wh*** in books, magazines, and newspapers, then use those words in a story of their own. Children can combine their stories in a book to put in the classroom library. You may wish to have them record their stories on audiotape for classmates to hear. Suggest that listeners try to keep track of how many words with **/hw/*wh*** they hear in each story.

Integrated Curriculum

WRITING CONNECTION

ALL ABOUT *Ww* Have children dictate sentences with **/hw/** and **/w/** words. Write the sentences on the chalkboard or on chart paper. Then have children circle the words with ***wh*** and underline the words with ***w***.

TECHNOLOGY CONNECTION

FUN WITH WIGGLEWORKS PLUS Have children use the Magnet Board to build words with ***wh***. Suggest that children begin with one of the following words: ***when, white,*** or ***wheel***.

Phonics Connection

Literacy Place: *Imagine That!*
Teacher's SourceBook, pp. T100–101; Literacy-at-Work Book, pp. 21–22

Big Book of Rhymes and Rhythms, 1A: "The Wheels on the Bus," pp. 6–7

Chapter Book: *The Puppet Club*, Chapter 6

Lesson 128

page 209

Build Words

QUICKCHECK ✔

Can children:

✔ build words?

If YES go to Read and Write.

TEACH

Review Sound-Spellings Review the sound-spelling relationships from the past few lessons. These include *s*-blends, /**ch**/*ch*, and /**hw**/*wh*.

Write ***chin*** on the chalkboard. Tell children to replace /**n**/, the third sound in ***chin***, with /**p**/ to make a new word. Ask children to name the letter that stands for /**p**/. Then replace the ***n*** in ***chin*** with ***p***. Blend the new word ***chip***. Continue by changing one letter or letter pair to build new words. You may wish to use this sequence: ***chin, chip, whip, which***.

READ AND WRITE

Build Words Distribute the following letter cards to children: ***c, h, i, n, o, p***. Have children build as many words as possible using the letter cards. Suggest that they record their words on paper. Continue building words with the following sets of cards: ***a, c, e, h, r, t*** and ***a, e, h, n, t, w***.

Then explain that you will give the beginning and ending words of a three- or four-word ladder. Children can change one letter or letter pair in each word to complete the word ladder. Use these words:

- teach • (beach) • beat
- sheet • (sheep) • shop
- wheel • (heel) • (feel) • feet

The words in parentheses are *possible* answers.

Complete Activity Page Read aloud the directions on page 209. Children can complete the page independently.

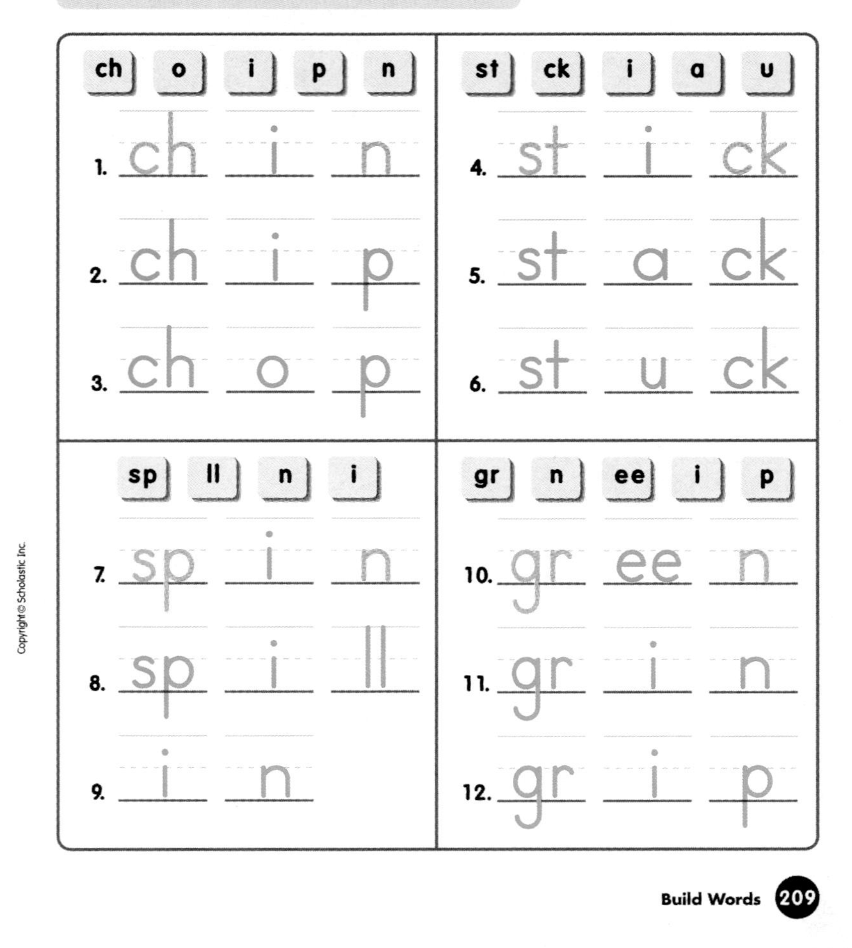

Supporting All Learners

KINESTHETIC LEARNERS

LEARNING CENTER To build awareness of consonant blends, set up Center 2, Activity 2, "Blend It Up!" from *Quick-and-Easy Learning Centers: Phonics*. This activity focuses on consonant blends that appear in utensil names.

VISUAL LEARNERS

MAP IT! Have children work together to make a map of the school neighborhood. Then have children use words with ***s***-blends, **/ch/ch**, and **/hw/wh** words to describe the map, give directions around the neighborhood, and rename streets or buildings.

CHALLENGE

Distribute to children the letter cards with digraphs, ***s***-blends, and all the familiar phonograms from the Phonics and Word Building Kit. Challenge children to build as many words as they can and to list these words. Then children can get together and read their words to one another.

POETRY

Read the poem. Finish each sentence using a word from the poem.

CHICKS

Can a chick snack?
Yes, with its beak.

Can a chick hide seeds?
Yes, in its cheek.

Can a chick speak?
Well, it can peep.

Can many chicks chat?
Yes! Cheep! cheep! cheep!

1. A chick can hide seeds.
2. It hides seeds with its beak.

210 Link to Spelling

Write and Read Words With *s*-blends, /ch/*ch*, /hw/*wh*

QUICKCHECK ✔

Can children:

✔ **spell words with *s*-blends, /ch/*ch*, /hw/*wh*?**

✔ **read a poem?**

If YES go to Read and Write.

TEACH

Link to Spelling Say one of the review sounds. Have a volunteer write on the chalkboard the spelling that stands for the sound. For example, the letters ***ch*** stand for /ch/. Continue with all the sounds. You may also wish to review ***r***-blends, ***l***-blends, and other ***s***-blends.

Phonemic Awareness Oral Segmentation
Say ***chat***. Have children orally segment the word. *(/ch/ /a/ /t/)* Ask them how many sounds the word contains. *(3)* Draw three connected boxes on the chalkboard, and have volunteers write the spelling that stands for each sound in ***chat*** in the appropriate box. Repeat with the words ***when*** and ***wheel***.

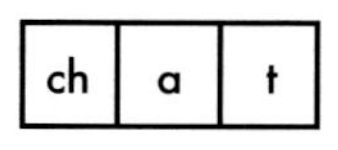

ch	a	t

READ AND WRITE

Dictate Dictate the following words and sentence. Have children write them on a sheet of paper. Then write the words and sentence on the chalkboard, and have children make any necessary corrections on their papers.

- **cheek** **when** **beach**
- **many** **which** **other**
- **We see many white whales.**

Complete Activity Page Read aloud the directions on page 210. Children can read the poem independently or with partners.

Supporting All Learners

KINESTHETIC LEARNERS

MATCH IT! GAME To practice and review words with consonant blends and digraphs, children may enjoy playing "Match It!" on pages 40–41 in *Quick-and-Easy Learning Games: Phonics*. Adapt the word list to include several words that begin with **/ch/** and **/hw/**, such as ***chop, chin, chest, cheek, whale, white, when,*** and ***which***.

CHALLENGE

Have children generate a list of words that end with **/ch/**. Record these words on the chalkboard. Then have children write their own stories using as many **/ch/** words on the chalkboard as they can. Suggest titles such as "The Beach" to help children. When completed, children can illustrate their stories and then share them with the class.

Lesson 130

pages 211-212

Read and Review Words With *s*-blends, /ch/*ch*, /hw/*wh*

QUICKCHECK ✔

Can children:

✓ **read words with *s*-blends, /ch/*ch*, /hw/*wh*?**

✓ **recognize high-frequency words?**

If YES go to Read and Write.

TEACH

Review Sound-Spellings Say one of the review sounds. Have a volunteer write on the chalkboard the spelling that stands for the sound. For example, the letters ***ch*** stand for /**ch**/. Continue with all the sounds.

Review High-Frequency Words Review the high-frequency words ***many, which, down, no,*** and ***out***. Write sentences on the chalkboard containing each word. Say one word, and have volunteers circle it. Invite children to generate additional sentences for each word.

READ AND WRITE

Build Words Distribute the following letter cards to children: ***a, e, k, n, s, p***. Provide time for children to build as many words as possible using the letter cards. Suggest that they record their words on paper. Continue with ***e, g, k, m, o, s; c, h, i, n, o, p;*** and ***e, n, h, i, l, t, w***.

Build Sentences Write the following words on note cards: ***I, they, reach, for, smell, when, down, sit, lunch, the, eat, times,*** and ***while***. Display the cards. Have children make sentences using the words.

Complete Activity Pages Read aloud the directions on pages 211–212. Children can complete the page independently.

Name

Check each word as you read it to a partner.
Circle any words you need to practice.

I can read!

☐ snap	☐ smile	☐ snake
☐ smoke	☐ smash	☐ snack
☐ chin	☐ teach	☐ chop
☐ chill	☐ cheek	☐ bench
☐ wheat	☐ white	☐ whale

Lookout Words!

☐ many	☐ which	☐ down	☐ no	☐ out
☐ about	☐ could	☐ when	☐ other	☐ was

Supporting All Learners

KINESTHETIC LEARNERS

TIC-TAC-TOE GAME To practice and review words with ***s***-blends, children may enjoy playing "Blend TIC-TAC-TOE" on pages 36–39 in *Quick-and-Easy Learning Games: Phonics*. Use the ***s***-blend game board, and introduce the ***sk*** blend with words such as ***skip*** and ***skate*** before children begin to play.

AUDITORY LEARNERS

RHYME IT Have children write or dictate a list of words with ***s***-blends, **/ch/*ch*,** and **/hw/*wh*,** with at least one word per list. Have children name and write a rhyming word for each word on their lists. Children can work together to figure out rhyming words, if they wish. Then have children find words that rhyme with each high-frequency word.

VISUAL/KINESTHETIC LEARNERS

PICTURE THIS Display pictures of objects whose names begin or end with ***sn, sm, ch,*** and ***wh***. Have children write the spelling that each picture name begins or ends with as the picture is displayed.

Fill in the bubble next to the sentence that tells about each picture.

1.	○ Look up at the plane. ● Look down at the steps.
2.	● The pigs sneak out of the pen. ○ Many chicks get in the pen.
3.	○ We will chill the snacks. ● Which peach will I eat?
4.	● She smells the cake. ○ I smell my lunch.
5.	○ A sick whale was on the beach. ● Can a whale chase a ship?
6.	○ I see a black and white dog. ● I teach the dog to sit down.
7.	○ The cats chase the mice. ● The mice ate the cheese.

Integrated Curriculum

SPELLING CONNECTION

SPELLING MIX UP Write the following scrambled words with *s*-blends, **/ch/*ch*,** and **/hw/*wh*** on the chalkboard: ***nacks, dilch, llems, hewn, hyw, lmsal, aisnl, cihck***. Ask children to write the scrambled words on sheets of paper. Then have children unscramble the words and write them beside the scrambled words. Encourage children to write or dictate sentences using these words.

MATH CONNECTION

WORD COUNT Have children look for words with ***s*-blends, /ch/*ch*,** and **/hw/*wh*** in books in the classroom. Then have children count these words and record the results of their counts in a chart such as the one shown.

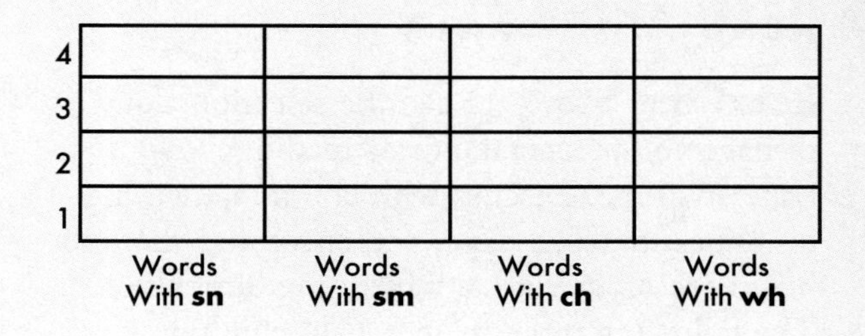

I CAN READ! OPTIONS

The I Can Read! page can be used for one or all of the following:

- **paired reading**
- **individual assessment**
- **choral reading**
- **homework practice**
- **program placement**

ESL Invite children acquiring English to play "Snatch and Match." Make duplicate word cards for several of the words being reviewed. (Or make one set of cards, and let children copy them to make the duplicate set.) Display one set of cards, and read each word with children. Then, as you hold up each duplicate word card, children can take turns snatching it from your hand, matching it, and reading aloud the pair of words.

EXTRA HELP

Children can write words with blends and digraphs in sand. Have children write a list of the words they write in the sand.

pages 213-214

Read Words in Context

TEACH

Assemble the Story Ask children to remove pages 213–214. Have them fold the pages in half to form the Take-Home Book.

Preview the Story Preview *Whales,* a nonfiction selection about whales. As you read aloud the title, you may wish to point out the *s* ending, meaning more than one whale. Invite children to browse through the first two pages of the selection and to comment on what they notice. Suggest that they point out any unfamiliar words. Read these words aloud as children repeat them. Then have children predict what they think they will learn from the selection.

READ AND WRITE

Read the Story Read the selection aloud, or have volunteers take turns reading aloud a page at a time. Discuss with children anything of interest on each page, and encourage them to help each other with any blending difficulties. The following prompts may help children who need extra support while reading:

- **What letter sounds do you know in the words?**
- **What word parts do you know?**

Reflect and Respond Have children share their reactions to the selection. What did they like best about it? What did they learn? Encourage children to write what else they would like to know about whales.

Develop Fluency Provide time for children to reread the selection on subsequent days to develop fluency and increase reading rate.

Invite children to share how they figure out unfamiliar words when reading other stories. Encourage children to look for words with *s*-blends, **/ch/*ch*,** and **/hw/*wh*** when they read. Children can apply what they have learned about these sound-spelling relationships to decode words. Continue to review sound-spelling relationships for children needing additional support.

A beach is no place for a whale.
A whale needs to be in the sea.
That is where a whale feels safe.

Circle the words in the story that tell what whales do.

4

Supporting All Learners

VISUAL LEARNERS

MY BOOKS For additional practice with *s*-blends and digraphs, have children read the following My Books: *Still Snoring, Where Is My Chick?* and *Six Shaggy Things.* For information on using the My Books on the computer, see the WiggleWorks Plus My Books Teaching Plan.

VISUAL/KINESTHETIC LEARNERS

WHALE ON THE WALL Draw the outline shape of a whale on mural paper. Invite children to take turns coming to the mural and writing sentences about whales inside the shape. Sentences may be original or from the selection. Then children can read their sentences aloud.

AUDITORY LEARNERS

WHALE FACTS Have children sit in a circle. Encourage each child to contribute a different fact about whales. Information can be from the selection or based on prior knowledge. Invite children to raise their hands each time someone uses a word with **/ch/** or **/hw/**.

Whales feed on sea life. Some whales eat seals and fish. Some eat plants. Some whales have teeth. Other whales do not.

2

Whales must come to the top of the sea to breathe. Some whales leap up and out of the sea. They seem to stop and wave.

Can whales speak? No, but they can click to each other. They can hum, too.

3

CHALLENGE

Have children write what they know about whales or some other animal. Children can address questions such as ***What do these animals eat? Where do they live? What makes them different from other animals?*** Have children illustrate their sentences and then share them with the class.

EXTRA HELP

Read aloud *Whales* with children. Then have children read aloud the book together. Children may also wish to take turns reading pages to each other.

Reflect on Reading

ASSESS COMPREHENSION

To assess their understanding of the selection, ask children questions such as:

- **What is a pack of whales called?** ***(a pod)***
- **What do whales eat?** ***(sea life such as seals, fish, and plants)***
- **Can whales speak?** ***(They do not speak like humans do, but they can click and hum.)***

HOME-SCHOOL CONNECTION

Send home *Whales*. Encourage children to read the book to a family member.

WRITING CONNECTION

WRITE ABOUT WHALES Have children write or dictate sentences about the selection *Whales*. What did they like about it? What do they want to learn more about? What questions do they have about the whales? What do they think about whales and the way they live? Provide books for children to find answers to their questions.

Phonics Connection

Phonics Readers:
#46, *Say It and Smile!*
#47, *Chuck's Lunch*
#48, *Whale of a Joke!*

Lesson 132

pages 215-216

Blend Words With /ā/ai, ay

QUICKCHECK ✔

Can children:

✔ **substitute sounds in words?**

✔ **identify /ā/?**

✔ **blend words with /ā/*ai, ay*?**

If YES go to Read and Write.

TEACH

Develop Phonemic Awareness

Phonemic Manipulation Explain to children that they are going to listen to words with /a/ as in *man*. Children are to replace the short *a* sound with the long *a* sound. For example, if you say *man*, children will say *main*. Use these and other words:

- pan
- pad
- mad
- ran

Connect Sound-Symbol Write the words *ran* and *rain* on the chalkboard as you read each word aloud. Ask children what sounds they hear in the middle of each word and how the words are different. Point out that /a/ in the middle of *ran* is spelled by *a*, and /ā/ in the middle of *rain* is spelled by *ai*. Write the word *ray* on the chalkboard. Explain that /ā/ can also be spelled by *ay*. Write and say other examples of *ay* words, such as *say, pay,* and *tray*.

READ AND WRITE

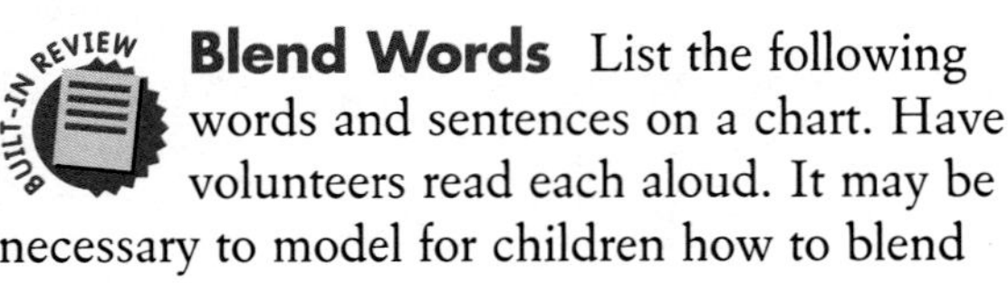

Blend Words List the following words and sentences on a chart. Have volunteers read each aloud. It may be necessary to model for children how to blend the words.

- rain   paid   main
- pay   gray   say
- May I play with clay?
- I paid for the clay.

Complete Activity Pages Read aloud the directions on pages 215–216. Review each picture name with children.

Name

Look at each picture. Write **a** or **ai** to finish the picture name.

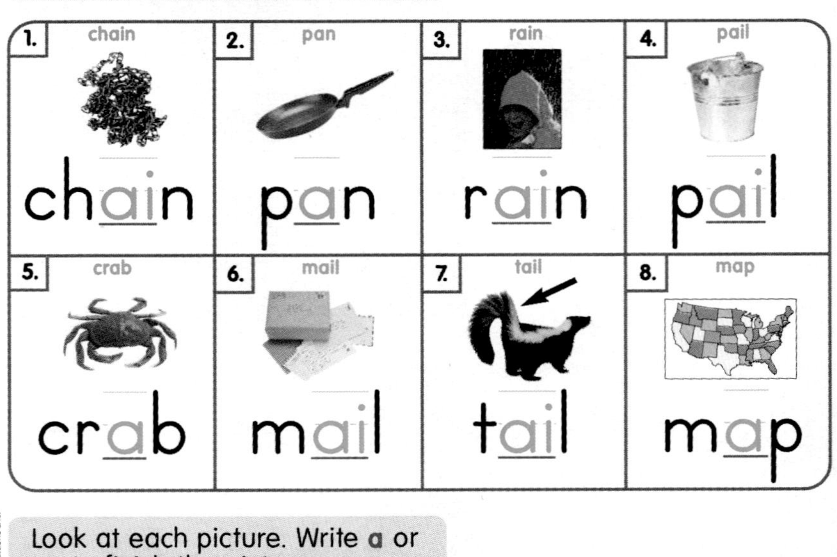

Look at each picture. Write **a** or **ay** to finish the picture name.

9. rat — r*a*t
10. pay — p*ay*
11. cat — c*a*t
12. tray — tr*ay*

Supporting All Learners

AUDITORY/VISUAL LEARNERS

WORD HUNT Read aloud "Rain, Rain, Go Away" on pages 14–15 in the *Big Book of Rhymes and Rhythms, 1B*. Have children find all the words with **/ā/** in the rhyme ***(rain, away, day, play)***. Write these words on the chalkboard. Invite volunteers to add other **/ā/** words to the list. Have volunteers circle the letters ***ai*** or ***ay*** in each word.

VISUAL LEARNERS

TIME LINE With children, brainstorm a list of events in a typical day, such as get up, get on the bus, go to school, play, and eat. Draw a time line on the chalkboard, and label it "Time Line of a Day." Place some times on one side of it. Have children decide what events to write for each time. Children can write or dictate these events on the appropriate places.

BLEND WORDS

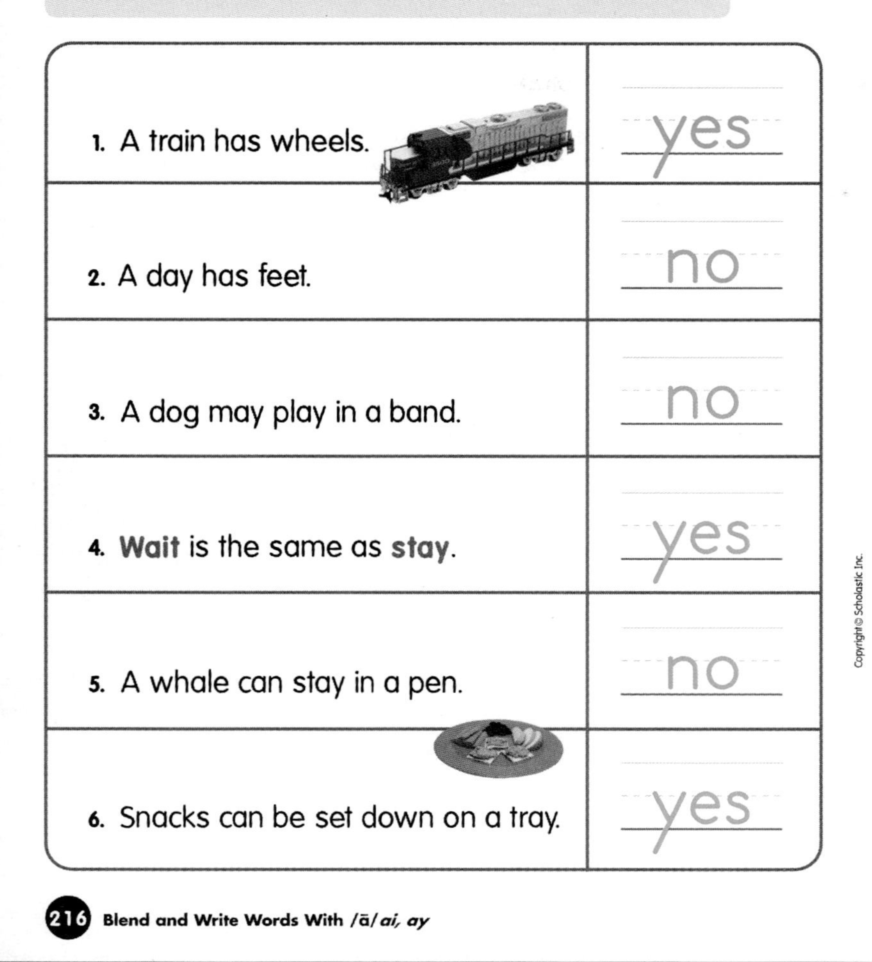
Read each sentence. If the sentence is true, write **yes** on the line. If it is not true, write **no** on the line.

1. A train has wheels.	yes
2. A day has feet.	no
3. A dog may play in a band.	no
4. **Wait** is the same as **stay**.	yes
5. A whale can stay in a pen.	no
6. Snacks can be set down on a tray.	yes

CHALLENGE

Make a two-column chart on the chalkboard. At the top of one column, write the word ***say,*** and underline ***ay***. At the top of the second column, write the word ***rain,*** and underline ***ai***. Challenge children to write as many words as they can with the same sound and spelling in each column.

Help children acquiring English become familiar with the ***ai*** and ***ay*** spelling patterns by completing long ***a*** words. Write ***ai*** and ***ay*** in different colors as headings on a chart, and in each column, write incomplete words such as these: ***st__ __n, p__ __d, r__ __n, t__ __l; d__ __, st__ __, pl__ __, m__ __***. For each column, have children use the matching color to fill in the missing letters. Then children can read the completed words.

Integrated Curriculum

SPELLING CONNECTION

DICTATION TIME Have children dictate sentences using words with **/ā/*ai, ay***. Write these sentences on the chalkboard. Children can point out the letters in each underlined word that stand for **/ā/**.

SOCIAL STUDIES CONNECTION

WORD SEARCH Have children look through books, magazines, and newspapers in the school or neighborhood library for words with **/ā/*ai, ay***.

Phonics Connection

Literacy Place: *Information Finders*
Teacher's SourceBook, pp. T54–55;
Literacy-at-Work Book, pp. 8–9

My Book: *Waiting for Suzy*

Big Book of Rhymes and Rhythms, 1B: "Rain, Rain, Go Away," pp. 14–15

Chapter Book: *Let's Go on a Museum Hunt,* Chapter 1

The Book Shop

- **Bringing the Rain to Kapiti Plain**
  by Verna Aardema
  This African folktale rhymes. **(/ā/*ai*)**
- **Moira's Birthday**
  by Robert Munsch
  Moira invites the school to her party. **(/ā/*ai*)**

Lesson 133

page 217

Recognize and Write Contractions

QUICKCHECK ✔

Can children:

✔ recognize contractions?

✔ write contractions?

If YES go to Read and Write.

TEACH

Introduce Contractions Write the word ***I'm*** on the chalkboard. Point out the apostrophe, and name it. Explain that the apostrophe in ***I'm*** shows that two words have been put together to form one word. The letter or letters left out have been replaced by the apostrophe. A word such as ***I'm*** is a contraction.

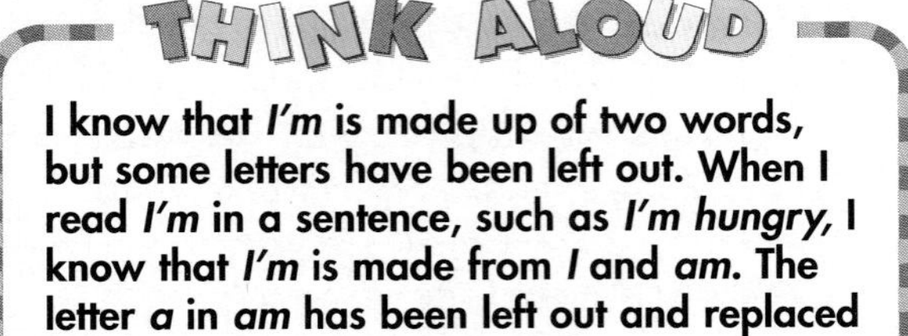

I know that *I'm* is made up of two words, but some letters have been left out. When I read *I'm* in a sentence, such as *I'm hungry,* I know that *I'm* is made from *I* and *am.* The letter *a* in *am* has been left out and replaced by an apostrophe to form *I'm.*

Write the following contractions on the chalkboard: ***isn't, she's, he's,*** and ***don't***. Have children name the two words that make up each contraction. Write them on the chalkboard next to the contraction. Then ask children which letters have been left out to form the contraction.

READ AND WRITE

Practice Write the following on the chalkboard. Ask children to tell what words the underlined word in each sentence stands for.

- **She's my cat Spot.**
- **I'll look for the hat.**
- **I don't see the lake.**
- **Isn't that your coat?**

Complete Activity Page Read aloud the directions on page 217. Children can complete the page independently.

Name

Match each contraction with the two words that make it up.

1. he's	can not	5. she's	I am
2. I'll	it is	6. don't	is not
3. can't	he is	7. I'm	do not
4. it's	I will	8. isn't	she is

Use one of the words from above to finish each sentence.

Remember the apostrophe (').

9. The cat isn't in her bed.
10. Yes, I'll look for my coat.
11. No, I don't have her hat.
12. The man can't fix the bike.

Supporting All Learners

VISUAL/KINESTHETIC LEARNERS

CONTRACTION ACTION Write the following words on note cards: ***he, she, it, I, can, do, is, not,*** and ***will***. Write ***is*** on three cards and ***not*** on two cards. Give partners the following cards: ***he, is; she, is; it, is; I, will; do, not; can, not***. Invite children to tape their two cards together, cross out letters on the card that do not belong, and add an apostrophe to form a contraction.

VISUAL LEARNERS

FUN WITH MY BOOKS For additional practice with contractions, have children read the following My Book: *Don't Be Afraid*. For information on using the My Books on the computer, see the WiggleWorks Plus My Books Teaching Plan.

EXTRA HELP

Write a list of contractions such as ***he'll, she'll, I'm,*** and ***don't*** on one side of a sheet of paper. Write ***he will, I am, she will,*** and ***do not*** on the other side of the paper. Help children to underline each contraction, circle each apostrophe, and draw a line from each contraction to the two words that make it up.

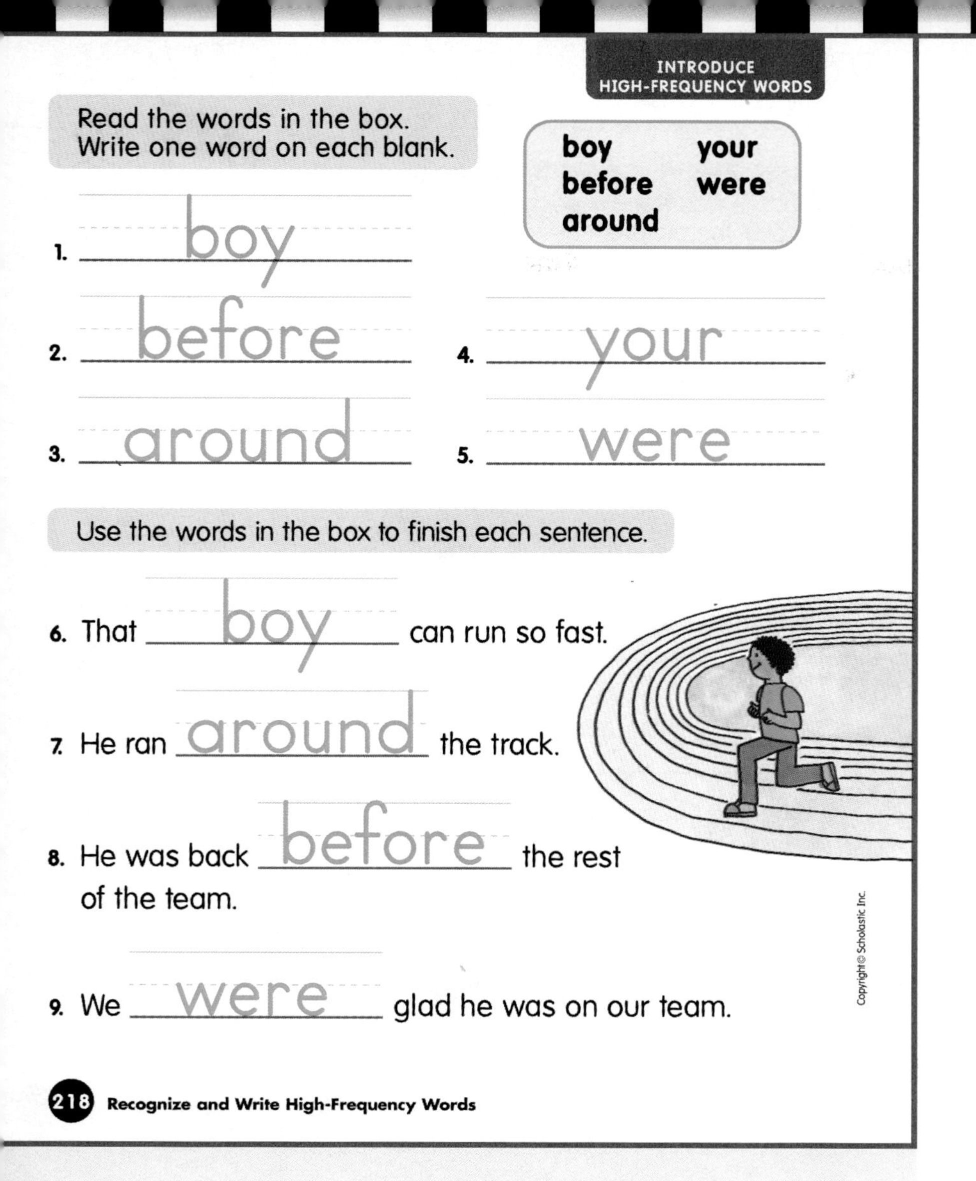

INTRODUCE HIGH-FREQUENCY WORDS

Read the words in the box.
Write one word on each blank.

boy	your
before	were
around	

1. boy
2. before
3. around
4. your
5. were

Use the words in the box to finish each sentence.

6. That boy can run so fast.
7. He ran around the track.
8. He was back before the rest of the team.
9. We were glad he was on our team.

Recognize High-Frequency Words

QUICKCHECK ✔

Can children:

✔ recognize and write the high-frequency words *boy, before, around, your, were*?

✔ complete sentences using the high-frequency words?

If YES go to Read and Write.

TEACH

Introduce the High-Frequency Words Write the words *boy, before, around, your,* and *were* in sentences on the chalkboard. Read aloud the sentences, underline the words, and ask children if they recognize them. You may wish to use the following sentences:

1. **The boy asks, "Where is the dog?"**
2. **I say, "Look around."**
3. **Some dogs were in the park before.**
4. **Was your dog with them?**

Ask volunteers to dictate sentences using the high-frequency words. Begin with the following sentence starters: *We were at the* ___ or *Where is your* ___*?*

READ AND WRITE

Practice Write each high-frequency word on a note card. Read each word aloud. Then do the following:

- **Mix the cards.**
- **Display one card at a time, and ask children to state each word aloud.**
- **Have children spell each word aloud, clapping on each letter.**
- **Ask children to write each word in the air as they state aloud each letter. Then have them write each word on a sheet of paper.**

Complete Activity Page Read aloud the directions on page 218. Have children complete the page independently.

Supporting All Learners

VISUAL/KINESTHETIC LEARNERS

WORD CALL Distribute words cards from the Phonics and Word Building Kit for the high-frequency words you have taught so far. Give each child one word card. Tell children that you will call out the words. Children should raise their hands when their word is called. Then children should copy their word on the chalkboard and use it in a sentence.

Write the sentences from the TEACH section of this lesson on sentence strips. Have children match the sentence strips to the correct sentences on the chalkboard. If you wish to simultaneously focus on word order, phonics, and meaning, you can also cut the words apart on each strip, and have children arrange them in sentences.

CHALLENGE

PORTFOLIO Have children use the high-frequency words to write sentences about their school or classroom.

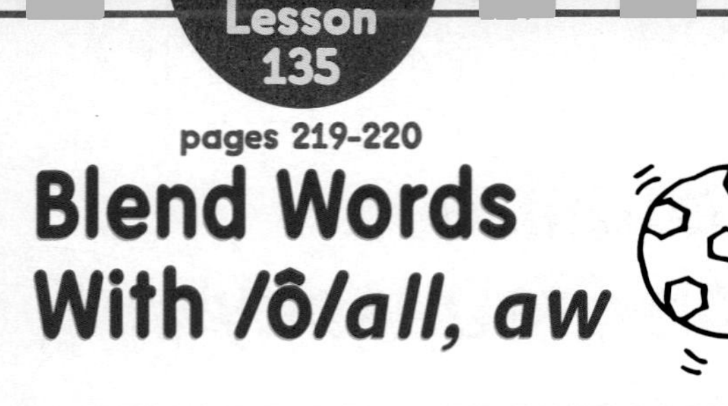

pages 219-220

Blend Words With /ô/all, aw

QUICKCHECK ✔

Can children:

- ✓ **substitute sounds in words?**
- ✓ **recognize /ô/?**
- ✓ **blend words with /ô/*all, aw*?**

If YES go to Read and Write.

TEACH

Develop Phonemic Awareness

Phonemic Manipulation Explain to children that they are going to replace the last sound in each word you say with /ô/. For example, if you say *see* children will say *saw*. Use these and other words:

pea **Lee** **fly** **so**

Connect Sound-Symbol Write the words *tap* and *tall* on the chalkboard as you read each aloud. Ask children what sounds they hear in the middle of each word and how the words are different. Point out that /a/ in the middle of *tap* is spelled with an *a,* and /ô/ in the middle of *tall* is spelled with an *a* followed by the letters *ll*. Continue with the words *rot* and *raw*. Ask children to suggest other words that contain /ô/. List these words on the chalkboard. Have volunteers circle the letters *aw* or *all* in each one.

READ AND WRITE

Blend Words List the following words and sentences on a chart. It may be necessary to model for children how to blend the words. Then have children blend the words.

- call tall fall
- jaw paw saw
- I saw a dog.
- I won't fall down.

Complete Activity Pages Read aloud the directions on pages 219–220. Review each picture name with children.

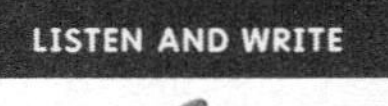

Name

saw

Look at each picture. If the picture name has the **aw** sound as in **saw**, color the picture.

1. ball	2. rainbow	3. wall
4. snail	5. hawk	6. paw
7. cheese	8. draw	9. saw

Write the letters **aw** or **all** to finish each word. Circle the word that names the picture.

10. cl aw

11. c all

12. sm all

Supporting All Learners

VISUAL LEARNERS

LIST WORDS Read aloud "Humpty Dumpty" on pages 16–17 in the *Big Book of Rhymes and Rhythms, 1B*. Have children find all the words with /ô/ in the rhyme. Write these words on the chalkboard. Invite volunteers to add other /ô/ words to the list. Encourage children to identify the letters that stand for /ô/ in each word.

VISUAL/KINESTHETIC LEARNERS

LEARNING CENTER To focus on word-ending clusters, vowel digraphs, or another target phonetic feature, use Center 7, Activity 3, "Coupon Collages," from *Quick-and-Easy Learning Centers: Phonics*. This activity uses coupons and grocery store circulars to build awareness of sound-spelling relationships.

Circle the word that best finishes each sentence. Then write the word on the line.

	Sentence	Choices
1.	The small cat can't get down.	smoke, (small), smile
2.	It is up in that tall tree.	tale, (tall), tap
3.	We can see its white spots and paws.	past, paid, (paws)
4.	We can see its claws.	chops, (claws), clay
5.	Deb will call to the cat.	(call), came, claw
6.	She will not let the cat fall.	fast, fan, (fall)

Integrated Curriculum

SCIENCE CONNECTION

ALL ABOUT FALL Invite children to make posters that tell about fall. Have children include drawings of fall clothes, fall activities, fall fruits and vegetables, and what the outdoors looks like at this time of year. Encourage children to write or dictate captions for their drawings.

Phonics Connection

Literacy Place: *Information Finders*
Teacher's SourceBook, pp. T96–97;
Literacy-at-Work Book, pp. 17–18

My Book: *The Awful Bug*

Big Book of Rhymes and Rhythms, 1B: "Humpty Dumpty," pp. 16–17

Chapter Book: *Let's Go on a Museum Hunt,* Chapter 2

EXTRA HELP

Children can use the magnetic letters from the Phonics and Word Building Kit to practice building words with **/ô/** spelled ***all*** or ***aw***. You may wish to supply words on cards for them to copy. After a word is formed, mix the letters, and have the child rebuild it.

CHALLENGE

Write ***all, aw,*** and ***au*** on different sections of the chalkboard. Model how to create words by adding consonants. Point out that the letter clusters all contain the same sound. Invite children to create as many words as they can in each list. They can use the words to write rhyming couplets.

Have children work together to write or dictate sentences with rhyming word pairs with **/ô/** spelled ***all*** or ***aw***.

- **Mr. Tall and Mr. Small**
  by Barbara Brenner
  A giraffe and a mouse are friends in this book about size. (**/ô/*all***)

Lesson 136

pages 221–222

Recognize and Write Words With *-ed*

QUICKCHECK ✔

Can children:

✓ recognize *-ed* verb endings?

✓ write *-ed* verb endings?

If YES go to Read and Write.

TEACH

Introduce *-ed* Explain to children that words that tell about actions that have already happened are often formed by adding *-ed* to the word. Write the word ***kick*** on the chalkboard. Ask a volunteer to read it. Use it in a sentence; for example, ***I can kick the ball***. Then write ***kicked*** on the chalkboard. Ask a volunteer to circle the word ***kick***, and ask another volunteer to underline the *-ed* ending.

I see a word I already know inside the word *kicked*. It is *kick*. I also know there is another word part in *kicked*. It is *-ed*. Both *kick* and *kicked* name an action. I can kick a ball today, or I kicked a ball yesterday. *Kicked* shows that I already did something.

Continue with ***smelled, rushed, locked, spilled, fixed, called***. Have children circle the word inside each word and underline the *-ed* verb ending. Ask children how the smaller word can help them to read the bigger word.

READ AND WRITE

Practice Write the following sentences on the chalkboard. Ask children to tell what is different about the underlined words in each.

- I fix the lock
  I fixed the lock.
- I call out to Sam.
  I called out to Sam.

Complete Activity Pages Read aloud the directions on pages 221–222. Have children complete the pages independently.

Name

Choose the word that best finishes each sentence. Then write the word on the line.

	Sentence	Choices
1.	Tom looked at me with a grin.	look / (looked)
2.	We had to clean the shed.	(clean) / cleaned
3.	I fixed his bike.	fix / (fixed)
4.	Jane and I like to play ball.	(play) / played
5.	We baked a cake last week.	bake / (baked)
6.	She kicked the ball.	kick / (kicked)

Supporting All Learners

AUDITORY/VISUAL LEARNERS

READ AND WRITE Read aloud "A Mouse in Her Room" on page 18 in the *Big Book of Rhymes and Rhythms, 1B*. Have children find all the words with ***-ed*** in the rhyme. Write these words on the chalkboard. Invite volunteers to add other ***-ed*** words to the list.

VISUAL LEARNERS

MY BOOKS TECHNOLOGY For additional practice with ***-ed*** endings, have children read *Don't Be Bored*. For information on using the My Books on the computer, see the WiggleWorks Plus My Books Teaching Plan.

TACTILE LEARNERS

ADD -ed Have children use clay or pipe cleaners to form the letters for action words such as ***kick, smell, rush, spill, fix,*** and ***call***. Then have children use clay or pipe cleaners to form the letters ***ed***. Children can put the letters together to build action words, then add the letters ***ed*** to each word to show that the action already happened.

Finish the story with words from the box. Write the words on the lines.

stayed	**played**
rained	**stacked**
fished	

The Trip to the Lake

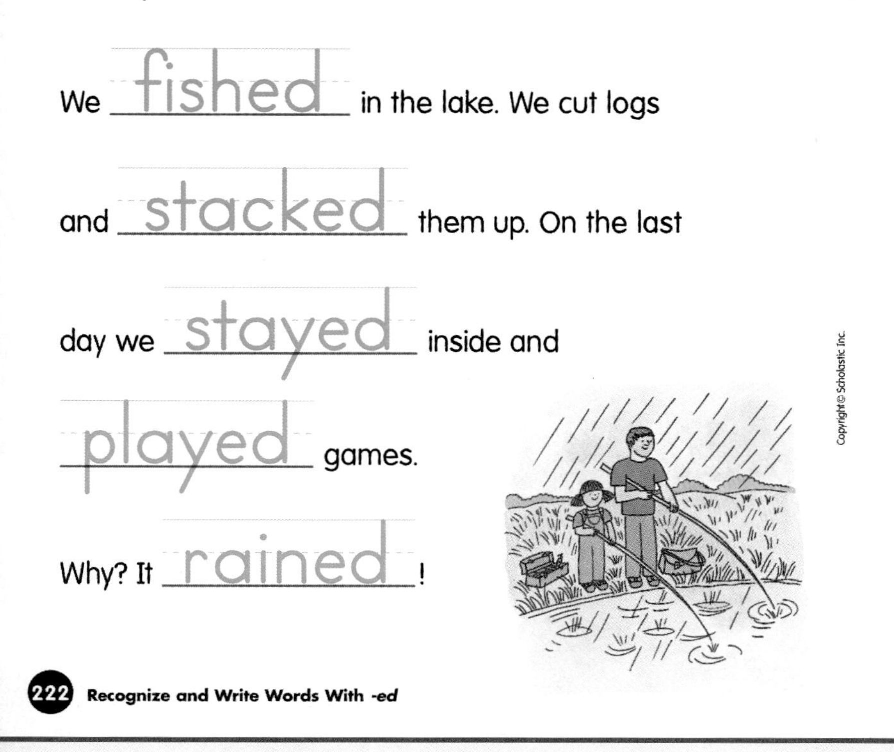

The trip to the lake was so much fun!

We fished in the lake. We cut logs

and stacked them up. On the last

day we stayed inside and

played games.

Why? It rained!

Integrated Curriculum

WRITING CONNECTION

ADDING *-ed* Write the following sentences on the chalkboard:

I walk the dog.
We jump rope.
I paint pictures.

Have children copy the sentences onto sheets of paper. Read the sentences aloud with children. Then have children add ***-ed*** to each underlined word and read each sentence aloud again.

SOCIAL STUDIES CONNECTION

SCHOOL NEIGHBORHOOD TOUR Have children walk around their school or neighborhood and name actions. Children should write or dictate a list of these actions. When children return to the classroom, have them use the verbs they collected in sentences that describe actions around the school or neighborhood. Ask children to identify the verbs that have ***-ed*** or could have ***-ed*** added to them.

ESL Write words such as ***packed, mixed, reached,*** and ***fished*** on note cards. Invite children to play charades by acting out the words, one at a time. Volunteers can guess the word and spell it.

EXTRA HELP

For additional practice with ***-ed,*** have children read the Phonics Readers #53, *Small Animals With Big Names,* and #54, *See the Sea.* You may wish to help children record the ***-ed*** words in one or both Phonics Readers.

CHALLENGE

Ask children to find and write more ***-ed*** words for the charades game. Children can write the new words on note cards, one word per card.

Phonics Connection

Literacy Place: *Information Finders*
Teacher's SourceBook, pp. T160–161;
Literacy-at-Work Book, pp. 29–30

My Book: *The Fixed-Up Park*

Big Book of Rhymes and Rhythms, 1B: "A Mouse in Her Room," p. 18

Chapter Book: *Let's Go on a Museum Hunt,* Chapter 3

Lesson 137

page 223

Write and Read Words With /ā/ai, ay, /ô/all, aw

QUICKCHECK ✔

Can children:

✔ **spell words with /ā/*ai, ay*, /ô/*all, aw*?**

✔ **read a poem?**

If YES go to Read and Write.

TEACH

Link to Spelling Review /ā/*ai, ay* and /ô/*all, aw*. Say either of these sounds, and have a volunteer write the spellings that stand for the sound. For example, the letters ***ai*** and ***ay*** stand for /ā/. You may also wish to review ***s***-blends, /**ch**/*ch*, and /*hw*/*wh*.

Oral Segmentation Say *paw* and have children segment it. (/**p**/ /ô/) Ask how many sounds the word contains and which sound is shown by two letters. Draw two connected boxes. Have a volunteer write the spelling that stands for each sound in **paw** in the appropriate box. Continue with *call* and *small*.

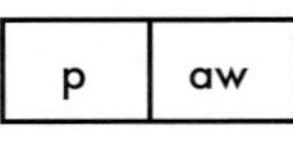

READ AND WRITE

Dictate Dictate these words and sentences. Have children write them on a sheet of paper. When they are finished, write the words and sentences on the chalkboard, and have children make any necessary corrections.

- saw call mall hall
- boy before out down
- Call that boy before he falls!
- We saw the boy at the mall.

Complete Activity Page Read aloud the directions on page 223. Children can read the poem independently or with partners.

POETRY

Name

Read the poem. Write a sentence telling about the boy's day.

I Play All Day

The rain can fall.
The sun can go away.
But I don't stop.
I play all day.

My feet can get wet.
The day can get gray.
But I don't stop.
I play all day.

Some boys go inside.
Others may not stay.
But I don't stop.
I play all day!

It's six o'clock
My mom will call.
But do I stop?
Yes! Now it's okay.

Answers will vary.

Supporting All Learners

VISUAL LEARNERS

WORD FIND Have children collect words with **/ā/*ai, ay*** and **/ô/*all, aw*** they see at home, in their neighborhood, or in school. Children can write these words in a list or journal.

CHALLENGE Have children generate a list of words that contain **/ô/**. Record these words on the chalkboard. Then have children create a story using as many **/ô/** words on the chalkboard as they can. Suggest a title such as "The Claw" to help children get started. Write the dictated story on chart paper for group and individual reading. Return to the story, rereading it in subsequent lessons.

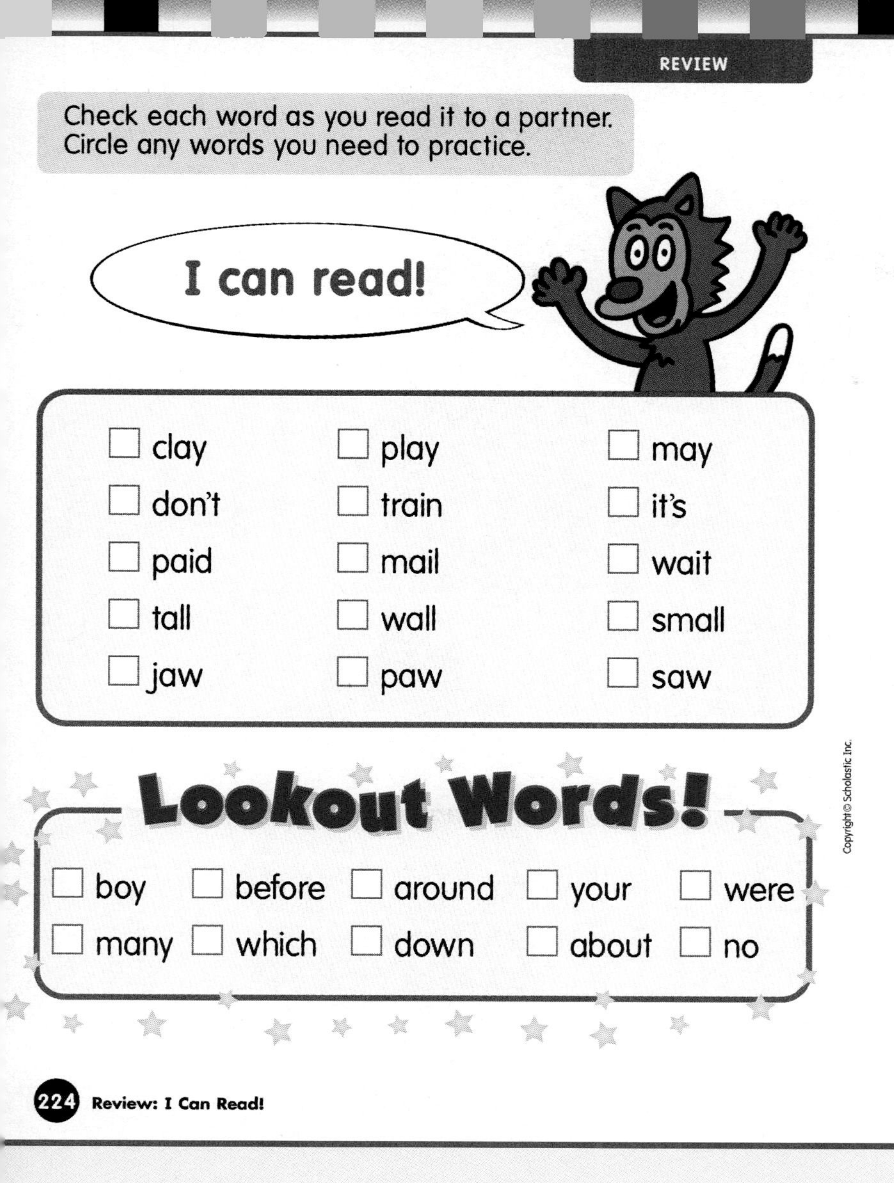

REVIEW

Check each word as you read it to a partner.
Circle any words you need to practice.

I can read!

☐ clay	☐ play	☐ may
☐ don't	☐ train	☐ it's
☐ paid	☐ mail	☐ wait
☐ tall	☐ wall	☐ small
☐ jaw	☐ paw	☐ saw

Lookout Words!

☐ boy	☐ before	☐ around	☐ your	☐ were
☐ many	☐ which	☐ down	☐ about	☐ no

Read and Review Words With /ā/ai, ay, /ô/all, aw

QUICKCHECK ✔

Can children:

✓ **read words with /ā/*ai, ay* and /ô/*all, aw*?**

✓ **recognize high-frequency words?**

If YES go to Read and Write.

TEACH

Review Sound-Spellings Review the following sound-spelling relationships: /ā/*ai, ay* and /ô/*all, aw*. Say one of these sounds, and ask a volunteer to write the spellings that stand for the sound. For example, the letters *aw* stand for /ô/. Then display pictures of objects whose names contain these sounds. Have children write the spelling pattern that each picture's name contains.

Review High-Frequency Words Review the high-frequency words ***boy, before, around, your,*** and ***were***. Write sentences with the words on the chalkboard. Say a word, and have a volunteer circle the word in each sentence. Then have children generate new sentences for each word.

READ AND WRITE

Build Words Distribute the following letter cards: ***a, d, m, p, l, y***. Allow time for children to build words using the letter cards. They can record their words on a separate sheet of paper. Continue building words with ***a, i, m, n, p, r; a, b, c, h, l, l;*** and ***a, l, p, r, s, w***.

Build Sentences Write these words on note cards: ***she, boy, mall, to, at, call, can, fall, stay, play, in, the, hall, saw***. Have children make sentences using the words. For example: ***The boy is in the hall.***

Complete Activity Page Read aloud the directions on page 224.

Supporting All Learners

KINESTHETIC/VISUAL LEARNERS

PICTURE SORT Have children make their own picture cards for words with /ā/***ai, ay*** and /ô/***all, aw***. Children can write or dictate the name of each picture on the back of the picture. Mix up the cards. Then have children sort them according to whether they hear /ā/ or /ô/ in each picture name.

CHALLENGE

Have children organize words that are of special interest to them by concept or category. For example, children may create a category called ***Words About Animals*** and list words such as ***paw, jaw,*** and ***tail*** in it, as well as words they have learned earlier (such as ***bat, mule, mole,*** and ***whale***) and other new words (such as ***claw***).

I CAN READ! OPTIONS

The I Can Read! page can be used for one or all of the following:

- paired reading
- individual assessment
- choral reading
- homework practice
- program placement

Lesson 139

pages 225–226

Read Words in Context

TEACH

Assemble the Story Ask children to remove pages 225–226. Have them fold the pages in half to form the Take-Home Book.

Preview the Story Preview *Bats at Home,* a nonfiction selection about bat habitats. Invite children to browse through the first two pages of the story and to point out any unfamiliar words. Read them aloud as children repeat them. Then have children predict what they think the selection might be about.

READ AND WRITE

Read the Story Have volunteers take turns reading aloud a page at a time. Discuss items of interest on each page, and encourage them to help each other with blending. The following prompts may help children while reading:

- **What letter sounds do you know in the word?**
- **Are there any word parts you know?**

Reflect and Respond Have children share their reactions to *Bats at Home.* What did they learn about bats? Did anything they read surprise them? Encourage children to share one interesting fact that they read.

Develop Fluency Reread the story as a choral reading, or have partners reread the story independently. Provide time for children to reread the story on subsequent days to develop fluency and increase reading rate.

When reading other stories, have children share how they figure out unfamiliar words and have them model blending. Encourage children to look for words with /ā/*ai, ay* and /ô/*aw, all* in other stories and to apply what they have learned about these sound-spelling relationships.

BATS AT HOME

Bats go home before the sun rises. Some bats like to rest on cave walls. Others make homes in trees and holes. Some bats make a home with each other. As many as ten bats can make a home in one small hole!

1

A bat can grab a bug with its claws. Sometimes it grabs one bug at a time. Other times it grabs a bunch of bugs. Sometimes a bat uses its tail to snap up a moth. Then the bat can take the moth home to eat it.

Underline your favorite bat fact. Write why you like it.

4

Supporting All Learners

VISUAL LEARNERS

BATS AT HOME Bats live in trees, caves, and boxes. Invite children to use shoe boxes to make dioramas showing bats at home. Children can write or dictate sentences about their dioramas on index cards and display them beside the dioramas. Create a classroom display of the children's pictures showing bats at home.

AUDITORY LEARNERS

BAT TALK Children can read *Bats at Home* aloud, then form a literature circle to talk about the selection. What did they like and dislike about it? What bat facts were the most interesting to them? What photographs did they like? Talk with children about the importance of respecting each other's opinions and giving everyone a chance to share their ideas.

A box can be a home for a bat. A bat box has no holes, but it has a small slit. Bats use the slit to get into the box. A bat stays safe inside a bat box.

2

Meet the tent bat. It makes a home in a big leaf. It can use the leaf as a nest. See its big claws? The bat uses its claws to grab onto the tree.

3

EXTRA HELP

For additional practice with ***/ā/ai, ay*** and ***/ô/aw, all,*** have children read the following My Books: *The Awful Bug* and *Small Daniel*. For information on using the My Books on the computer, see the WiggleWorks Plus My Books Teaching Plan.

CHALLENGE

Have children take turns reading *Bats at Home* to each other and to children from other classes. You may also wish to suggest that children point out the photographs and describe them as they read *Bats at Home*.

Reflect on Reading

ASSESS COMPREHENSION

ONGOING ASSESSMENT

To assess their understanding of the story, ask children questions such as:

- **When do bats go home?** ***(before the sun rises)***
- **What can you put in your backyard as a home for bats?** ***(a bat box)***
- **How can a bat use its tail to eat?** ***(It can use its tail to catch bugs.)***

HOME-SCHOOL CONNECTION

HOME/SCHOOL CONNECTION

Send home *Bats at Home*. Encourage children to read the story to a family member. You may also wish to encourage children to tell family members about the letters, sounds, and words they are learning. Children may point out some of these letters, sounds, or high-frequency words in print that they find in messages, grocery lists, greeting cards, and other writing in their home.

WRITING CONNECTION

BAT CAPTIONS Have children draw a picture of bats doing something based on what they read in *Bats at Home*. Then have children write or dictate captions for their pictures. If possible, display books about bats to help children generate ideas.

Phonics Connection

Phonics Readers:
#50, *A Rain Forest Day*
#51, *Dinosaur Hall*
#53, *Small Animals With Big Names*

Lesson 140

pages 227-228

Blend Words With /ō/o, ow

QUICKCHECK ✔

Can children:

✓ orally blend word parts?

✓ recognize /ō/?

✓ blend words with /ō/*o, ow*?

If YES go to Read and Write.

TEACH

Develop Phonemic Awareness

Oral Blending Say aloud the following word parts, and ask children to oraly blend them. For example, say the sounds /**sh**/ /ō/. Guide children to say the whole word, ***show***. Continue with the following word parts:

/g/ /ō/	**/n/ /ō/**	**/l/ /ō/**
/gr/ /ō/	**/bl/ /ō/**	**/sl/ /ō/**

Connect Sound-Symbol Write the words *log* and *low* on the chalkboard. Read each word aloud. Ask children what vowel sounds they hear in each word and how the words are different. Point out that /o/ in the middle of ***log*** is spelled with an ***o***, and /ō/ at the end of ***low*** is spelled ***ow***. Continue with the words ***got*** and ***go***. Point out that /ō/ at the end of ***go*** is spelled ***o***.

Ask children to suggest other words that contain /ō/. List the words on the chalkboard. Have volunteers circle the letters ***o*** or ***ow*** in each word that contains one of these spellings.

READ AND WRITE

Blend Words List the following words and sentences on a chart. Have volunteers read each aloud.

- go slow no low
- grow blow so show
- No, the pot is not hot.
- They go to play in the snow.

Complete Activity Pages Read aloud the directions on pages 227–228. Review each picture name with children.

Name

Look at each picture. If the picture name has the **long o** sound as in **go**, color the picture.

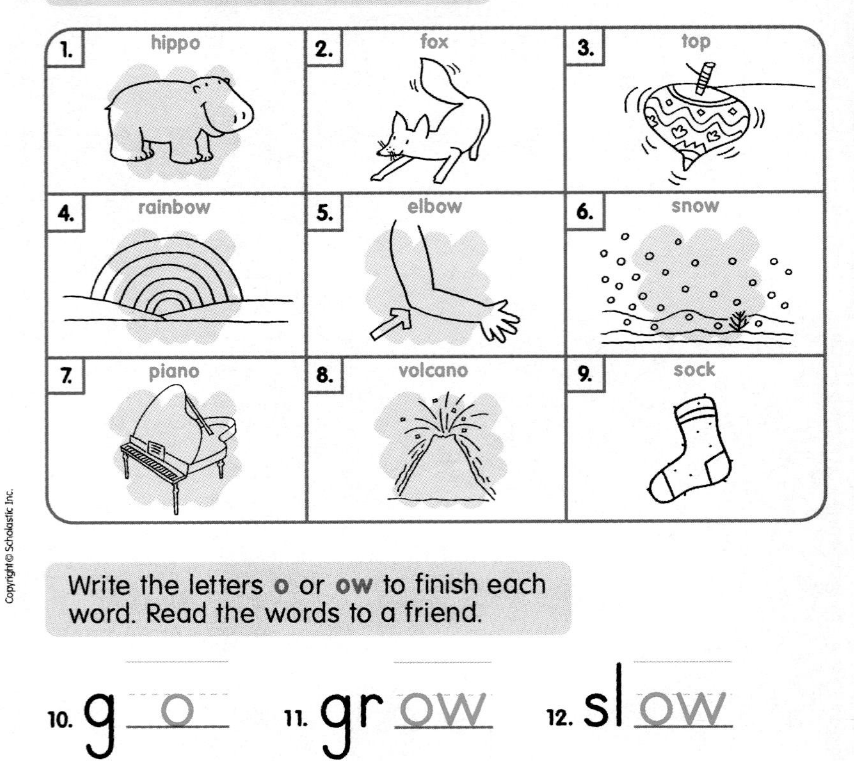

Write the letters **o** or **ow** to finish each word. Read the words to a friend.

10. g o 11. gr ow 12. sl ow

Supporting All Learners

AUDITORY/VISUAL LEARNERS

FIND WORDS Read aloud "Mary Had a Little Lamb" on page 19 in the *Big Book of Rhymes and Rhythms, 1B*. Have children find all the words with **/ō/** in the rhyme. Write these words on the chalkboard. Invite volunteers to add other **/ō/** words to the list.

TACTILE/AUDITORY LEARNERS

TIME TO BUILD Provide children with the magnetic letters from the Phonics and Word Building Kit. Have children build words with **/ō/*o, ow***. Then have children say each word and spell it aloud as they trace its letters.

Look at each picture. Write the picture name on the line. Then find the name in the word search.

snow hello blow rainbow

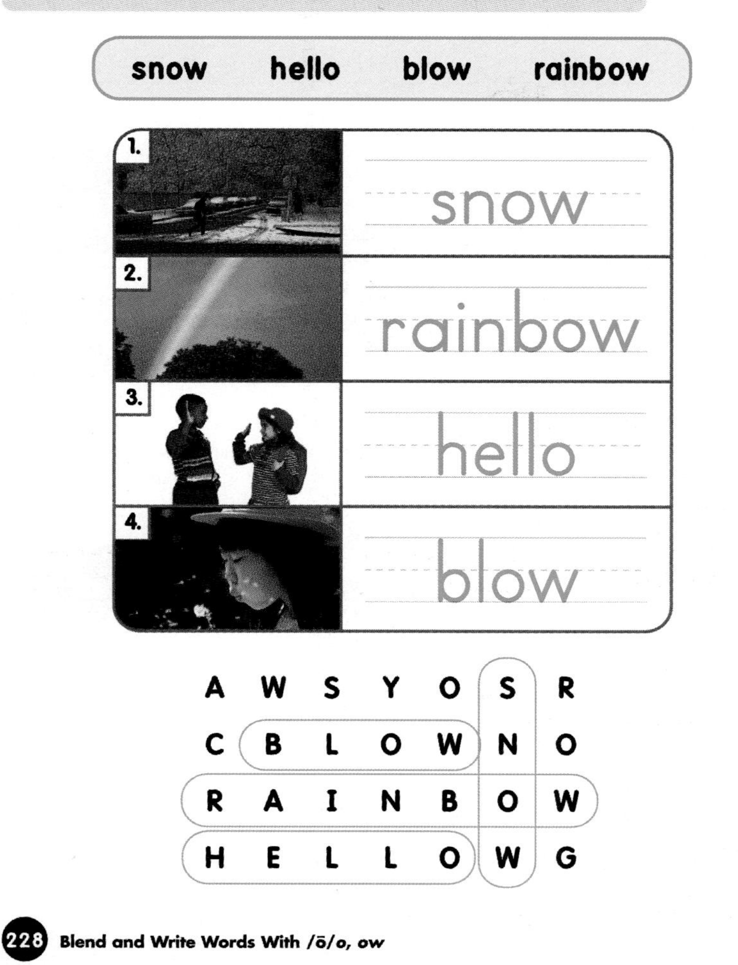

Integrated Curriculum

SPELLING CONNECTION

O-GAME Write the following word parts on index cards, one word part per card: ***g__, sn __, l__, s__, gr__, sl__, n__, sh__, bl__***. Hide the cards around the room while children close their eyes. Then have children look for the cards. Each time children find a card, they should add the letters ***o*** or ***ow*** to make a word. Then have all children say the word formed.

Phonics Connection

Literacy Place: *Information Finders*
Teacher's SourceBook, pp. T202–203; Literacy-at-Work Book, pp. 38–39

My Book: *The Pet Show*

Chapter Book: *Let's Go on a Museum Hunt,* Chapter 4

ESL For children acquiring English, provide a context for some new words by making a two-column chart with headings such as "Go Fast" and "Go Slow." Then list words children already know that name things or animals that go fast, such as a jet, and things or animals that go slow, such as a mole.

EXTRA HELP

Write context sentences such as the following for **/ō/** words:

- **The train is slow.**
- **Let's all go.**
- **Plants need sun to grow.**

Help children track print as they read the sentences. As needed, frame words and target word elements that present difficulty. Then have children reread the sentences with you and point out words with ***/ō/o, ow***.

The Book Shop

- **Bob the Snowman**
  by Sylvia Laretan
  Bob decides to go South.
  (/ō/o, ow)
- **Snowsong Whistling**
  by Karen Lotz
  This book is a rhyming account of typical activities as the first snowfall approaches.
  (/ō/o, ow)

Lesson 141

page 229

Review Long Vowel Sounds

QUICKCHECK ✔

Can children:

- ✔ **blend words with /ā/, /ē/, /ō/, and final *e*?**
- ✔ **write words with /ā/, /ē/, /ō/, and final *e*?**

If YES go to Read and Write.

TEACH

Link to Spelling Review the following sound-spelling relationships: /ā/*ai, ay;* /ē/*ea, ee, y, ey;* /ō/*o, ow; a-e; i-e; o-e;* and *u-e*. Say one of these sounds, and have a volunteer write the spelling or spellings that can stand for the sounds. For example, the letters *ai* can stand for /ā/. Continue with all the sounds.

Phonemic Awareness Oral Segmentation
Say the word ***tail***. Ask children to say the word slowly and to tap lightly on their desk for each sound in the word. Draw three connected boxes on the chalkboard, and have a volunteer write the spelling that stands for each sound in the word in the appropriate box.

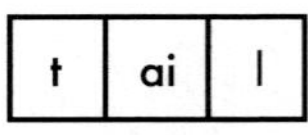

t	ai	l

Continue with other long vowel sounds. For example, *ee* as in ***sheep*** and *ow* as in ***blow***. Emphasize the long vowel sound in each word.

READ AND WRITE

Practice Distribute three counters and one copy of the Segmentation Reproducible Master on page T29 to each child. Read aloud a word. Ask children to count the sounds in the word and to place one counter in each box for each sound they hear. For example, if you say ***no***, the children will place one counter on each of the first two boxes. Continue with: ***no, not, note; net, neat; ran, ray, rain; row, rot, rode.***

Complete Activity Page Read aloud the directions on page 229. Review each picture name with children.

Supporting All Learners

TACTILE LEARNERS

PLAY WITH CLAY Have children use clay to form words with ***/ā/ai, ay; /ē/ea, ee, y, ey; /ō/o, ow;*** and ***a-e, i-e, o-e, u-e.*** After children form the words, they can add them to a list of words with long vowel sounds.

KINESTHETIC LEARNERS

WORD SORT Write the following sets of words on word cards: Set 1: ***rat, hat, bag, day, ham, pain, tail, pail;*** and Set 2: ***no, slow, grow, low, hot, log, fog, frog.*** Mix the cards, and have children sort each set into two groups according to short and long vowel sounds.

EXTRA HELP

Have children listen for words with long vowel sounds in the following songs on the Sounds of Phonics Audiocassette (Teamwork): "Little Green Frog", "Hokey Pokey", "If I Could Play", and "Mary Had a Little Lamb". Help children identify all the words they hear that contain ***/ā/, /ē/,*** and ***/ō/*** in each song. Write these words on the chalkboard. Read the lists aloud with children, emphasizing the long vowel sound in each.

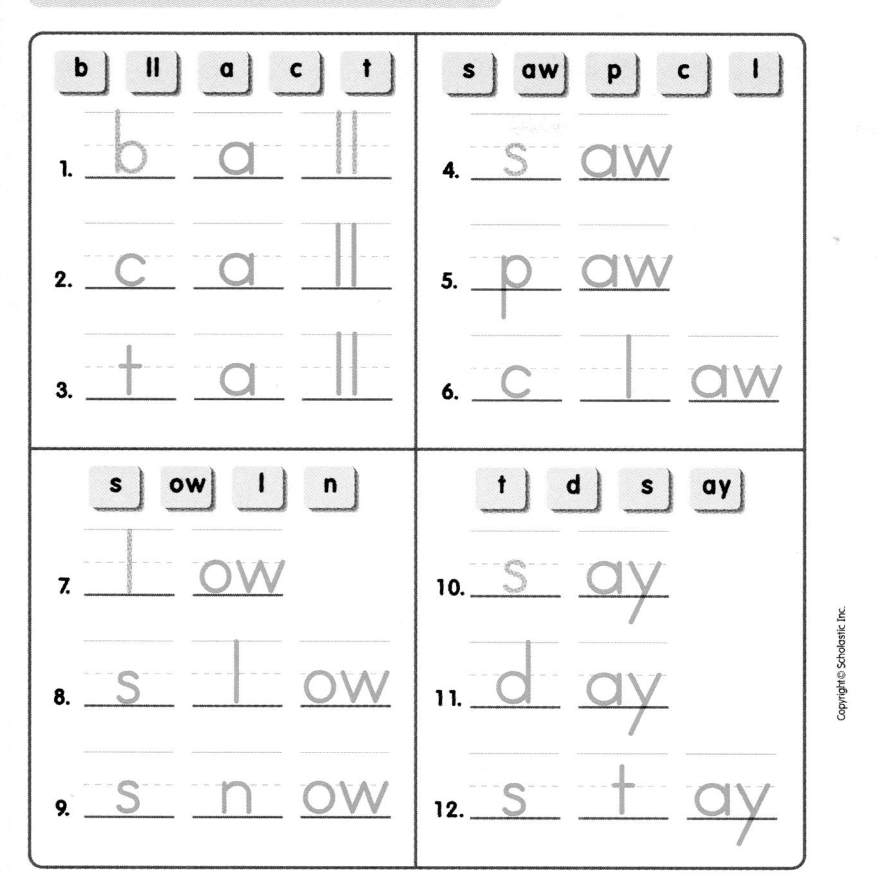

Use the letter tiles to make words.

b | ll | a | c | t

1. b a ll
2. c a ll
3. t a ll

s | aw | p | c | l

4. s aw
5. p aw
6. c l aw

s | ow | l | n

7. l ow
8. s l ow
9. s n ow

t | d | s | ay

10. s ay
11. d ay
12. s t ay

page 230

Build Words

QUICKCHECK ✔
Can children:
✔ build words?
If YES go to Read and Write.

TEACH

Review Sound-Spellings Review the sound-spelling patterns from the past few lessons including /ā/*ai, ay;* /ô/*aw, all,* and /ō/*o, ow*.

Write the word ***show*** on the chalkboard. Ask children to replace /**sh**/ with /**l**/ to make a new word. Replace the letters ***sh*** in ***show*** with the letter ***l***, and blend the new word formed. Continue changing one letter or letter pair to build new words.

low
grow
gray
clay
claw

READ AND WRITE

Build Words Distribute the following letter cards: ***g, l, n, o, r, w***. Provide time for children to build words using the letter cards. Ask them to record their words. Continue building words with: ***a, c, l, p, w, y; a, i, l, p, r, t;*** and ***a, b, c, h, l, l.***

Complete Activity Page Read aloud the directions on page 230. Children can complete the page independently.

Supporting All Learners

VISUAL LEARNERS

CLAWS AND PAWS Have children write or dictate a list of animals that have claws and/or paws. Then children can draw pictures of these animals. After children have completed their pictures, they can write or dictate captions using the words ***claws*** and/or ***paws,*** depending on what each animal has.

KINESTHETIC/AUDITORY LEARNERS

CENTER TIME For additional practice with vowels, set up Center 1, Activity 1, in *Quick-and-Easy Learning Centers: Phonics*. Adapt the activity so that children are sorting names according to short and long vowel sounds, or review other phonics skills, such as sorting by beginning sounds, ending sounds, or number of syllables.

CHALLENGE

Children can make a flip book featuring **/ō/*ow***. The top word may be ***blow***. Under the ***bl,*** children can put ***sl, sn, sh,*** and ***gr***. Children can take turns reading the words in their flip books and can make up sentences using the words. You may also have children make flip books for **/ā/*ai, ay*** and **/ô/*aw, all***.

pages 231-232

Blend Words With /ē/ey, y

QUICKCHECK ✔

Can children:

- ✔ orally segment words?
- ✔ recognize /ē/?
- ✔ blend words with /ē/ey, y?

If YES go to Read and Write.

TEACH

Develop Phonemic Awareness

Oral Segmentation Explain to children that you are going to say words sound by sound. Ask children to listen to the sounds and to say the whole word. For example, say /sh/ /ē/. Guide children to say the whole word *she*.

/k/ /ē/	**(key)**
/p/ /ō/ /n/ /ē/	**(pony)**
/r/ /e/ /d/ /ē/	**(ready)**
/h/ /u/ /n/ /ē/	**(honey)**

Connect Sound-Symbol Write *pony* on the chalkboard, and read it aloud. Ask children what sound they hear at the end of *pony*. Point out that /ē/ at the end of *pony* is spelled with *y*. Then write *monkey* on the chalkboard, and read it aloud. Ask children what sound they hear at the end of the word. Point out that /ē/ at the end of *monkey* is spelled with *ey*.

Ask children to suggest other words that end with /ē/. List them on the chalkboard, and have volunteers circle *y* or *ey* in each word.

READ AND WRITE

Blend Words List the following words and sentence on a chart. Have volunteers read each aloud.

- **key monkey pony**
- **We met a pony called Fred.**

Complete Activity Pages Read aloud the directions on pages 231–232. Review the art with children.

LISTEN AND WRITE

Name

Look at each picture. If the picture name has the **long e** sound as in **baby**, write the letter **y** on the line.

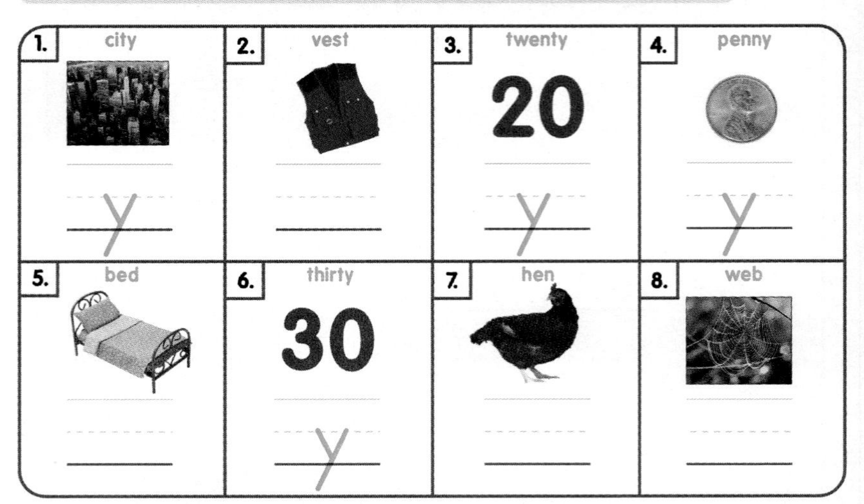

Look at each picture. If the picture name has the **long e** sound as in **monkey**, write the letters **ey** on the line.

9. key — ey
10. men
11. chimney — ey
12. turkey — ey

Supporting All Learners

AUDITORY/VISUAL LEARNERS

FUN WITH RHYMES AND RHYTHMS Read aloud "Three Little Monkeys" on pages 20–21 in the *Big Book of Rhymes and Rhythms, 1B*. Have children find all the words with **/ē/** in the rhyme. Write these words on the chalkboard. Invite volunteers to add other **/ē/** words to the list.

VISUAL LEARNERS

WORD SEARCH GAME Divide the class into two groups. Have each group look through books, magazines, and newspapers for words with **/ē/y, ey**. Ask them to list the words they find on a sheet of paper. Then have children compare their word lists and blend each word. Which group found the most words with **/ē/y, ey**?

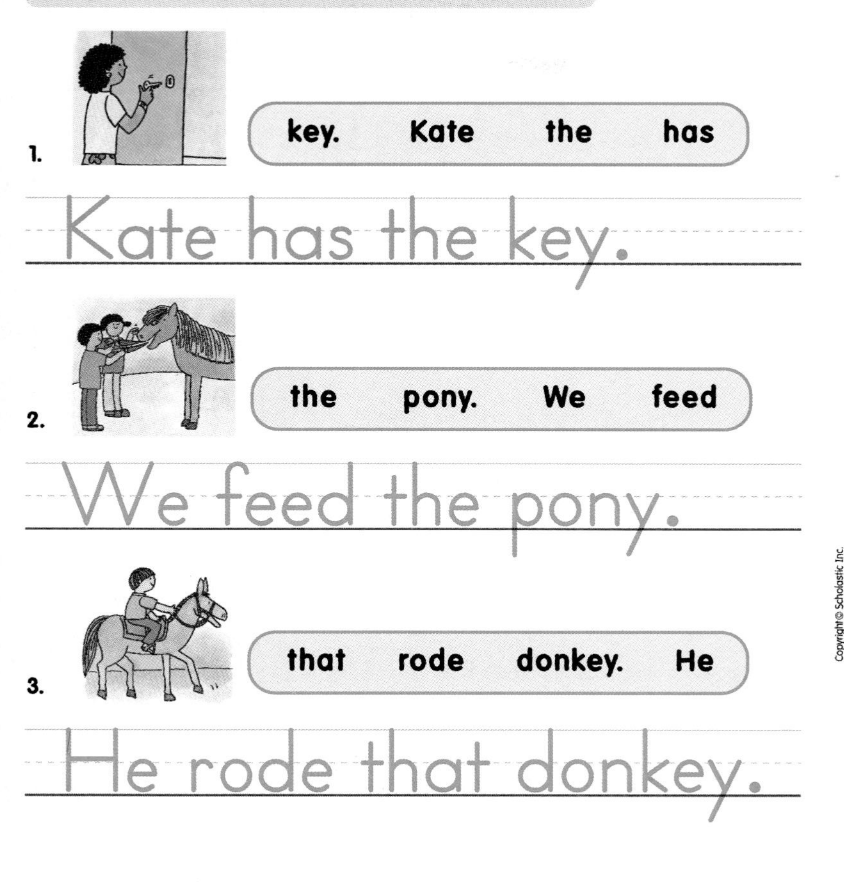

Look at the picture. Then unscramble the words to make a sentence that tells about the picture. Write the sentence on the line.

1. key. Kate the has

Kate has the key.

2. the pony. We feed

We feed the pony.

3. that rode donkey. He

He rode that donkey.

Integrated Curriculum

SOCIAL STUDIES CONNECTION

ENVIRONMENTAL PRINT WALL

Create an environmental print wall. Have children bring in items to post, including cereal boxes, wrappings for toys, and place mats from restaurants. Invite children to find words with **/ē/y, ey** and other target sounds and letters in these labels. Then children can post these labels on the environmental print wall.

Phonics Connection

Literacy Place: *Information Finders*
Teacher's SourceBook, pp. T260–261;
Literacy-at-Work Book, pp. 50–51

My Book: *Monkey See, Monkey Do*

Chapter Book: *Let's Go on a Museum Hunt,* Chapter 5

The Book Shop

- **Clifford's Puppy Days**
  by Norman Bridwell
  An oversized, red dog is born! **(/ē/ey, y)**
- **Arthur's Funny Money**
  by Lilian Hoban
  Arthur and his sister start a business. **(/ē/ey, y)**

ESL Have children find or draw two or more silly monkeys. Encourage children to label their pictures and to write a sentence that tells about the monkeys. Post the pictures in the classroom.

EXTRA HELP

Guide children to use the pocket ABC cards and pocket chart from the Phonics and Word Building Kit to build words with **/ē/y, ey**. Write the words children build on the chalkboard. Then have children underline the ***y*** or ***ey*** in each word.

CHALLENGE

Write the following word parts on index cards, one word part per card: ***k__, hon__, pon___, read__, monk__***. Place the cards in a bag. Have children take turns picking one card. Ask children to write the word parts on sheets of paper, and add ***ey*** or ***y*** to each. Then have children take turns saying the words they made. Repeat the game by returning the cards to the bag.

Lesson 144

page 233

Recognize and Write Plurals

QUICKCHECK ✔

Can children:

✔ recognize plurals?

✔ write plurals?

If YES go to Read and Write.

TEACH

Introduce Plurals Remind children that many words that tell more than one are formed by adding *s*. Write the word ***donkey*** on the chalkboard. Ask a volunteer to read it aloud. Use the word ***donkey*** in a sentence such as ***I see a donkey***. Then write ***donkeys*** on the chalkboard, and use it in a sentence such as ***I see two donkeys***. Point out that the letter *s* makes the word mean "more than one."

Create a two-column chart on the chalkboard. Label the first column ***One***. Label the second column ***More Than One***. Write the following words under ***One*** as you say them: ***claw, paw, key, monkey***. Add an *s* to each word, and say the new word. Then write the plural form of each word under ***More Than One***.

READ AND WRITE

Practice Write the following sentence pairs on the chalkboard. Ask children to tell what is the same and what is different about the underlined words in each pair.

- **I have the key.**
  I have the keys.
- **Jill saw the monkeys.**
  Jill saw the monkey.

Complete Activity Page Read aloud the directions on page 233. Have children complete the page independently.

Name

Read the word next to each sentence. Add an **s** to the word. Then write the new word on the line. Read each sentence to a friend.

1.	**leg**	That bug has six legs.
2.	**egg**	The hens lay eggs.
3.	**shell**	He picks up many clam shells.
4.	**Seal**	Seals swim in the sea.
5.	**monkey**	We like to see the monkeys play.
6.	**cat**	The cats like to run and jump.

Supporting All Learners

KINESTHETIC LEARNERS

LABELS WITH AND WITHOUT *S* Have children create labels on self-sticking notes for various classroom objects or groups of objects such as one ball, two pens, one desk, two note pads, and some pins. Read each label with children, and note that the letter ***s*** added to the end of a word means more than one.

VISUAL LEARNERS

HOW MANY? Children can look for pictures that show one and more than one thing or animal, or draw pictures of one and more than one thing or animal. After children finish their drawings, have them write or dictate captions describing the pictures or drawings. For example, a caption might read ***This picture shows two cats.***

EXTRA HELP

For additional practice with plurals, have children read aloud the following My Book: *Lots of Oranges*. For information on using the My Books on the computer, see the WiggleWorks Plus My Books Teaching Plan.

INTRODUCE HIGH-FREQUENCY WORDS

Read the words in the box.
Write one word on each blank.

girl	more
school	there
their	

1. girl
2. school
3. their
4. more
5. there

Use the words in the box to finish each sentence.

6. That girl is late for the bus.
7. She needs to ride the bus to her school.
8. Can she get there on time?

Supporting All Learners

VISUAL LEARNERS

LEARNING CENTER For additional practice and review, use Center 8, Activity 1, "Junk Mail Collection" in *Quick-and-Easy Learning Centers: Phonics*. This activity fosters awareness of high-frequency words.

KINESTHETIC LEARNERS

BE A SENTENCE Write the words from the sentences in the Teach section on note cards, one word per card. Give the words in the first sentence to a group of five children, the words in the second sentence to a group of six children, and so on. Have these groups of children arrange themselves to form the sentences. Then have other children read the sentences.

CHALLENGE

Have children write sentences using the high-frequency words ***girl, school, their, more,*** and ***there***. Ask children to cover the high-frequency word or words with sticky notes. Then children can exchange sentences and challenge each other to figure out the words under the sticky notes.

page 234

Recognize High-Frequency Words

QUICKCHECK ✔

Can children:

✔ **recognize and write the high-frequency words *girl, school, their, more, there*?**

✔ **complete sentences using the high-frequency words?**

If YES go to Read and Write.

TEACH

Introduce the High-Frequency Words

Write the high-frequency words ***girl, school, their, more,*** and ***there*** in sentences on the chalkboard. Read aloud the sentences, and underline the words. Ask children if they recognize them. If necessary, read the sentences again. You may wish to use the following sentences:

1. **The girl runs to school.**
2. **The boy runs there with her.**
3. **Their school is close to their home.**
4. **It is no more than a hop away!**

Then ask volunteers to dictate sentences using high-frequency words. You may wish to begin with the following sentence starters: ***The school is*** ___ or ***There is my*** ___.

READ AND WRITE

Practice Write each high-frequency word on a note card. Read each word aloud as you display the cards. Then do the following:

- **Mix the cards.**
- **Display one card at a time, and ask children to state each word aloud.**
- **Have children spell each word aloud, clapping on each letter.**
- **Ask children to write each word in the air as they state aloud each letter. Then have them write each word on a sheet of paper.**

Complete Activity Page Read aloud the directions on page 234. Have children complete the page independently.

Lesson 146

pages 235-236

Blend Words With /ō/oa

QUICKCHECK ✔

Can children:

✔ **orally segment words?**

✔ **recognize /ō/?**

✔ **blend words with /ō/*oa*?**

If YES go to Read and Write.

TEACH

Develop Phonemic Awareness

Oral Segmentation Explain to children that you are going to say words. Children will tell you how many sounds they hear in each word. For example, say the word ***coat***, then say it sound by sound—/k/ /ō/ /t/. Guide children to hear that the word ***coat*** contains three separate sounds. Continue with the following words.

- soap
- goat
- toe
- road
- so
- toad

Connect Sound-Symbol Write the words ***got*** and ***goat*** on the chalkboard as you read each aloud. Ask children what sound they hear in the middle of each word. Ask children how the sounds are different. Point out that the short ***o*** sound in ***got*** is spelled with an ***o***, and the long ***o*** sound in ***goat*** in spelled with ***oa***.

READ AND WRITE

BUILT-IN REVIEW

Blend Words List the following words and sentences on a chart. Have volunteers read each aloud. Model for children how to blend the words. Then have children blend the words independently.

- cot coat got goat
- coach soap load groan
- The boat is big.
- The coach ran down the road.

Complete Activity Pages Read aloud the directions on pages 235–236. Review each picture name with children.

Name

Look at each picture. If the picture name has the **long o** sound as in **boat**, write the letters **oa** on the line. If it has the **short o** sound as in **top**, write **o**.

1. coat — oa	2. log — o	3. pot — o	4. goat — oa
5. dot — o	6. road — oa	7. top — o	8. box — o
9. soap — oa	10. mop — o	11. float — oa	12. toast — oa

Supporting All Learners

VISUAL/AUDITORY LEARNERS

MISS LUCY Read aloud "Miss Lucy" on pages 22–23 in the *Big Book of Rhymes and Rhythms, 1B*. Have children find all the words with **/ō/ *oa*** in the rhyme. Write these words on the chalkboard. Invite volunteers to add other **/ō/*oa*** words to the list.

AUDITORY LEARNERS

GAME TIME To review words with long vowel sounds, children may enjoy playing "Raceway" on pages 30–32 in *Quick-and-Easy Learning Games: Phonics*. You may wish to adapt the game slightly to avoid the few words with vowel spellings such as ***ie*** that children have not yet learned.

VISUAL/AUDITORY LEARNERS

CIRCLE IT Ask children to suggest words that contain the long ***o*** sound. List these words on the chalkboard. Read them aloud. Have children write them. Then have volunteers come to the chalkboard and circle those words that contain ***oa***.

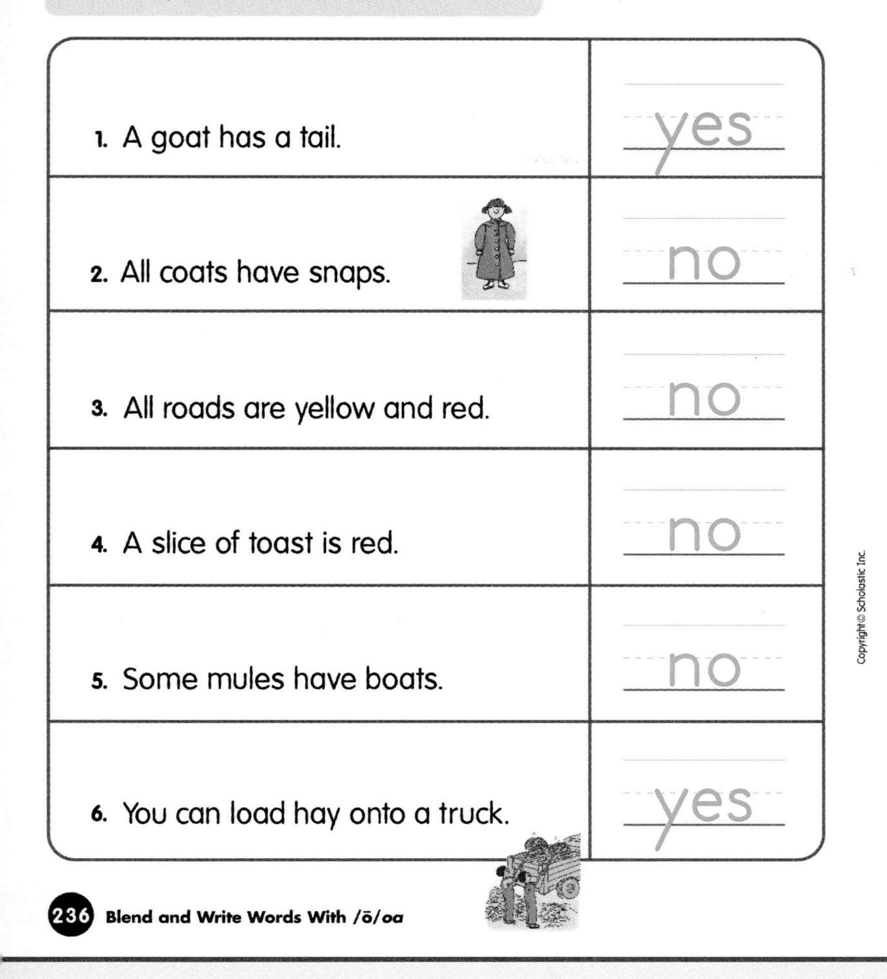
BLEND WORDS

Read each sentence. If the sentence is true, write **yes** on the line. If it is not true, write **no** on the line.

1. A goat has a tail.	yes
2. All coats have snaps.	no
3. All roads are yellow and red.	no
4. A slice of toast is red.	no
5. Some mules have boats.	no
6. You can load hay onto a truck.	yes

236 **Blend and Write Words With /ō/oa**

EXTRA HELP

Write the following words on index cards: ***cot, cob, throat, soap, hot, jog, boat, goal, mop, toad, box,*** and ***groan***. Have children sort them according to short and long vowel sounds.

CHALLENGE

Have children look through books, magazines, and newspapers for words with ***oa***. Have children list the words on a sheet of paper. Children can use these words to write sentences such as ***The goat ate my coat!***

Integrated Curriculum

SPELLING CONNECTION

COATS AND MORE COATS Have children draw and cut out outlines of coats from colored construction paper. Then have children write the following on the coats: ***g_t, s_p, b_t, t_d, g_l***. Say ***goat, soap, boat, toad,*** and ***goal***. Have children fill in ***oa*** in the spaces on their coats to make the words.

SCIENCE CONNECTION

GOATS, TOADS, AND FOALS Have children find or draw pictures of goats, toads, and foals. Then have children draw pictures of settings or environments where each animal could be found. Children can match each animal to each environment.

Phonics Connection

Literacy Place: *Information Finders*
Teacher's SourceBook, pp. T314–315;
Literacy-at-Work Book, pp. 60–61

My Book: *Rabbit's New Coat*

- **A New Coat for Anna**
  by Harry Ziefert
  Thanks to her mother's careful planning, Anna gets a new coat. **(/ō/oa)**
- **Boats**
  by Anne Rockwell
  Pictures and text show where and how boats go. **(/ō/oa)**

Lesson 147

page 237

Recognize and Write Compound Words

QUICKCHECK ✓

Can children:

✓ recognize compound words?

✓ write compound words?

If YES go to Read and Write.

TEACH

Introduce Compound Words Explain to children that a compound word is made up of two smaller words. Often the two smaller words provide clues to the meaning of the compound word. Write the compound word ***doghouse*** on the chalkboard. Ask a volunteer to circle the two smaller words in ***doghouse***. Point out that a ***doghouse*** is a house for a dog.

The word *doghouse* is a compound word, so I need to think about the two little words *dog* and *house* it contains. I know that a dog is an animal and a house is a place to live. So I bet a *doghouse* is a place where a dog lives.

Write the following words on the chalkboard: ***sidewalk, apple, rabbit, someone, running, treehouse, lunchbox***. Have children circle the compound words. Then have them underline the two smaller words that make up each compound word. Ask children to state the meaning of each compound word.

READ AND WRITE

Practice Write the following words on the chalkboard: ***ball, cook, light, sun, foot, hot, dog, book***. Have volunteers combine two of the words to form a compound word. Then have them use the compound word in a sentence.

Complete Activity Page Read aloud the directions on page 237. Children can complete the page independently.

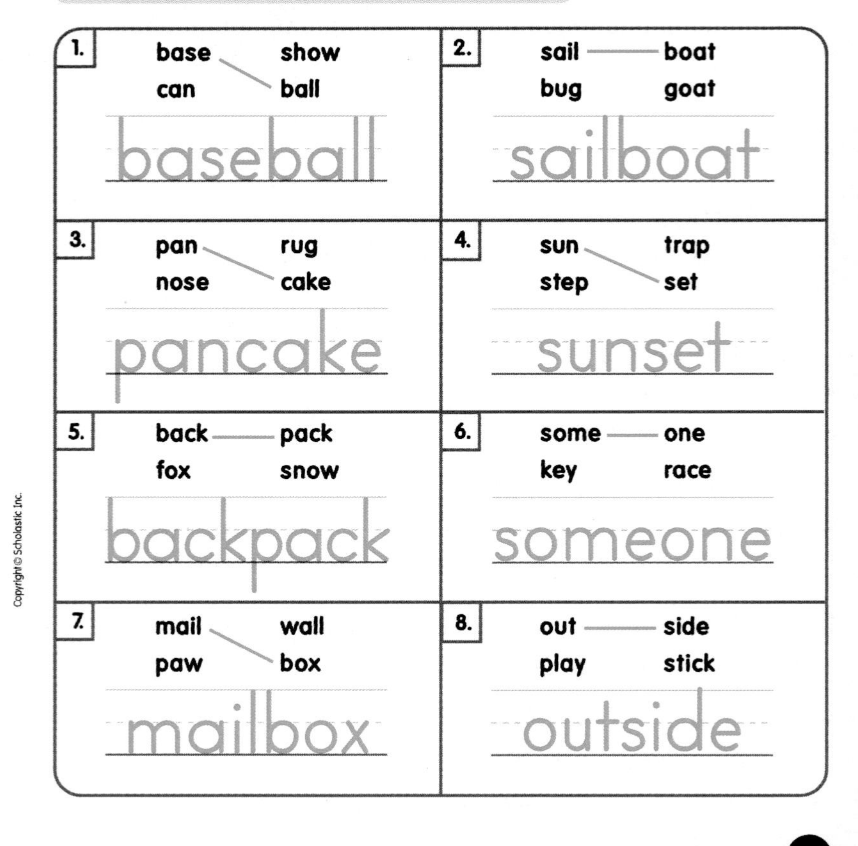

COMPOUND WORDS

Name

Draw a line to connect the two words in each box that make a compound word. Then write the word on the line.

1. base show / can ball — baseball	2. sail boat / bug goat — sailboat
3. pan rug / nose cake — pancake	4. sun trap / step set — sunset
5. back pack / fox snow — backpack	6. some one / key race — someone
7. mail wall / paw box — mailbox	8. out side / play stick — outside

Supporting All Learners

KINESTHETIC LEARNERS

WORD CARD CONCENTRATION Write the following words on note cards: ***house, ball, base, boat, sun, walk, dog, print, hand, light, sail, tree, side, foot***. Have partners play Concentration using the cards. Children should place the cards facedown on a desk or on the floor. One player at a time should then turn over two cards. If the cards form a compound word, the player should state the word and use it in a sentence. The player then gets a point, but must turn the cards facedown again. Only one point can be earned for each word. The player with the most points at the end of the game wins.

Select compound words such as ***lunchbox*** and ***raincoat*** to provide children acquiring English contextual practice. Have partners tell the meaning of each word using the sentence pattern ***A ____ is a ____. (A lunchbox is a box for keeping one's lunch.)*** Adjust the sentence as needed to fit the meaning of the compound word.

Find the word in the box that answers each clue. Then write the word in the puzzle.

DOWN

1. run into someone
2. beat the other team
3. home to whales

see	**sea**
meet	**meat**
week	**weak**
one	**won**
sun	**son**
road	**rode**

ACROSS

2. it has 7 days
3. it shines
4. a place to drive

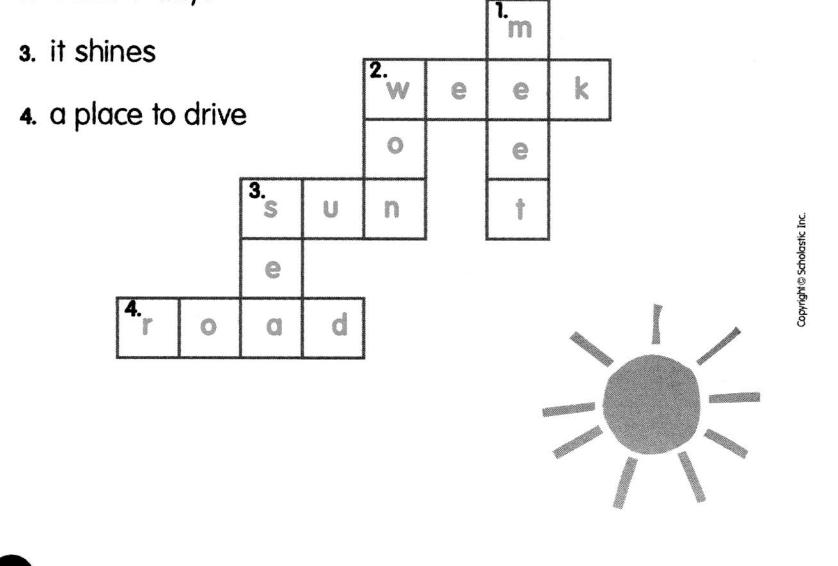

Supporting All Learners

VISUAL/AUDITORY LEARNERS

HOMOPHONE MATCH Have children draw and label pictures of an ant, an eye, a flower, and the sail on a boat. Write these on the chalkboard:

- **My aunt is my mother's sister.**
- **I like my cat.**
- **The car is for sale.**
- **I used flour to make the cake.**

Read each sentence aloud. Have children hold up the pictures they drew that have names that sound like each underlined word. Talk about how the spellings and meanings are different.

EXTRA HELP

For practice with homophones, have children read the My Book *Going to Sea*. For information on using the My Books on the computer, see the WiggleWorks Plus My Books Teaching Plan.

ASSESSMENT

ONGOING ASSESSMENT

Some children may more quickly differentiate between pairs of homophones that name at least one concrete object, such as ***plane*** and ***plain,*** than between more abstract homophone pairs such as ***there*** and ***their***. Many children will need additional practice with such abstract pairs.

page 238

Recognize and Write Homophones

QUICKCHECK ✔

Can children:

✔ recognize homophones?

✔ write homophones?

If YES go to Read and Write.

TEACH

Introduce Homophones Write the word ***there*** on the chalkboard. Ask a volunteer to read it. Use the word ***there*** in a sentence such as ***There is my cat***. Then write ***their*** on the chalkboard, and use it in a sentence such as ***I see their cat***. Point out the difference in spelling between these two words that sound the same. Ask children how the words differ in meaning.

Explain that sometimes words sound the same but have different meanings and spellings. When two words sound alike, children can tell what the meaning is by using two kinds of clues. One clue is spelling. Another clue is the surrounding words in the sentence. These give clues to the word's meaning.

READ AND WRITE

Practice Write the following sentence pairs on the chalkboard. Ask children to tell what is the same and what is different about the underlined words in each pair.

- **I see the beach.**
  I swim in the sea.
- **Sam rode the donkey.**
  Gail met Sam on the road.

Complete Activity Page Read aloud the directions on page 238. Have children complete the page independently.

Lesson 149

page 239

Blend and Build Words With Phonograms

QUICKCHECK ✔

Can children:

✔ blend words with phonograms *-ake, -oke*?

✔ build words with *-ake, -oke*?

If YES go to Read and Write.

TEACH

Introduce the Phonograms Write the phonograms *-ake* and *-oke* on the chalkboard. Point out the sounds these phonograms stand for. Add the letter *m* to the beginning of *-ake*. Model how to blend the word formed. Have children repeat the word ***make*** aloud as you blend the word again. Then add *w* to the beginning of *-oke,* and model for children how to blend the word.

READ AND WRITE

Build Words Write *-ake* on the chalkboard. (If available, use a pocket chart and letter cards.) Have children add a letter (or pair of letters) to the beginning of the phonogram to make a new word. Continue by having children replace the initial consonant, blend, or digraph in each word to build a new word. For example, children may build ***wake*** from ***bake,*** then ***take*** from ***wake***. Continue the activity by building words with the word part *-oke*.

Complete Activity Page Read aloud the directions on page 239. Have children complete the page independently.

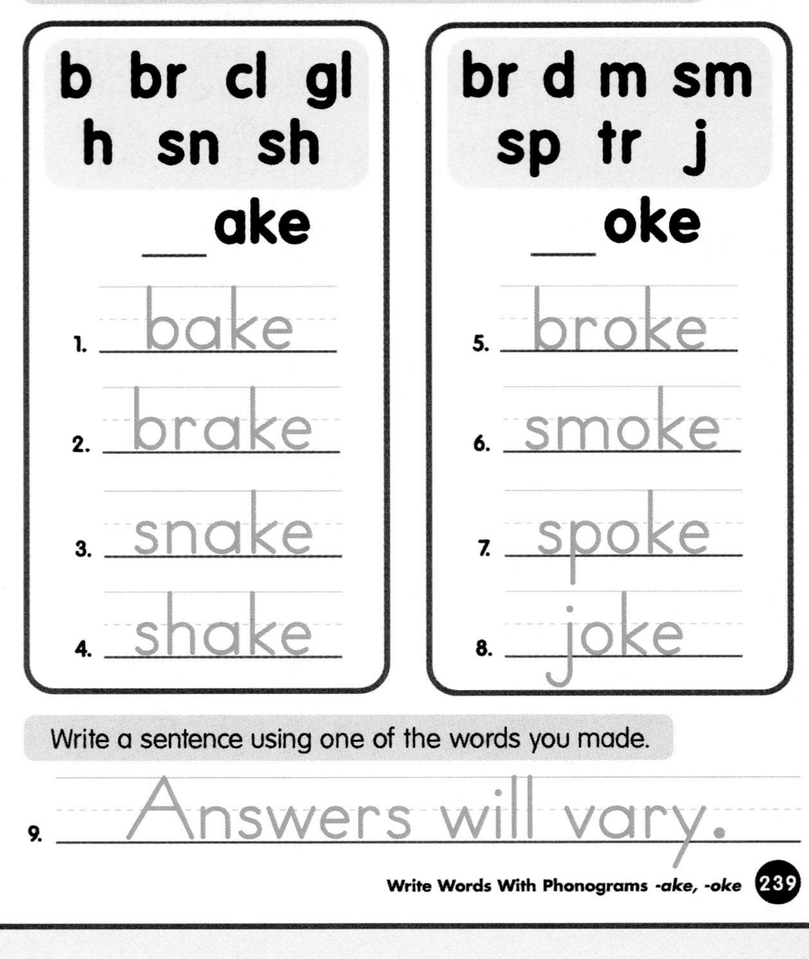

Supporting All Learners

VISUAL LEARNERS

WORD WALL Have children write ***-ake*** and ***-oke*** words on large cards. Write the beginning consonant, consonant blend, or digraph in black. Write the phonograms ***-ake*** and ***-oke*** in red. Children can add the cards to the Word Wall.

VISUAL/AUDITORY LEARNERS

TIME TO RHYME Ask children to suggest words that rhyme with ***make*** and ***woke***. List these words on the chalkboard in separate columns. Have volunteers underline the phonogram ***-ake*** or ***-oke*** in each. Be sure to include the words ***cake, lake, bake, take, joke, poke, spoke,*** and ***broke***. Blend these words for children as they repeat them aloud. Encourage children to look for the word parts ***-ake*** and ***-oke*** as they read.

EXTRA HELP

For additional practice, have children build ***-ake*** and ***-oke*** words on the Magnet Board. Then suggest that they listen as the word is pronounced. See the WiggleWorks Plus My Books Teaching Plan for information on this and other Magnet Board activities.

POETRY

Read the poem. Write a sentence telling about the bike.

Sad, Sad Bike

The wheels can't spin.
The seat shakes.
The chain has rust.
And the bike needs brakes!

Just last week
The kickstand broke.
And one wheel
Lost one spoke.

"Sad, sad bike!" said the girl.
"I'm on your side.
I'll fix you up.
And then we'll ride!"

Answers will vary.

Supporting All Learners

VISUAL LEARNERS

PICTURE & CAPTION ACTION Have children draw a picture showing a detail in the poem. Have them label their drawing with the appropriate line or lines from the poem. For fun, children can take turns displaying their picture with the caption covered up so that others can guess what it says.

CHALLENGE

Have children generate a list of words with long ***/ō/oa*** and long ***/ē/y, ey***. Record these words on the chalkboard. Then have children create a story, using as many long ***o*** and ***e*** words on the chalkboard as they can. Write the dictated story on chart paper for group and individual reading. Return to the story, rereading it in subsequent lessons.

Write and Read Words With Long Vowel Sounds, Plurals, and Contractions

QUICKCHECK ✔

Can children:

- ✔ **spell words with long vowel sounds, plurals, and contractions?**
- ✔ **read a poem?**

If YES go to Read and Write.

TEACH

Link to Spelling Review long vowel sound spellings, plurals, and contractions. Say one of the long vowel sounds. Have a volunteer write the spelling that stands for the sound on the chalkboard. For example, the spelling *ee* stands for /ē/. Continue with all the long vowel sounds and their spellings.

Phonemic Awareness Oral Segmentation
Say ***chain***. Have children orally segment the word. (/***ch***/ /ā/ /***n***/) Draw three connected boxes on the chalkboard. Have a volunteer write the spelling that stands for each sound in the word ***chain*** in the appropriate box. Continue with ***wheel*** and ***spoke***.

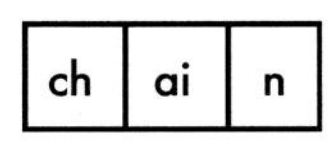

ch	ai	n

READ AND WRITE

Dictate Dictate the following. Have children write the words and sentences. Then write them on the chalkboard. Have children make any corrections.

- **seed** **key** **pony** **seat**
- **goat** **brake** **stay** **snow**
- **The pony can eat hay.**
- **I need the key to the van.**

Complete Activity Page Read aloud the directions on page 240. Children can read the poem independently or with partners.

Lesson 151

pages 241-242

Read and Review Words With /ō/o, ow, oa; /ē/ey, y

QUICKCHECK ✔

Can children:

✔ read words with /ō/o, ow, oa; /ē/ey, y?

✔ recognize high-frequency words?

If YES go to Read and Write.

TEACH

Review Sound-Spellings Review with children the sound-spelling relationships from the past few lessons. These include /ō/*o, ow, oa* and /ē/*ey, y*. State aloud one of these sounds. Have a volunteer write on the chalkboard the spellings that stand for the sound. For example, the letters *o* and *ow* stand for /ō/. Continue with the other sound. Then display pictures of objects whose names contain one of these sounds. Have children write the letters that each picture's name contains as the picture is displayed.

Review High-Frequency Words Review the high-frequency words ***girl, school, their, more,*** and ***there***. Write sentences containing each high-frequency word on the chalkboard. Say one word, and have volunteers circle the word in the sentences. Continue by having children generate sentences for each word.

READ AND WRITE

Build Words Distribute the following letter cards to children: ***e, k, n, o, p, y***. If children have their own set of cards, have them locate the letter set. Provide time for children to build as many words as possible using the letter cards. Suggest that they record their words on a separate sheet of paper. Then have partners compare their lists. Continue building words with the following sets of letter cards: ***a, d, g, n, o, r*** and ***a, b, c, g, o, t***.

Complete Activity Pages Read aloud the directions on pages 241–242. Review the art with children.

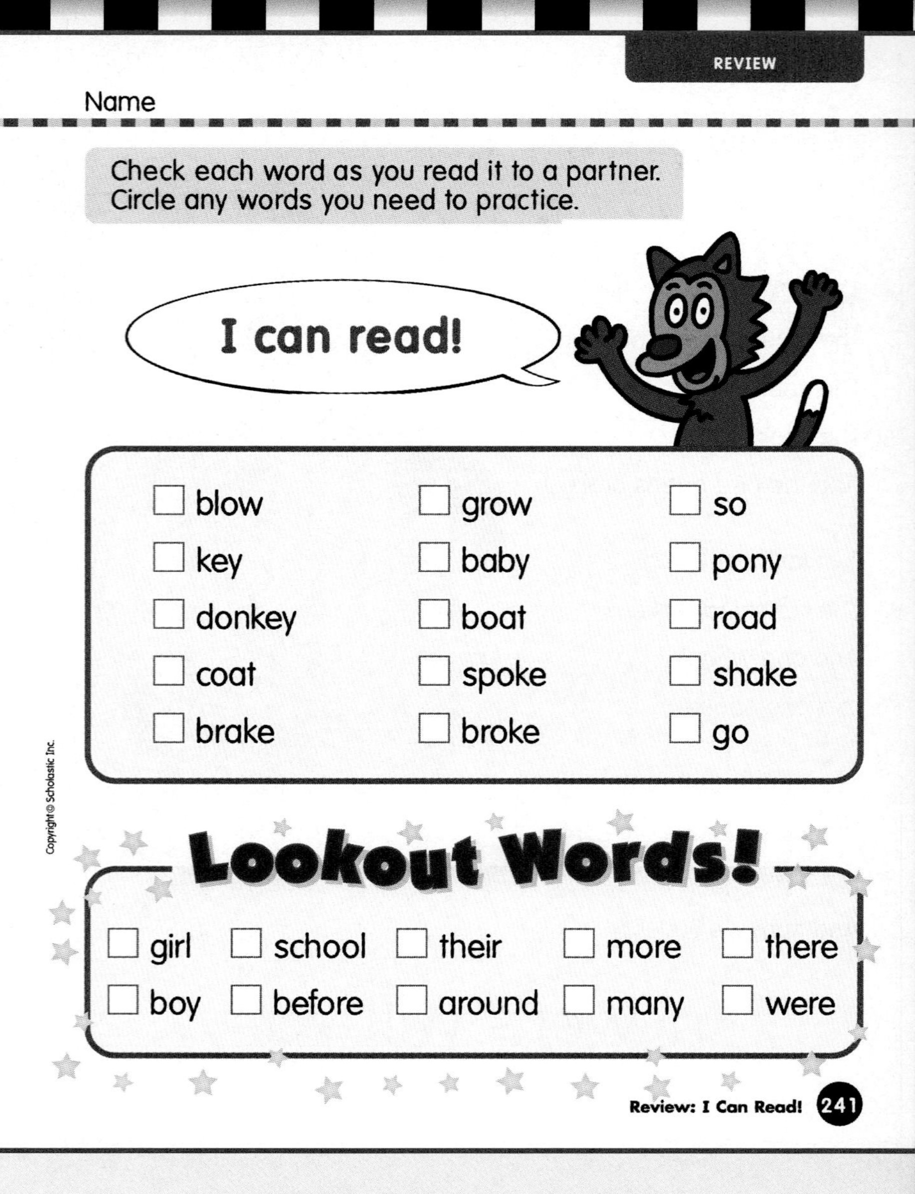
REVIEW

Name

Check each word as you read it to a partner.
Circle any words you need to practice.

☐ blow	☐ grow	☐ so
☐ key	☐ baby	☐ pony
☐ donkey	☐ boat	☐ road
☐ coat	☐ spoke	☐ shake
☐ brake	☐ broke	☐ go

Lookout Words!

☐ girl	☐ school	☐ their	☐ more	☐ there
☐ boy	☐ before	☐ around	☐ many	☐ were

Supporting All Learners

VISUAL LEARNERS

LIST RHYMES Invite small groups of three or four children to write rhyming words for ***boat*** and ***grow***. Have each group list as many words as they can. When finished, help them to notice and, as needed, correct the differences in spelling for the same vowel sounds.

KINESTHETIC/AUDITORY LEARNERS

FUN WITH WORDS Have partners take turns using the pocket ABC cards and pocket chart from the Phonics and Word Building Kit to make words with **/ō/o, ow, oa** and **/ē/ey, y**. Each time a child makes a word, his or her partner should use the word in a sentence.

KINESTHETIC LEARNERS

BUILD SENTENCES Write the following words on note cards: ***I, they, got, a, coat, boat, had, pony, key, the***. Display the cards, and have children make sentences using the words. To get them started, suggest this sentence: ***They had the key***.

TEST YOURSELF

Fill in the bubble next to the sentence that tells about each picture.

1. ● The girl has her coat on.
○ The girl plays in the rain.

2. ○ The plant will grow in the sun.
● I let the cat go outside.

3. ● The boy leads a goat.
○ The boy calls out to the goat.

4. ● The men jog around the lake.
○ We row the boat on the lake.

5. ○ The lady has a nice dress.
● That lady showed us her baby.

6. ○ Trains have wheels and brakes.
● We were lucky to see a rainbow.

Wow!
You have finished.
What a great job!

Integrated Curriculum

WRITING CONNECTION

TWISTERS! Have children work together to write or dictate a list of words with ***/ō/o, ow*** and ***/ē/ey, y***. Then help children to write or dictate tongue-twister sentences such as ***Slow crows fly low in snow***. Display the sentences, then challenge children to say them two or three times as fast as possible without making a mistake.

TECHNOLOGY CONNECTION

TECHNOLOGY TIME Have children use the Magnet Board to build words with ***/ō/oa, o, ow***. Suggest that they begin with one of these words: ***coat, go,*** or ***slow***.

I CAN READ! OPTIONS

The I Can Read! page can be used for one or all of the following:

- paired reading
- individual assessment
- choral reading
- homework practice
- program placement

CHALLENGE

Challenge children to classify the I Can Read! words into these categories: Words That Name People and Animals; Words That Name Actions; Words That Name Places and Things; and All Other Words. Draw a large four-column chart on the chalkboard. Invite children to write the words in the appropriate columns.

ESL To help children acquiring English participate with greater confidence in the Review activities, go over the key sound-spellings, decodable words, and high-frequency words beforehand. For the Build Words and Build Sentences activities, you may wish to pair these children with those whose primary language is English, then let the same partners work together to complete the pages.

Lesson 152

pages 243-244

Read Words in Context

TEACH

Assemble the Story Ask children to remove pages 243–244. Have them fold the pages in half to form the Take-Home Book.

Preview the Story Preview *All About Seeds,* a nonfiction selection about how plants grow. Invite children to browse through the first two pages of the selection and to comment on anything they notice. Suggest that they point out any unfamiliar words. Read these words aloud. Have children repeat them. Then have children predict what the selection might be about.

READ AND WRITE

Read the Story Read the story aloud, or have volunteers take turns reading aloud a page at a time. Discuss anything of interest on each page, and encourage children to help each other with any blending difficulties. The following prompts may help children who need extra support while reading:

- **What letter sounds do you know in the word?**
- **Are there any word parts you know?**

Reflect and Respond Have children share their reactions to the story. What did they learn about seeds? What did they already know that the story told them? Encourage children to think of one more thing they would like to learn about seeds and plants.

Develop Fluency You may wish to reread the story as a choral reading or have partners reread the story independently. Provide time for children to reread the story on subsequent days to develop fluency and increase reading rate.

Have children share how they figure out unfamiliar words. Continue to review challenging or confusing sound-spelling relationships for children needing additional support.

ALL ABOUT SEEDS

Look at the grapes on that vine! The vine and grapes began as a seed.

1

Copyright © Scholastic Inc.

Many plants come from seeds. Look at the plants growing in rows. They were seeds at one time. They will make more seeds. That way, more plants can grow.

Why are seeds important?

Many plants come from seeds.

4

Supporting All Learners

KINESTHETIC LEARNERS

WORD SCRAMBLE Use the pocket ABC cards and pocket chart from the Phonics and Word Building Kit to build the following scrambled words: ***og, yek, wosh, toba, yonp, nomyek, oact, os,*** and ***wol.*** Have children unscramble the words to make words with ***/ō/o, ow, oa;*** and ***/ē/ey, y.***

AUDITORY LEARNERS

CLIMB THE LADDER! Draw a large ladder on chart paper. Invite children to "climb" the ladder by naming words that begin with ***/ō/o, ow, oa*** or ***/ē/ey, y.*** Write the words on the ladder, one word per rung, beginning with the bottom rung, and working toward the top. Remind children to look for words with ***/ō/o, ow, oa*** and ***/ē/ey, y*** when they read. Make a second word ladder on which to add the words children find.

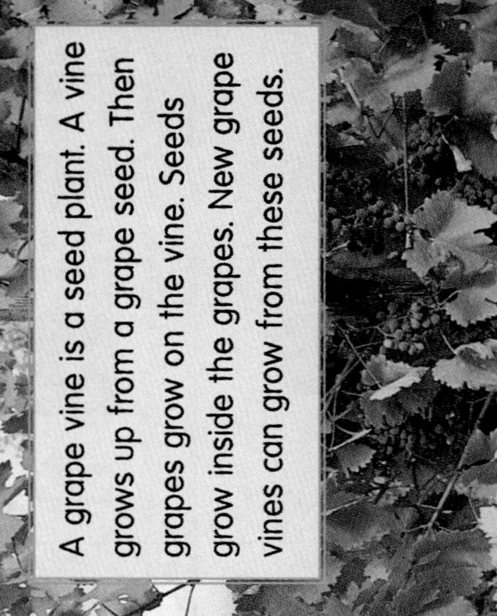

Seeds can be many sizes and shapes. Some seeds grow big and get fluffy tops. They can blow away in the wind. The seeds have shells that help them float to a place to grow.

2

Reflect On Reading

ASSESS COMPREHENSION

ONGOING ASSESSMENT To assess their understanding of the story, ask children questions such as:

- **Why do some seeds get fluffy tops?** ***(The fluffy tops help the seeds float. They float to new places to grow.)***
- **Where does a grape seed grow?** ***(inside the grape)***
- **Where do new seeds come from?** ***(Plants make their own seeds.)***

HOME-SCHOOL CONNECTION

HOME/SCHOOL CONNECTION Send home *All About Seeds*. Encourage children to read the story to a family member.

WRITING CONNECTION

WRITING EXTENSION Have children generate a list of **/ō/o, ow, oa** and **/ē/ey, y** words. Record these words on the chalkboard. Then have children create a story using as many of these words as they can. To begin, suggest a title such as "The Boat Ride" or "The Monkey." Write the dictated story on chart paper for group and individual reading. Return to the story, rereading it in subsequent lessons.

EXTRA HELP

For additional practice with **/ō/o, ow, oa** and **/ē/ey, y** have children read the following My Books: *Don't Go So Slow!*; *Monkey See, Monkey Do*; *Rabbit's New Coat*; and *The Pet Show*. For information on using the My Books on the computer, see the WiggleWorks Plus My Books Teaching Plan.

CHALLENGE

Have children find words with **/ō/o, ow, oa** and **/ē/ey, y** in the classroom or school. Children can count how many words they found and record the results in a chart such as the one shown.

4			
3			
2			
1			
	Words With ***oa***	Words With ***o, ow***	Words With ***ey, y***

Phonics Connection

Phonics Readers:
#55, *Follow It!*
#56, *Bo's Bows*
#57, *Baby Pig at School*
#58, *Donkey and Monkey*
#59, *The Hungry Toad*
#60, *Goat's Book*

Lesson 153

pages 245-246

Blend Words With /o͝o/oo

QUICKCHECK ✔

Can children:

✔ substitute sounds in words?

✔ recognize /o͝o/?

✔ blend words with /o͝o/oo?

If YES go to Read and Write.

TEACH

Develop Phonemic Awareness

Phonemic Manipulation Explain to children that they are going to listen to nonsense words with /o/ as in ***bok***. Children are to replace the short *o* sound with /o͝o/. For example, if you say ***bok***, children will say ***book***. Use these and other nonsense words: ***hok, fot, wod, hod, lok, tok, shok, stod***.

Connect Sound-Symbol Write ***hot*** and ***hoof*** on the chalkboard as you read each aloud. Ask children what sounds they hear in the middle of ***hot*** and ***hoof*** and how each sound is different. Point out that /o/ in the middle of ***hot*** is spelled with an *o*, and /o͝o/ in the middle of ***hoof*** is spelled *oo*.

Ask children to suggest other words that contain /o͝o/. List these words on the chalkboard. Have volunteers circle the letters *oo* in each word that contains this spelling.

READ AND WRITE

Blend Words List the following words and sentences. Have volunteers read each aloud.

- look stood book took
- foot wood good shook
- Look at the big pile of wood.
- She stood in back of the cook.

Complete Activity Pages Read aloud the directions on pages 245–246. Review each picture name with children.

LISTEN AND WRITE

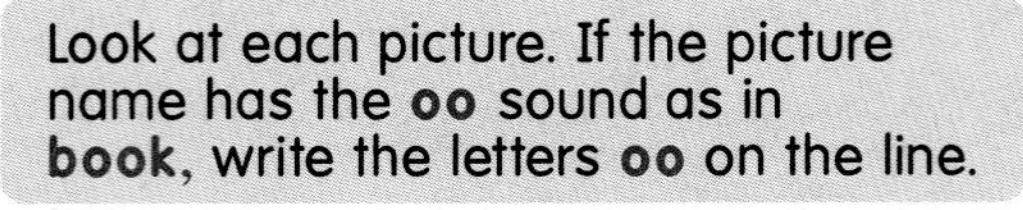

Look at each picture. If the picture name has the **oo** sound as in **book**, write the letters **oo** on the line.

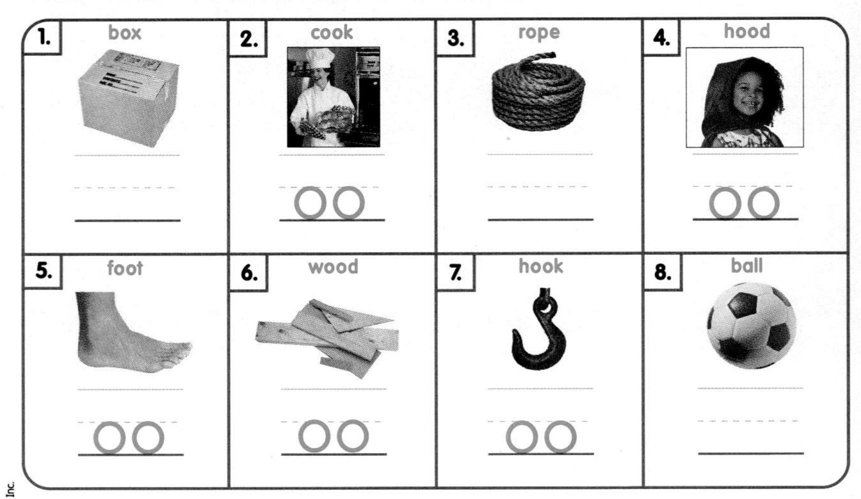

Write the letters **oo** to finish each word. Circle the word that tells about the picture.

Supporting All Learners

AUDITORY/VISUAL LEARNERS

LOOK FOR WORDS Read aloud "Teddy Bear Picnic" on pages 24–25 in the *Big Book of Rhymes and Rhythms, 1B*. Have children find all the words with **/o͝o/**. Write these words on the chalkboard. Invite volunteers to add other **/o͝o/** words to the list.

VISUAL LEARNERS

WEBS To extend meaning and to provide practice with familiar words, create webs on the chalkboard. For example, write the word ***cook*** in the center of chart paper. Have children suggest things people cook or things cooks use. Write these suggestions around ***cook*** to make a web. Repeat the activity with the word ***look*** by having children suggest places to look for something that is lost or times to look and listen for safety.

Integrated Curriculum

WRITING CONNECTION

COOKBOOKS Have children write sentences using the word ***cook*** and other words with ***/o͞o/oo,*** one sentence per page. For example, children might write ***I like to cook***. Then children can illustrate their sentences.

SOCIAL STUDIES CONNECTION

WORD SEARCH Have children look through books, magazines, and newspapers for words with ***/o͞o/oo***. Have them list the words they find on a sheet of paper. Then have children categorize the words—animals, objects, people, actions.

Phonics Connection

Literacy Place: *Hometowns*
Teacher's SourceBook, pp. T60–61;
Literacy-at-Work Book, pp. 9–10

My Book: *Little Red Riding Hood and the Good Wolf*

EXTRA HELP

Draw and cut out a large foot shape. Write the letters ***oo*** and the word ***foot*** on the shape. Ask children to draw and cut out their own foot shapes. Have partners suggest words with ***/o͞o/oo*** and write them on foot shapes, one word per shape. Remind children to look for words with ***/o͞o/oo*** during their reading. Add these words to the foot shapes. You may wish to display the feet "walking" around the room.

Invite children acquiring English to draw pictures and label them for any of the following words: ***wood, foot, cook, hook, hood, book***. Children may also write a sentence telling about their pictures.

- **Everybody Cooks Rice**
  by Norah Dooley
  Carrie's neighbors cook rice. **(/o͞o/oo)**
- **Good News**
  by Barbara Brenner
  Good news spreads fast. **(/o͞o/oo)**

Lesson 154

pages 247-248

Blend Words With /o͞o/oo

QUICKCHECK ✔

Can children:

✔ **orally blend word parts?**

✔ **recognize /o͞o/?**

✔ **blend words with /o͞o/oo?**

If YES go to Read and Write.

TEACH

Develop Phonemic Awareness

Oral Blending Explain to children that you are going to say words by saying their parts. Children are to listen to the parts and then say the whole word. For example, say the sounds /**r**/ /o͞o/ /**m**/. Guide children to say the whole word ***room***. Continue with the following word parts.

/k/ /o͞o/ /l/	**/m/ /o͞o/ /n/**
/br/ /o͞o/ /m/	**/t/ /o͞o/ /l/**
/h/ /o͞o/ /t/	**/z/ /o͞o/ /m/**

Connect Sound-Symbol Write the words ***hop*** and ***hoop*** on the chalkboard as you read each aloud. Ask children what sounds they hear in the middle of each word and how the words are different. Point out that /**o**/ in the middle of ***hop*** is spelled with an ***o***, and /o͞o/ in the middle of ***hoop*** is spelled ***oo***.

Ask children to suggest other words that contain /o͞o/. List these words on the chalkboard. Have volunteers circle the letters *oo* in each.

READ AND WRITE

Blend Words List the following words and sentences on a chart. Have volunteers read each aloud.

- **room** **bloom** **moon** **spoon**
- **toot** **pool** **school** **tooth**
- **We hope to go in the pool soon.**
- **She lost her tooth at school.**

Complete Activity Pages Read aloud the directions on pages 247–248. Review each picture name with children.

LISTEN AND WRITE

Name

Look at each picture. If the picture name has the **oo** sound as in **spoon**, write the letters **oo** on the line.

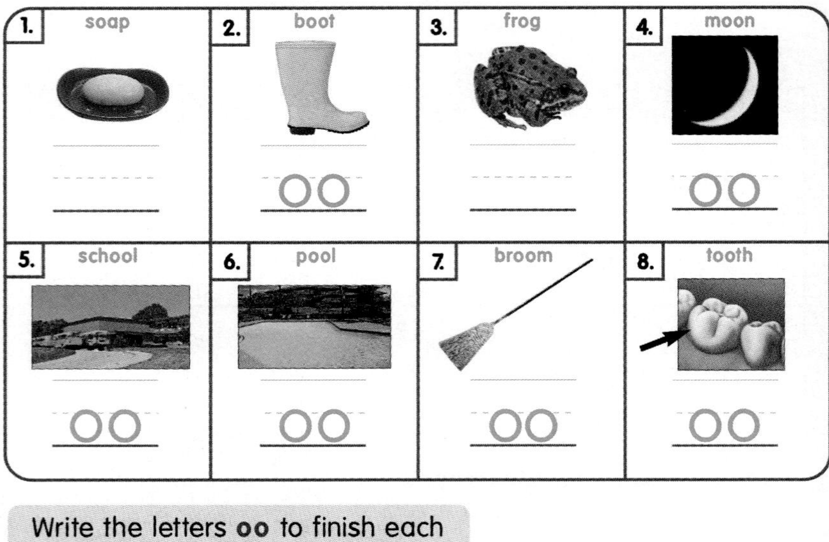

Write the letters **oo** to finish each word. Read the words to a friend.

9. z oo m 10. r oo m 11. r oo t

Supporting All Learners

VISUAL/AUDITORY LEARNERS

WORD HUNT Read aloud "The Teddy Bear Picnic" on pages 24–25 in the *Big Book of Rhymes and Rhythms, 1B*. Have children find all the words with **/o͞o/** in the rhyme. Write these words on the chalkboard. Invite volunteers to add other **/o͞o/** words to the list.

AUDITORY LEARNERS

TIME TO LISTEN Have children listen for vowel sounds in "Who Took the Cookie From the Cookie Jar?" on the Sounds of Phonics Audiocassette (Personal Voice). Invite children to identify all the words they hear that contain **/o͞o/** and **/ē/**, as well as other vowel sounds they hear in the various first names in the song. Write the words that children name, and then read the list, emphasizing the vowel sounds in each.

Choose a word from the box to answer each riddle. Write the words on the lines.

pool	tool
broom	bedroom
tooth	spoon

What am I?

1. You sweep with me.	broom
2. You swim in me.	pool
3. You use me to pick up food.	spoon
4. You can fix a clock with me.	tool
5. You bite with me.	tooth
6. You sleep in me.	bedroom

Integrated Curriculum

SPELLING CONNECTION

DICTATE! Have children dictate sentences using words with **/o͞o/oo**. Write these sentences, and have children underline the words with **/o͞o/oo**.

SOCIAL STUDIES CONNECTION

Have children draw pictures of their favorite room at home. Have them label things in the rooms and then write a caption using words with **/o͞o/oo**.

Phonics Connection

Literacy Place: *Hometowns*
Teacher's SourceBook, pp. T60–61;
Literacy-at-Work Book, pp. 9–10

My Book: *In Kalamazoo*

Big Book of Rhymes and Rhythms, 1B: "The Teddy Bear Picnic," pp. 24–25

Chapter Book: *A Bag of Tricks*, Chapter 1

CHALLENGE

Have children make up rhyming sentences using words with **/o͞o/oo**. For example: ***We go to the pool after school***. Then children can trade their sentences and challenge each other to find the rhyming words.

EXTRA HELP

For additional review of words with **/o͞o/oo** and **/o͝o/oo,** children may enjoy playing "Vowel Concentration" on pages 44-45 in *Quick-and-Easy Learning Games: Phonics.*

- **Sally's Room**
  by M. K. Brown
  A girl's messy room fights back. **(/o͞o/oo)**
- **Too Many Babas**
  by Carolyn Croll
  A folktale from Russia. **(/o͞o/oo)**

Lesson 155

page 249

Blend and Build Words With Phonograms

QUICKCHECK ✔

Can children:

✔ **blend words with phonograms *-ink, -ing*?**

✔ **build words with *-ink, -ing*?**

If YES go to Read and Write.

TEACH

Introduce the Phonograms Write *-ink* and *-ing* on the chalkboard. Point out the sounds these phonograms stand for. Add *s* to the beginning of *-ink,* and model how to blend the word formed. Have children repeat ***sink*** aloud as you blend the word again. Then add ***r*** to the beginning of *-ing* and model how to blend ***ring***.

Ask children to suggest words that rhyme with ***sink*** and ***ring***. List these words on the chalkboard in separate columns. Have volunteers underline *-ink* or *-ing* in each. Be sure to include ***think, wink, drink, pink, sing, king, bring,*** and ***wing***. Blend these words for children as they repeat them aloud. Encourage children to look for the word parts *-ink* and *-ing* as they read.

READ AND WRITE

Build Words Write *-ink* and *-ing* on the chalkboard. (If available, use a pocket chart and letter cards.) Have children add a letter to the beginning of each phonogram to make a new word. Continue by having children replace the initial consonant in the first word to build a second word. Children may build ***think*** from ***sink,*** and ***thing*** from ***think***.

Complete Activity Page Read aloud the directions on page 249. Have children complete the page independently.

Name

Add each letter or letters to the word part below it. Blend the word. If it is a real word, write it on the line.

s f g k h br w	b p h dr w th y
__ing	__ink
1. sing	5. pink
2. king	6. drink
3. bring	7. wink
4. wing	8. think

Write a sentence using one of the words you made.

9. Answers will vary.

Supporting All Learners

VISUAL/AUDITORY LEARNERS

FLIP BOOKS Have children make flip books with phonograms, consonants, and blends. A flip book for the phonogram ***-ing*** can begin with the word ***ring***. Under the ***r***, children can put ***s, k, th, br, w, st,*** and perhaps ***str***. The ***-ing*** remains constant. Children may take turns reading the words in their flip books and then make up sentences using these words.

TACTILE LEARNERS

MOLDING WORDS Have children form the phonograms ***-ink*** and ***-ing*** from clay or pipe cleaners. Children can display these word parts on a table or desk top. Then have children form letters to add to these phonograms to make words. Write a list of the words children make. Have children trace each word as they say it aloud.

EXTRA HELP

Children who need additional practice with the phonograms ***-ink*** and ***-ing*** can read Phonics Reader #65, *Clay Things, Play Things!* aloud in small groups or with partners.

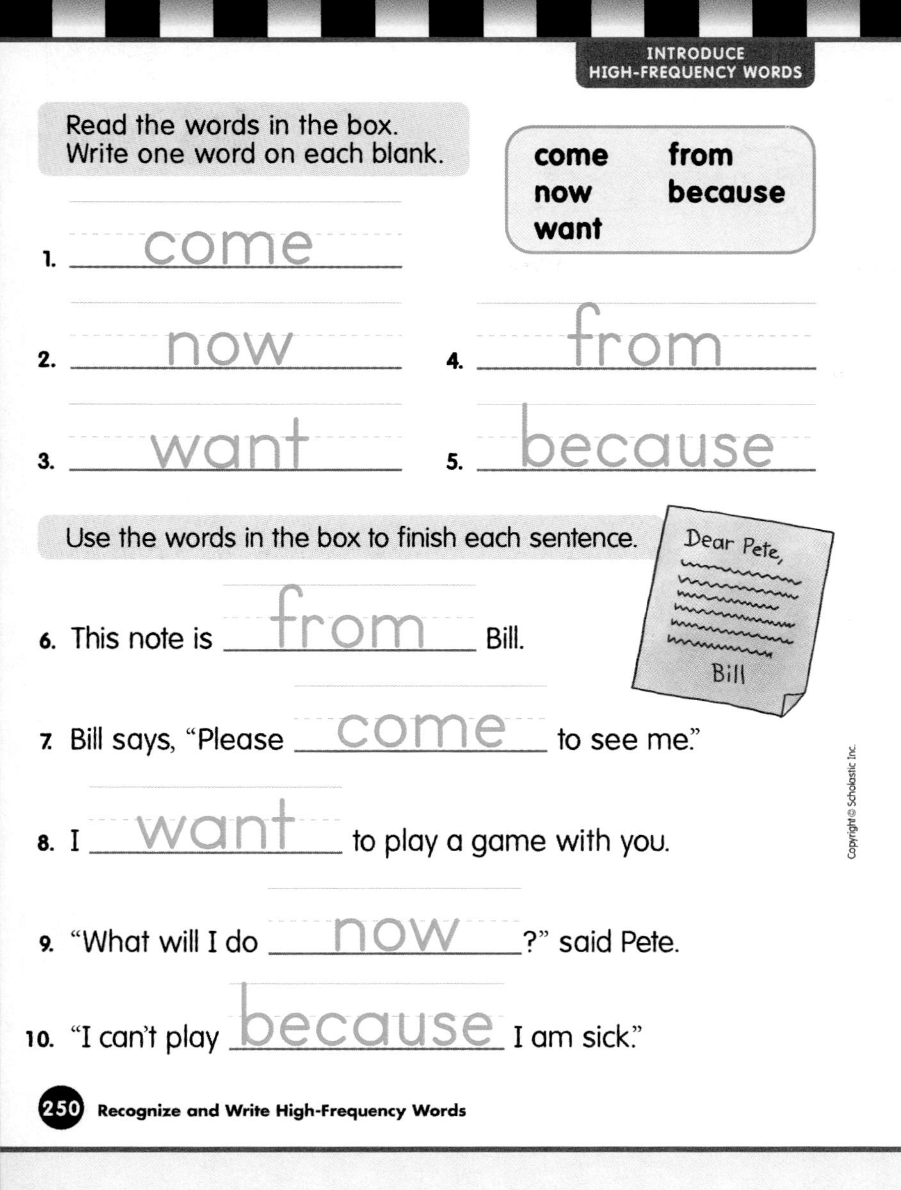
INTRODUCE HIGH-FREQUENCY WORDS

Read the words in the box. Write one word on each blank.

come	from
now	because
want	

1. come
2. now
3. want
4. from
5. because

Use the words in the box to finish each sentence.

6. This note is from Bill.
7. Bill says, "Please come to see me."
8. I want to play a game with you.
9. "What will I do now?" said Pete.
10. "I can't play because I am sick."

Supporting All Learners

KINESTHETIC/VISUAL LEARNERS

LEARNING CENTER For additional practice with high-frequency words, set up Center 8, Activity 2, in *Quick-and-Easy Learning Centers: Phonics*. This activity focuses on writing.

VISUAL/AUDITORY LEARNERS

SENTENCE FUN Ask children to dictate sentences using high-frequency words. You may wish to begin with the following sentence starters: ***They want a*** ___ or ***Can you come*** ___**?** Write children's sentences on the chalkboard, and have a volunteer circle the high-frequency words in each.

EXTRA HELP

Write the high-frequency words you have taught so far on index cards, one word per card. Have partners use the cards as flash cards. Have children read each word, use it in a sentence, and then, without looking at it, spell it.

page 250

Recognize High-Frequency Words

QUICKCHECK ✔

Can children:

✔ **recognize and write the high-frequency words *come, now, want, because, from*?**

✔ **complete sentences using the high-frequency words?**

If YES go to Read and Write.

TEACH

Introduce the High-Frequency Words

Write the high-frequency words *come, now, want, because,* and *from* in sentences. Read aloud the sentences, underline the words, and ask children if they recognize them. If necessary, read the sentences again. You can use these sentences:

1. **Where do all the weeds come from?**
2. **We want to get rid of them now.**
3. **We pick them because it will help other plants grow.**

READ AND WRITE

Practice Write each high-frequency word on a note card. Read each word aloud as you display the cards. Then do the following:

- **Mix the cards.**
- **Display one card at a time, and ask children to state each word aloud.**
- **Have children spell each word aloud, clapping on each letter.**
- **Ask children to write each word in the air as they state aloud each letter. Then have them write each word on a sheet of paper.**

Complete Activity Page Read aloud the directions on page 250. Have children complete the page independently.

Lesson 157

pages 251-252

Blend Words With /ou/ou, ow

QUICKCHECK ✔

Can children:

- ✓ **orally segment words?**
- ✓ **recognize /ou/?**
- ✓ **blend words with /ou/ou, ow?**

If YES go to Read and Write.

TEACH

Develop Phonemic Awareness

Oral Segmentation Explain to children that you are going to say a word. Children will tell you how many sounds they hear in the word. For example, say the word ***town***. Then say the word sound by sound: **/t/ /ou/ /n/**. Guide children to understand that the word ***town*** contains three separate sounds. Continue with the following words:

- now
- down
- brown
- mouth
- clown
- crown
- how
- loud
- shout

Connect Sound-Symbol Write the words ***dot*** and ***down*** on the chalkboard as you read each word aloud. Ask children what sound they hear in the middle of each word and how the words are different. Point out that **/o/** in the middle of ***dot*** is spelled with an ***o***, and **/ou/** in the middle of ***down*** is spelled ***ow***. Continue with the words ***shot*** and ***shout***. Point out that **/ou/** in the middle of ***shout*** is spelled ***ou***.

READ AND WRITE

BUILT-IN REVIEW

Blend Words List the following words and sentences on the chalkboard. Have volunteers read each aloud.

- now how brown town
- out loud shout cloud
- How do we get down these steps?
- A clown does not make me frown!

Complete Activity Pages Read aloud the directions on pages 251–252. Review each picture name with children.

LISTEN AND WRITE

Name

house

If the picture name has the **ou** sound as in **cow**, write the letters **ow** on the line.

1. clown	2. coat	3. brown	4. owl
ow		ow	ow

5. flower	6. cone	7. crown	8. rose
ow		ow	

If the picture name has the **ou** sound as in **house**, write the letters **ou** on the line.

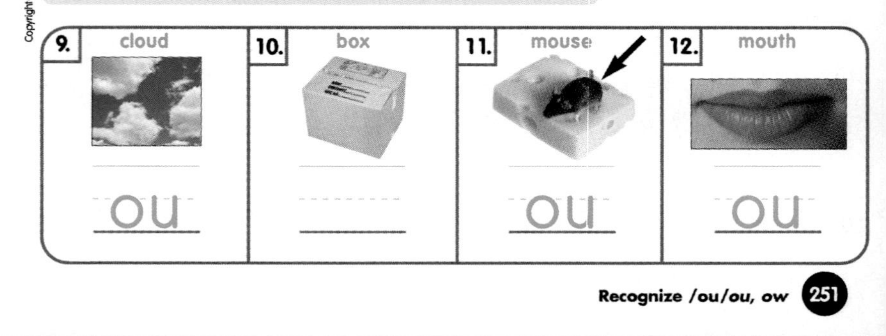

Supporting All Learners

VISUAL/AUDITORY LEARNERS

WORD HUNT! Read aloud "Eensy Teensy Spider" on page 26 in the *Big Book of Rhymes and Rhythms, 1B*. Have children find all the words with **/ou/** in the rhyme. Write these words on the chalkboard. Invite volunteers to add other **/ou/** words to the list.

KINESTHETIC LEARNERS

WORD CARD FUN Write the following words on word cards: ***cow, brown, sound, round, town, crown, out, shout, toot, how, pool, shoot, zoom, room, school,*** and ***tool***. Mix up the cards, and have partners sort the cards into two piles according to their vowel sounds. Then challenge children to sort the cards into rhyming pairs or groups.

VISUAL LEARNERS

MAKE A LIST Ask children to suggest other words that contain **/ou/**. List these words on the chalkboard. Have volunteers circle the letters ***ow*** or ***ou*** in each.

BLEND WORDS

Find a word in the box that rhymes with each word. Then write the word on the line.

house shout down cloud how found

1. cow how
2. loud cloud
3. out shout
4. mouse house
5. round found
6. brown down

Use one of the words from above to finish each sentence.

7. Was the dog black or brown?
8. Who found the dime that I lost?
9. Did you fall down?

Integrated Curriculum

SOCIAL STUDIES CONNECTION

AROUND TOWN Help children to make a map of their town on large sheets of paper. Children can color places where there might be grass or ponds and use collage materials to add details to the map. Have children label the buildings, streets, and other areas of the map.

Phonics Connection

Literacy Place: *Hometowns*
Teacher's SourceBook, pp. T100–101;
Literacy-at-Work Book, pp. 19–20

My Book: *How Fox Got Lost*

Big Book of Rhymes and Rhythms, 1B: "Eensy Teensy Spider" p. 26

Chapter Book: *A Bag of Tricks*, Chapter 2

ESL To help children acquiring English learn new words, create relationships with known words. For example, if children are familiar (at least orally) with the words ***up, soft, later,*** and ***whisper,*** these words can be employed to help create a semantic map for their opposites—***down, hard, now,*** and ***shout***. Acting out pairs of words—such as ***shout*** and ***whisper, hard*** and ***soft,*** and ***up*** and ***down***—will also help children learn the words in a meaningful way.

CHALLENGE

Have children look through books, magazines, and newspapers for words with ***/ou/ou, ow***. Have children list the words they find on a sheet of paper. Then have children use these words to write or dictate a story. When the stories are finished, give children an opportunity to share them.

EXTRA HELP

Write these word parts on the chalkboard: ***d__n, l__d, cl__d, n__, r__nd, c___, h__, s__nd***. Have volunteers add the letters ***ou*** or ***ow*** to each. Ask children to say each word and write it. Then have children dictate sentences using the words.

- **Baseball Ballerina**
  by Kathryn Cristaldi
  This little girl prefers baseball to ballet.
  (/ou/ou, ow)

- **The Cow Who Wouldn't Come Down**
  by Paul Brett Johnson
  Miss Rosemary's cow learns to fly.
  (/ou/ou, ow)

Lesson 158

page 253

Write and Read Words With /o͞o/oo, /o͝o/oo, /ou/ou, ow

QUICKCHECK ✔

Can children:

✓ **spell words with /o͞o/oo, /o͝o/oo, /ou/ou, ow?**

✓ **read a poem?**

If YES go to Read and Write.

TEACH

Link to Spelling Review with children the following sound-spelling relationships: */o͞o/oo, /o͝o/oo, /ou/ou, ow*. Say /o͞o/ and /o͝o/. Have a volunteer write on the chalkboard the spelling that stands for these sounds.

Phonemic Awareness Oral Segmentation

Say ***root***. Have children orally segment the word. (/r/ /o͞o/ /t/) Ask them how many sounds the word contains. *(3)* Draw three connected boxes on the chalkboard, and have a volunteer write the spelling that stands for each sound in the appropriate box. Continue with ***loud*** and ***down***.

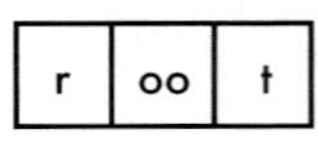

r	oo	t

READ AND WRITE

Dictate Dictate the following words and sentence. Have children write them. Then write the words and sentence on the chalkboard. Have children make any corrections on their papers.

- **how** **mouth** **tooth** **tool**
- **they** **from** **come** **could**
- **Where did all that good food come from?**

Complete Activity Page Read aloud the directions on page 253. Have children read the page independently or with partners.

POETRY

Name

Read the poem. Finish each sentence using a word from the poem.

Look Around

Brown Cow stood up.
Pink Pig sat down.

Brown Cow ran to school.
Pink Pig ran to town.

Brown Cow met a king.
Pink Pig met a clown.

Did you see Brown Cow?
Did you see Pink Pig?

Look around.
Look around.

1. Brown cow ran to school.

2. Pink pig met a clown.

Link to Spelling 253

Supporting All Learners

AUDITORY/VISUAL LEARNERS

GAME TIME To practice and review words with ***/o͞o/oo, /o͝o/oo,*** and ***/ou/ou, ow,*** children may enjoy playing "Frog Hop" on pages 16–20 in *Quick-and-Easy Learning Games: Phonics*. As suggested, you can adapt the game for play with different vowel sounds. Among the many words you may use are ***crown, cow, now, shout, mouth, cloud, tooth, tool, broom, wood, hook,*** and ***foot***.

CHALLENGE

Have children generate a list of words that have ***/ou/ou, ow*** in the middle. Record these words on the chalkboard. Then have children create a story using as many ***/ou/*** words on the chalkboard as they can. To begin, suggest a title such as "The Brown Cow," "The Clown," or "My House." Write the dictated story on chart paper for group and individual reading. Return to the story, rereading it in subsequent lessons.

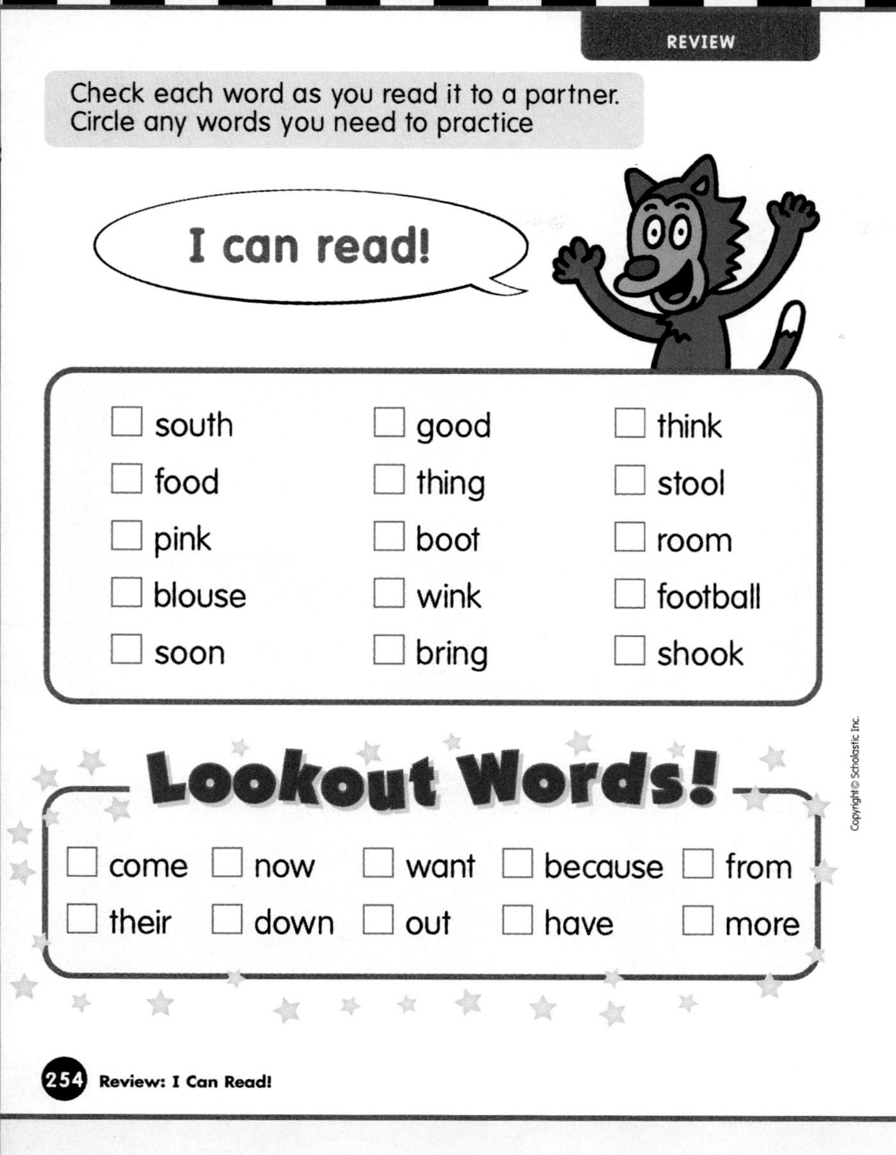

Supporting All Learners

AUDITORY LEARNERS

RHYMING WORDS Invite children to generate other words that rhyme with each I Can Read! word. For example, they may list ***mouth,*** which rhymes with ***south,*** and ***stood,*** which rhymes with ***good***. Have children read their lists to one another.

EXTRA HELP

Children can work together to use the magnetic letters from the Phonics and Word Building Kit to build words with ***/o͞o/oo, /o͝o/oo,*** and ***/ou/ou, ow***.

I CAN READ! OPTIONS

The I Can Read! page can be used for one or all of the following:

- **paired reading**
- **individual assessment**
- **choral reading**
- **homework practice**
- **program placement**

Read and Review Words With /o͞o/oo, /o͝o/oo, /ou/ou, ow

QUICKCHECK ✓

Can children:

✓ **read words with /o͞o/oo, /o͝o/oo, /ou/ou, ow?**

✓ **recognize high-frequency words?**

If YES go to Read and Write.

TEACH

Review Sound-Spellings Review these sound-spelling relationships: /o͞o/*oo*, /o͝o/*oo*, and /ou/*ou, ow*. Say one of these sounds. Have a volunteer write the spelling or spellings that stand for the sound. Then display pictures of objects whose names contain these sounds. Have children write the spelling pattern that each picture's name contains.

Review High-Frequency Words Review the high-frequency words ***come, now, want, because,*** and ***from***. Write sentences containing each high-frequency word. Say one word, and have volunteers circle it in the sentences. Have children generate sentences for each word.

READ AND WRITE

Build Words Distribute these letter cards: ***b, c, k, l, o, o***. Have children build as many words as possible using the letter cards and record their words. Then have children compare their lists. Continue building words with ***b, m, o, o, r, t; c, d, l, o, u;*** and ***b, c, n, o, r, w***.

Build Sentences Write these words on note cards: ***I, cloud, clown, go, room, pool, to, a, brown, cow, see, the***. Have children make sentences using them.

Complete Activity Page Read aloud the directions on page 254. Children can complete the page independently.

Lesson 160

pages 255-256

Read Words in Context

TEACH

Assemble the Story Ask children to remove pages 255–256. Have them fold the pages in half to form the Take-Home Book.

Preview the Story Preview *How to Make a Cake,* a story about Cow's humorous attempts at baking a cake. Invite children to browse through the first two pages of the story and to comment on anything they notice. Suggest that they point out any unfamiliar words. Read these words aloud as children repeat them. Then have children predict what they think the selection might be about.

READ AND WRITE

Read the Story Read the story aloud, or have volunteers take turns reading aloud a page at a time. Discuss with children anything of interest on each page, and encourage them to help each other with any blending difficulties. The following prompts may help children who need extra support while reading:

- **What letter sounds do you know in the word?**
- **What do the pictures tell you about the story?**

Reflect and Respond Have children share their reactions to *How to Make a Cake.* Does it remind them of other stories they have read? What did they laugh at in the story? Encourage children to think of advice to give to Cow and Mouse for the next time they bake.

Develop Fluency Reread the story as a choral reading, or have partners reread the story. Provide time for children to reread the story on subsequent days to develop fluency and increase reading rate.

Have children share how they figure out unfamiliar words when reading other stories, and model how they blend words. Continue to review confusing or challenging sound-spelling relationships for children needing additional support.

Supporting All Learners

KINESTHETIC LEARNERS

READERS THEATER Invite partners to act out the story. You may provide simple props such as a measuring cup, bowl, spoon, cake pan, and pot holders. Children may memorize some of the dialogue or read it from the book so that they say it exactly as it is given in the story. Some children may enjoy creating new dialogue to add to the story.

AUDITORY LEARNERS

READ ALOUD TIME Have children take turns reading aloud *How to Make a Cake*. Then have them talk about Cow and Mouse and how they solved the problems in the story. What did they do right? What should they have done differently? Have children also look at the art and talk about how it helped them to understand the story.

But Cow had the cookbook upside down. She could not read it. So Cow made it up!

Cow said, "Get five cups of mud. Get some nuts, too. Drop in some sticks. Mix well with a big spoon. Now bake. Then let it cool."

2

Cow put all of the mix into a big pan. Then Cow put the pan in to bake.

Cow and Mouse sat down to wait. "I like good food," said Cow. "I hope the cake comes out well."

3

Reflect On Reading

ASSESS COMPREHENSION

To assess their understanding of the story, ask children questions such as:

- **Why can't Cow read the cookbook?** ***(She holds it upside down.)***
- **Who makes up the recipe?** ***(Cow)***
- **What does the cake look like when it is done?** ***(a mud house)***

HOME-SCHOOL CONNECTION

Send home *How to Make a Cake.* Encourage children to read the story to a family member.

WRITING CONNECTION

MORE ABOUT COW AND MOUSE

Suggest that children write a story about Cow and Mouse preparing lunch or dinner for friends. Have children talk about what Cow and Mouse will cook and how they will cook it. Then have children take turns writing or dictating sentences to add to the story. You may wish to begin the story with a sentence such as ***Cow and Mouse wanted to make dinner for Pig and Horse.*** When the story is completed, read it aloud with children. Give them an opportunity to revise the story, if necessary.

EXTRA HELP

For additional practice with **/o͞o/oo, /o͝o/oo,** and **/ou/ou, ow,** have children read the following My Books: *Kalamazoo, Sleeping Out,* and *How Fox Got Lost*. For information on using the My Books on the computer, see the WiggleWorks Plus My Books Teaching Plan.

CHALLENGE

Have children write or dictate another story about Cow and Mouse. You may wish to suggest that children write about how Cow and Mouse clean Cow's room or how Cow and Mouse make an art project. Encourage children to illustrate their stories.

Phonics Connection

Phonics Readers:

#61, *Pizza Cook*
#62, *Root for the Team*
#63, *All Around the Farm*
#64, *The Proud Mayor*

Lesson 161

pages 257-258

Recognize and Write Words With *-ed*

QUICKCHECK ✔

Can children:

✔ recognize *-ed* verb endings?

✔ write *-ed* verb endings?

If YES go to Read and Write.

TEACH

Introduce *-ed* Remind children that words that tell about actions that already happened sometimes end with *-ed*. Often these words are formed by just adding *-ed* to a word that names an action. Write the word ***fixed*** on the chalkboard. Have a volunteer underline its two parts: the word ***fix*** and the ending *-ed*.

Explain to children that sometimes when you add *-ed* to a word, it creates a new syllable. Say the word ***fixed*** again. Note that it has one syllable. Then say the word ***painted***. Write the word on the chalkboard. Have a volunteer underline its two parts: the word ***paint*** and the ending *-ed*. Note that this word has two syllables.

Then write the following words on the chalkboard: ***licked, loaded, filled, planted, stayed, hooted, toasted, needed,*** and ***picked***. Read each word, and have children clap the number of syllables they hear in each. Note **/t/** and **/d/** that precede *-ed* in the words that have two syllables.

READ AND WRITE

Practice Ask children to add *-ed* to the following words: ***rest, play, rain, snow, work, rent, wait, roast, shout,*** and ***look***. Have children write the words and then read them to a friend. Then work as a class to add *-ed* to each word and to read the words.

Complete Activity Pages Read aloud the directions on pages 257–258. Children can complete the pages independently.

WORD ENDINGS

Name

Finish the letter with words from the box. Write the words on the lines.

planted **rained** **stayed** **played** **painted**

To Grandma,

We had fun outside. Dad painted the house. I played ball with Pam. Mom planted an oak tree. We stayed outside until it rained!

Bill

Supporting All Learners

AUDITORY/VISUAL LEARNERS

READ AND FIND Read aloud "There Was a Crooked Man" on page 27 in the *Big Book of Rhymes and Rhythms, 1B*. Have children find all the words with ***-ed*** in the rhyme. Write these words on the chalkboard. Invite volunteers to add other ***-ed*** words to the list.

VISUAL LEARNERS

GAME TIME To provide additional practice with ***-ed*** and ***-s,*** set up Center 3, Activity 1, "Sign Collection" on page 24 in *Quick-and-Easy Learning Centers: Phonics*. This center focuses on spelling patterns and word endings.

KINESTHETIC LEARNERS

MAKING *-ed* WORDS To review ***-ed,*** distribute the pocket ABC cards and pocket chart from the Phonics and Word Building Kit. Help children form ***-ed*** words. Write the following words on the chalkboard: ***need, pick, plant, rent, snow, play, wait, fix, paint,*** and ***stay***. Have children place letter cards in front of ***-ed*** to make the words on the chalkboard.

Add **-ed** to the words in the box. Write the words on the lines. Then read the sentences to a friend.

1.	look	We looked for the lost cat.
2.	help	Al and Rose helped Mom to cook.
3.	wait	All of us waited for the bus.
4.	shout	Bill shouted, "Get a home run!"
5.	train	We trained the dog to sit up.

Write a sentence using one of the words you wrote.

6. Answers will vary.

Integrated Curriculum

SPELLING CONNECTION

DICTATION TIME! Have children dictate sentences with and without words with ***-ed*** verb endings. Write the sentences on the chalkboard or on chart paper. Then have children underline each ***-ed*** ending. Give children an opportunity to discuss how the ***-ed*** verb ending changes the meaning of each word.

TECHNOLOGY CONNECTION

MY BOOKS/TECHNOLOGY Have children read the My Book *The Fixed-Up Park* to get additional practice with ***-ed*** endings. For information on using the My Books on the computer, see the WiggleWorks Plus My Books Teaching Plan.

Phonics Connection

Literacy Place: *Hometowns*
Teacher's SourceBook, pp. T198–199; Literacy-at-Work Book, pp. 39–40

My Book: *The Fixed-Up Park*

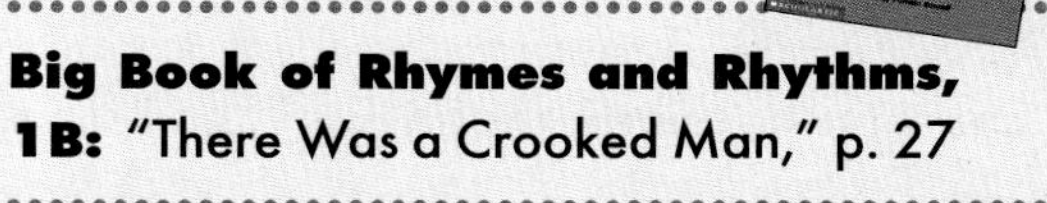

Big Book of Rhymes and Rhythms, 1B: "There Was a Crooked Man," p. 27

Chapter Book: *Let's Go on a Museum Hunt*, Chapter 3

ESL For additional practice with ***-ed*** endings, have children acquiring English get together with English-proficient partners to read the following My Book: *The Fixed-Up Park*. For information on using the My Books on the computer, see the WiggleWorks Plus My Books Teaching Plan.

CHALLENGE

Children can make two or more word equations such as the following:

fix + ed = fixed
snowed – ed = snow

When children have completed their word equations, they can use the words in each equation in sentences. For example: ***I will fix the window. I fixed the window yesterday***. You may also wish to have children place sticky notes over the answers to their equations. Then children can exchange equations with others, figure out the answer to each, and use each of the words in sentences.

Lesson 162

page 259

Blend and Build Words With Phonograms

QUICKCHECK ✔

Can children:

- ✓ **blend words with phonograms *-ank, -unk*?**
- ✓ **build words with *-ank, -unk*?**

If YES go to Read and Write.

TEACH

Introduce the Phonograms Write the phonograms *-ank* and *-unk* on the chalkboard. Point out the sounds these phonograms stand for. Add the letter *s* to the beginning of *-ank,* and model for children how to blend the word formed. Have children repeat the word *sank* aloud as you blend the word again. Then add *s* to the beginning of *-unk,* and model for children how to blend the word *sunk*.

Ask children to suggest words that rhyme with *sank* and *sunk*. List these words on the chalkboard in separate columns. Have volunteers underline the phonogram *-ank* or *-unk* in each. Be sure to include the words *bank, tank, thank, dunk, junk,* and *skunk*. Blend these words as children repeat them aloud. Encourage children to look for the word parts *-ank* and *-unk* as they read.

READ AND WRITE

Build Words Write the word parts *-ank* and *-unk* on the chalkboard. (If available, use a pocket chart and letter cards.) Have children add a letter or a letter pair to the beginning of each phonogram to make a new word. Continue by having children replace the initial consonant in the first word to build a second word. Children may build *hunk* from *junk* and *trunk* from *hunk*.

Complete Activity Page Read aloud the directions on page 259. Have children complete the page independently.

Name ____________________

Add each letter or letters to the word part below it. Blend the word. If it is a real word, write it on the line.

b f s t n th __ank	d s l j r sk __unk
1. bank	5. dunk
2. sank	6. sunk
3. tank	7. junk
4. thank	8. skunk

Write a sentence using one of the words you made.

9. Answers will vary.

Supporting All Learners

AUDITORY/VISUAL LEARNERS

PING! POW! THUNK! Some words spell out sounds. At this point, children should be able to encode a number of these words. You may dictate ***ping, pow, thunk, plunk, wham, vroom, zoom, clank,*** and other sound words. After children spell them, and you review them as a class, invite children to illustrate one of their favorite sounds.

KINESTHETIC/VISUAL LEARNERS

WORD SORT Write the following words on index cards: ***bank, tank, sank, thank, junk, dunk, hunk, skunk***. Have children sort them according to the word parts ***-ank*** and ***-unk***.

ESL To help children learn new words, provide sentence frames with familiar context clues. For example, you may write these: ***The fish swim in the ___*** and ***The ship ___ when it hit a rock***. Have children copy the entire sentence. Then help children to complete the sentences with the words ***tank*** and ***sunk***. Provide visual clues as needed.

INTRODUCE HIGH-FREQUENCY WORDS

Read the words in the box. Write one word on each blank.

you	very
first	does
after	

1. you
2. first
3. after
4. very
5. does

Use the words in the box to finish each sentence.

6. We will play baseball with you.
7. Baseball is a very good game.
8. Sam will be first to bat the ball.
9. Let me run after the ball.
10. It does feel good to hit a home run!

Copyright © Scholastic Inc.

260 Recognize and Write High-Frequency Words

Supporting All Learners

AUDITORY LEARNERS

RIDDLES Challenge children to work in small groups to make up riddles for each of the high-frequency words. Riddles may use both sound clues, such as the number of syllables and kinds of sounds, as well as meaning clues, such as ***opposite of***. For example, children may say, ***I begin with /f/. I'm the opposite of*** **last. (first)** Have groups test their riddles on each other.

KINESTHETIC/VISUAL LEARNERS

SPEED SEARCH The word ***you*** is a very common and useful high-frequency word. Have small groups or partners race to find as many examples of ***you*** as they can in classroom materials. Give children a three-minute time limit. Children can mark instances of ***you*** with self-sticking notes. Children may also search for the word ***very***. When the race is over, these words can be read in context.

EXTRA HELP

To give children additional help in writing the high-frequency words, have children write the words on construction paper, trace the letters with glue, then pour sand over the glue. When the glue dries, children can trace the words with their fingers as they say the words.

Lesson 163

page 260

Recognize High-Frequency Words

QUICKCHECK ✔

Can children:

- ✔ **recognize and write the high-frequency words *you, first, after, very, does*?**
- ✔ **complete sentences using the high-frequency words?**

If YES go to Read and Write.

TEACH

Introduce the High-Frequency Words

Write the high-frequency words ***you, first, after, very,*** and ***does*** in sentences on the chalkboard. Read aloud the sentences, and underline the words. You may wish to use these sentences:

1. **First you get a bat.**
2. **After that, you hit the ball.**
3. **Then you run very fast.**
4. **Does that sound like fun?**

Then ask volunteers to dictate sentences using the high-frequency words. You may wish to begin with the following sentence starters: ***First we will*** ___ or ***When do you*** ___?

READ AND WRITE

Practice Write each high-frequency word on a note card. Read each word aloud as you display the cards. Then do the following:

- **Mix the cards.**
- **Display one card at a time, and ask children to state each word aloud.**
- **Have children spell each word aloud, clapping on each letter.**
- **Ask children to write each word in the air as they state aloud each letter. Then have children write each word on a sheet of paper.**

Complete Activity Page Read aloud the directions on page 260. Have children complete the page independently.

Lesson 164

pages 261-262

Blend Words With /ī/igh, y

QUICKCHECK ✔

Can children:

✔ substitute sounds in words?

✔ identify /ī/?

✔ blend words with /ī/*igh, y*?

If YES go to Read and Write.

TEACH

Develop Phonemic Awareness

Phonemic Manipulation Explain to children that they are going to listen to words with /i/ as in *pig*. Children are to replace the middle sound in each word with a long *i* sound. For example, read the word *lit*. Ask children to identify the middle sound, /i/. Guide children to replace /i/ with /ī/ and to say the new word *light*. Continue with the following words: *sit, fit, knit*.

Connect Sound-Symbol Write the words *lit* and *light* on the chalkboard as you read each aloud. Ask children what sound they hear in the middle of each word and how the words are different. Point out that /i/ in the middle of *lit* is spelled by *i*, and /ī/ in the middle of *light* is spelled by *igh*. Then write the words *fly* and *by*. Point out that /ī/ at the end of these words is spelled by *y*.

Ask children to suggest other words that contain /ī/. List these words on the chalkboard. Have volunteers circle the letters *igh* or *y* in each word that contains one of these spellings.

READ AND WRITE

BUILT-IN REVIEW **Blend Words** List the following words and sentences on a chart. Have volunteers read each aloud.

- **night** **right** **flight** **bright**
- **why** **try** **my** **by**
- **Why can't you come at night?**
- **My night light is very bright.**

Complete Activity Pages Read aloud the directions on pages 261–262. Review each picture name with children.

LISTEN AND WRITE

light

Name

Look at each picture. If the picture name has the long i sound as in fry, color the picture.

1. night 2. bone 3. gate 4. fly
5. flashlight 6. sky 7. light 8. six
9. cake 10. butterfly 11. goat 12. pig

Write the letters igh or y to finish each word. Read the words to a friend.

13. by 14. my 15. right

Supporting All Learners

AUDITORY/VISUAL LEARNERS

WORD HUNT Read aloud "Wynken, Blynken, and Nod" on pages 28–29 in the *Big Book of Rhymes and Rhythms, 1B*. Have children find all the words with **/ī/*igh, y*** in the rhyme. Write these words on the chalkboard. Invite volunteers to add other **/ī/*igh, y*** words to the list.

VISUAL LEARNERS

HANGMAN Play Hangman on the chalkboard with children, using words with ***igh*** and ***y***. After you have played a few games as a sample, invite partners to play the game with each other.

AUDITORY LEARNERS

PHONEMIC AWARENESS KIT For children needing additional phonemic awareness training, see the Scholastic Phonemic Awareness Kit. The oral segmentation exercises will help children to break apart words sound by sound. This is necessary for children to be able to encode, or spell, words while writing.

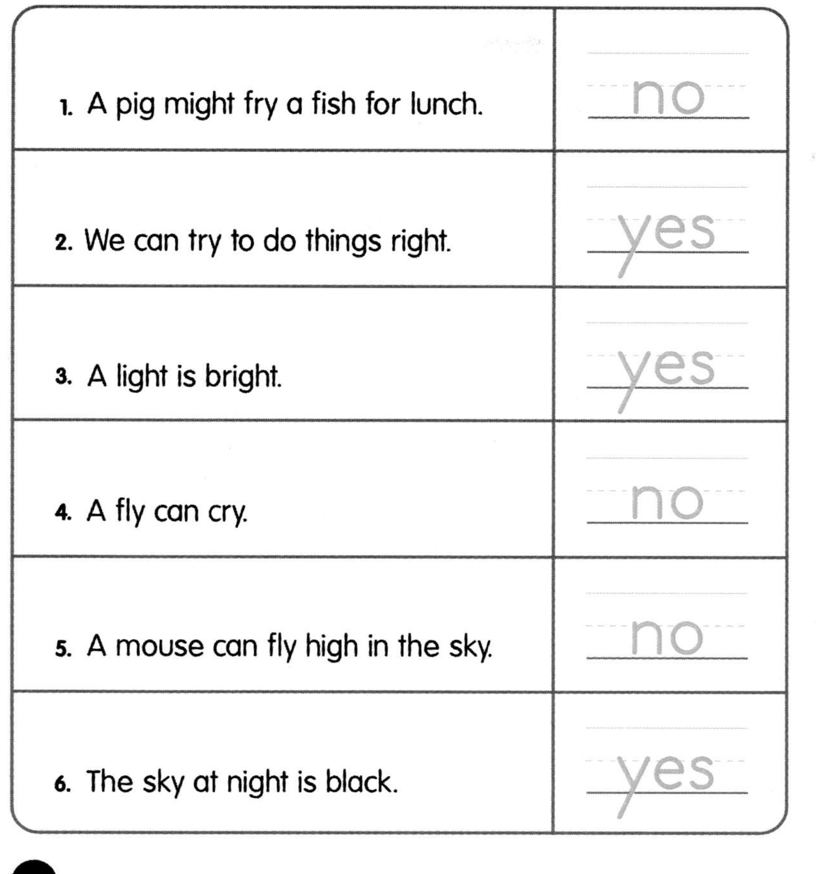

BLEND WORDS

Read each sentence. If the sentence is true, write **yes** on the line. If it is not true, write **no** on the line.

1. A pig might fry a fish for lunch.	no
2. We can try to do things right.	yes
3. A light is bright.	yes
4. A fly can cry.	no
5. A mouse can fly high in the sky.	no
6. The sky at night is black.	yes

262 **Blend and Write Words With /ī/*igh, y***

Integrated Curriculum

SPELLING CONNECTION

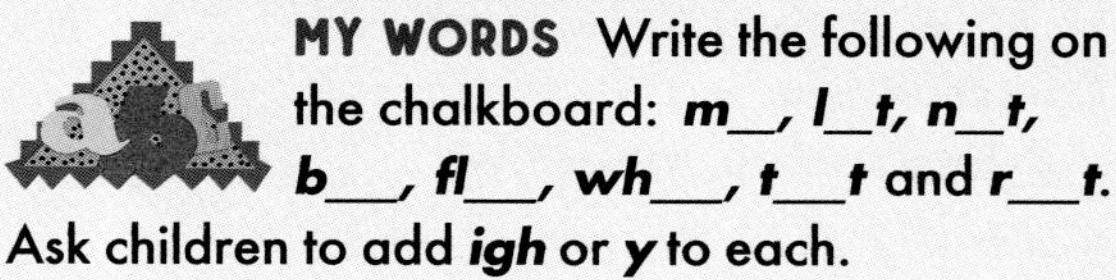

MY WORDS Write the following on the chalkboard: ***m__, l__t, n__t, b__, fl__, wh__, t__t*** and ***r__t***. Ask children to add ***igh*** or ***y*** to each.

SOCIAL STUDIES CONNECTION

WORD SEARCH Have children look through books, magazines, and newspapers for words with ***/ī/igh, y***. Have children list the words they find.

Phonics Connection

Literacy Place: *Hometowns*
Teacher's SourceBook, pp. T200–201;
Literacy-at-Work Book, pp. 41–42

My Book: *There Was Clyde, Right by My Side*

Big Book of Rhymes and Rhythms, 1B: "Wynken, Blynken, and Nod," pp. 28-29

Chapter Book: *A Bag of Tricks*, Chapter 5

ESL Make a set of duplicate cards for ***igh*** and ***y*** words. To help children acquiring English become familiar with the words, display each word, and read it with them chorally. Then children can work with English-proficient partners to sort the words by spelling pattern, mix and match pairs, play Concentration or Go Fish, and find matching pictures.

CHALLENGE

Have partners play tic-tac-toe with ***igh*** and ***y*** words. Draw a large grid on the chalkboard or on chart paper. Have one child be ***igh*** and the other child be ***y***. Then have children play tic-tac-toe using ***igh*** and ***y*** words instead of ***x***'s and ***o***'s. When children get three words in a row, challenge them to use each word in a sentence.

- **Night Sounds, Morning Colors**
  by Rosemary Wells
  A child enjoys colors and sounds. **(/ī/*igh*)**
- **No Fighting, No Biting**
  by Else A. Minarak
  Two little alligators stop fighting. **(/ī/*igh*)**

Lesson 165

pages 263–264

Blend Words With /ī/-*ild*, -*ind*

QUICKCHECK ✔

Can children:

✔ delete sounds in words?

✔ identify /ī/?

✔ blend words with /ī/-*ild*, -*ind*?

If YES go to Read and Write.

TEACH

Develop Phonemic Awareness

Phonemic Manipulation Ask children to listen to words that end with *-ind* and *-ild*. Say each word, and ask children to repeat it without its first sound. For example, if you say ***kind***, children will say ***ind***. Use these and other words.

- find • wild • mild • child
- wind • grind • blind • hind

Connect Sound-Symbol Write ***grand*** and ***grind*** on the chalkboard, and ask a volunteer how the words are different. Point out that the ending sounds of ***grand*** are spelled by ***-and***, and the ending sounds of ***grind*** are spelled by ***-ind***. Write the word ***mild*** on the chalkboard. Point out the ending ***-ild***.

Write these word parts on the chalkboard: ***ch_***, ***k_***, ***gr_***, ***w_***, and have volunteers add the word endings ***-ind*** or ***-ild*** to each. Model how to blend each word.

READ AND WRITE

Blend Words List these words and sentences on a chart, and have volunteers read them aloud.

- wild child find mind
- kind blind mild wind
- She is kind to the child.
- We may find some wild plants.

Complete Activity Pages Read aloud the directions on pages 263–264. Have children complete the pages independently.

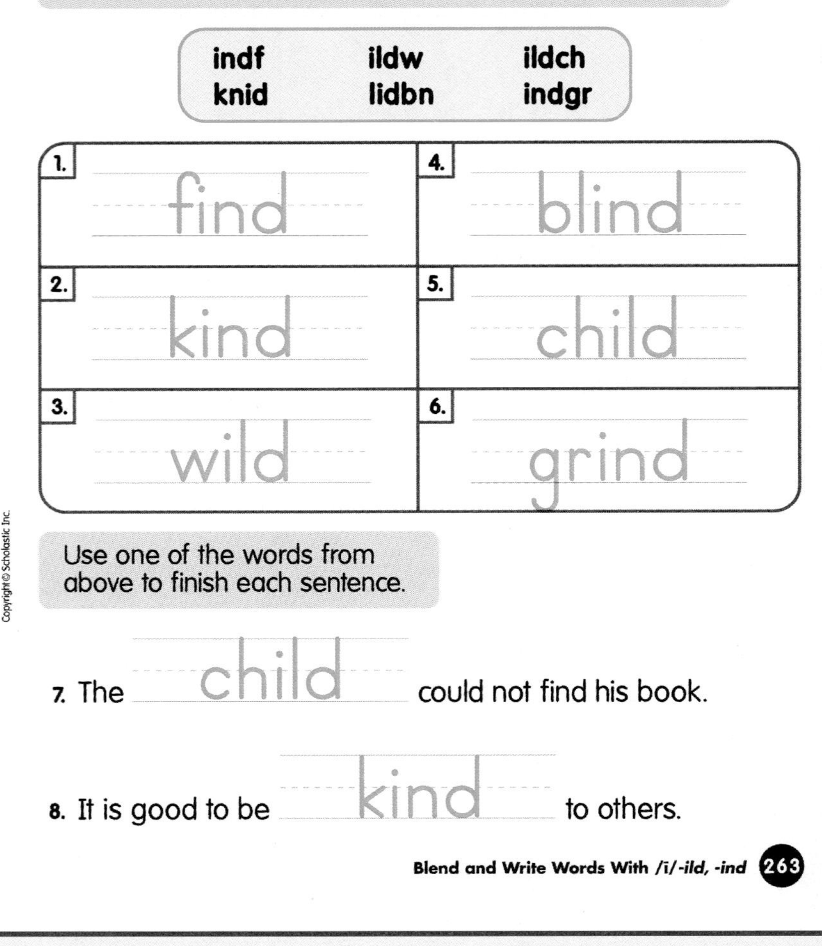

BLEND WORDS

Name

Unscramble the letters to make words that rhyme with **mild** or **mind**. Then write the words on the lines.

indf	**ildw**	**ildch**
knid	**lidbn**	**indgr**

1. find
2. kind
3. wild
4. blind
5. child
6. grind

Use one of the words from above to finish each sentence.

7. The child could not find his book.

8. It is good to be kind to others.

Supporting All Learners

AUDITORY/VISUAL LEARNERS

FIND /ī/! Read aloud "Ten Fingers" on pages 30–31 in the *Big Book of Rhymes and Rhythms, 1B*. Have children find all the words with /ī/ in the rhyme. Write these words on the chalkboard. Invite volunteers to add other /ī/ words to the list.

KINESTHETIC LEARNERS

SORT WORDS Make word cards with these words: ***wild, child, mild, mile, while, bike, kind, find, grind, hit, lip,*** and ***six***. Mix up the cards. Challenge children to sort them into two piles: one pile with long ***i*** words and the other with short ***i*** words. Then ask children to sort the long ***i*** cards according to whether the sound is represented by the ***i-e, -ind,*** or ***-ild*** spelling pattern.

BLEND WORDS

Circle the word that best finishes each sentence. Then write the word on the line.

1. The child is named Tom.	wild, (child), chin
2. We can wind up the clock.	blind, find, (wind)
3. The wild cat can run fast.	win, (wild), child
4. Say yes or no. Make up your mind.	mild, (mind), find
5. Thank you. You are very kind!	grind, king, (kind)
6. Where did you find that book?	mind, (find), fin

Integrated Curriculum

SPELLING CONNECTION

WILD ABOUT /ī/ Have children dictate sentences with ***/ī/-ild, -ind*** words. Write them on a chart.

SCIENCE CONNECTION

WILD ANIMALS Have children make posters of animals that live in the wild. Children can cut out pictures from magazines and newspapers or draw their own pictures.

Phonics Connection

Literacy Place: *Hometowns*
Teacher's SourceBook, pp. T314–315; Literacy-at-Work Book, pp. 62–63

My Book: *Wild Animals*

Big Book of Rhymes and Rhythms, 1B: "Ten Fingers," pp. 30–31

Chapter Book: *A Bag of Tricks*, Chapter 6

EXTRA HELP

To review words with long vowel sounds, children may enjoy playing "Sound Bingo" on pages 28–29 in *Quick-and-Easy Learning Games: Phonics*. Use Game 2.

CHALLENGE Have children go on a word hunt in your school. Children should look through books, magazines, and newspapers as well as on signs and posters for words with ***/ī/-ild, -ind***. Children can keep a list of the words they find and where they find them. Then children can present the word lists to the class.

The Book Shop

- **Jamaica's Find**
  by Juanita Havill
  Jamaica finds a stuffed dog in the park. **(/ī/-*ild*)**
- **Wild, Wild Sunflower Child Anna**
  by Nancy White Carlstrom
  A girl enjoys nature. **(/ī/-*ild*)**

Lesson 166

pages 265-266

Recognize and Write Possessives

QUICKCHECK ✔

Can children:

✔ recognize possessives?

✔ write possessives?

If YES go to Read and Write.

TEACH

Introduce Possessives Write ***the girl's cap*** on the chalkboard. Underline ***girl's***. Then circle the apostrophe and name it. Explain that sometimes the apostrophe shows that two words have been combined to make a contraction. Other times the apostrophe shows that someone owns something.

I know that there is an apostrophe in *girl's*. But when I read the sentence, it doesn't make sense that *girl's* could stand for two words. Instead, I know that the apostrophe is there because it shows that the girl owns the cap.

Write these phrases on the chalkboard: ***the boy's bike, the dog's bone, the man's soap***. Have children circle each apostrophe. Then have them decide what belongs to whom.

READ AND WRITE

Practice Write the following phrases on the chalkboard. Ask children to rewrite each using a possessive form. For example, in the first item, the word ***boy*** should be made possessive.

- **the football that belongs to the boy**
- **the notebook that belongs to Scott**
- **the pen that belongs to Katy**

Complete Activity Pages Read aloud the directions on pages 265–266. Review the art with children.

Name ____________________

Read the first sentence. Then finish the second sentence with the word that tells who owns the object in the picture.

1. The dog has a [bone]. It is the dog's bone.
2. Jill has a [hat]. It is Jill's hat.
3. The king has a [ring]. It is the king's ring.
4. The hen has a [chick]. It is the hen's chick.

Now make up one of your own.

5. The ____________ has a [ ].

It is the ____________ ____________.

Supporting All Learners

VISUAL/AUDITORY LEARNERS

CLASSROOM POSSESSIONS Ask children to think of five classmates and five things that belong to them and to write each person's name and the object. For example, children may write ***Ben*** and ***book***. Then have children write a phrase that shows ownership—for example, ***Ben's book***.

TACTILE/VISUAL LEARNERS

WORD CHECKERS To review possessives and contractions, put a self-sticking note on each square of a checkerboard. On each note, write a contraction or the possessive form of a noun or name. Have children play checkers using the board. Children must read the words on the squares they want to jump over or land on, or else find a different move.

Read the word next to each sentence. Make the word show that the person or thing owns something. Then write the new word on the line. Read the sentence.

1.	man	The man's coat is brown.
2.	pig	Is the pig's food in the pen?
3.	Mom	Mom's job is in this town.
4.	Dad	Dad's tools are in the shed.
5.	queen	Is the queen's gown red?
6.	duck	How big are the duck's feet?

Integrated Curriculum

WRITING CONNECTION

WHO HAS WHAT? Write each child's name in the class on an index card, one name per card. Place the cards in a bag. Write the name of the same number of objects on index cards, one object per card. Place these cards in a separate bag. Have children take turns picking a name and an object. Children should write the name and the object on a sheet of paper, using an apostrophe to show ownership. For example, if a child picks ***Sandy*** and ***truck,*** the child should write ***Sandy's truck***. Give children an opportunity to share their writing.

WIGGLEWORKS TECHNOLOGY

MY BOOKS/TECHNOLOGY For additional practice with possessives, have children read the following My Book: *Maggie's Bad Day*. For information on using the My Books on the computer, see the WiggleWorks Plus My Books Teaching Plan.

Phonics Connection

Literacy Place: *Hometowns*
Teacher's SourceBook, pp. T62–63;
Literacy-at-Work Book, pp. 11–12

My Book: *Maggie's Bad Day*

EXTRA HELP Have children dictate sentences with possessives. Write the sentences on the chalkboard. Give children a few examples such as ***I see Pam's dog, I play with Kim's train,*** and ***He rides Dan's bike***. After children have finished dictating, have them point out the apostrophe in each sentence. Ask children what is owned and by whom. Then have children replace the words with apostrophes with other possessive words. For example, ***Pam's dog*** might change to ***Don's dog***.

Write the following sentences on the chalkboard without apostrophes: ***I see Pams dog, I play with Kims train, He rides Dans bike***. Read each sentence with children. Model adding the first apostrophe, then have children add the other two apostrophes. You may want to have children copy the sentences onto strips to use for oral reading practice.

Lesson 167

page 267

Build Words

QUICKCHECK ✔

Can children:

✔ **build words?**

If YES go to Read and Write.

TEACH

Review Sound-Spellings Review with children the sound-spelling relationships from the past few lessons. These include /ī/ *igh, y* and the phonograms *-ind* and *-ild*.

Then write the word ***kind*** on the chalkboard. Ask children to replace /**k**/, the first sound in ***kind,*** with /**f**/ to make a new word. Ask children what letter stands for /**f**/. Then replace the letter ***k*** in ***kind*** with the letter ***f,*** and blend the new word formed. Continue by changing one letter (or pair of letters) to build new words.

find
mind
bind
blind
grind

READ AND WRITE

Build Words Distribute the following letter cards: ***g, h, i, l, m, t***. Provide time for children to build as many words as possible using the letter cards. Suggest that they record their words on a separate sheet of paper. Continue building words with the following sets of letter cards: ***b, c, m, r, t, y; d, f, i, k, m, n;*** and ***c, d, h, i, l, m***.

Complete Activity Page Read aloud the directions on page 267. Have children complete the page independently.

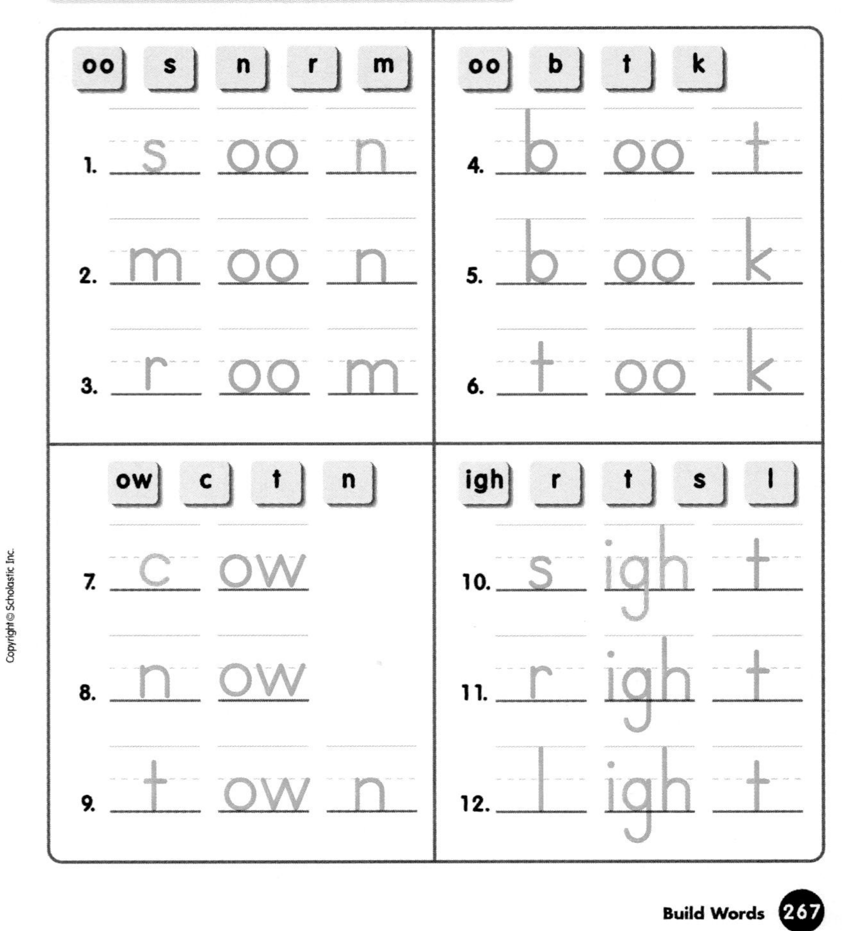

BUILD WORDS

Name

Use the letter tiles to make words.

oo s n r m

1. s oo n
2. m oo n
3. r oo m

oo b t k

4. b oo t
5. b oo k
6. t oo k

ow c t n

7. c ow
8. n ow
9. t ow n

igh r t s l

10. s igh t
11. r igh t
12. l igh t

Build Words 267

Supporting All Learners

KINESTHETIC LEARNERS

GAME FUN To practice word-building skills, children may enjoy playing "Build a House" on pages 21–24 in *Quick-and-Easy Learning Games: Phonics*. You may wish to introduce any unfamiliar blends, such as ***sw,*** or word parts, such as ***or,*** before children begin.

EXTRA HELP

Children who need extra practice can use the Magnet Board to build words with ***y*** and ***igh***. Suggest that they begin with one of the following words: ***try, fly, tight***.

POETRY

Read the poem. Write a new line for the poem. Begin with the word **where**.

Where can pigs fly in the sky?

Where can hens sing all night?

Where can cows cry?

Where can cats drive by?

Where can bats light a bright light?

Where can all this be right?

In your mind when you dream at night.

Answers will vary.

Supporting All Learners

AUDITORY/VISUAL LEARNERS

STORY TIME Have children generate a list of words that contain ***-ind*** and ***-ild***. Record these words on the chalkboard. Then have children create a story, using as many ***-ind*** and ***-ild*** words on the chalkboard as they can. To begin, suggest a title such as "The Kind Child." Write the dictated story on chart paper for group and individual reading. Return to the story, rereading it in subsequent lessons.

Introduce the concept of silent letters. Explain that English has many silent letters such as ***gh***. Help children acquiring English understand this concept by writing the words ***fight, right,*** and ***tight,*** pronouncing the words, and asking volunteers to cross out the letters they do not hear as individual sounds.

Write and Read Words With /ī/igh, y, -ild, -ind

QUICKCHECK ✔

Can children:

✔ spell words with /ī/*igh, y, -ild, -ind*?

✔ read a poem?

If YES go to Read and Write.

TEACH

Link to Spelling Review the sound-spelling relationship /ī/***igh, y***. Have a volunteer write the many spellings that stand for /ī/. Continue with other sounds. ***(-ild, -ind)*** You may also wish to review /o͝o/*oo*, /o͞o/*oo* and /ou/*ou, ow*.

Phonemic Awareness **Oral Segmentation** Say the word ***right,*** and have children orally segment it. (/r/ /ī/ /t/) Draw three connected boxes on the chalkboard, and have a volunteer write the spelling that stands for each sound in ***right***. Continue with the word ***try***.

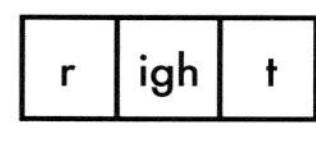

READ AND WRITE

Dictate Dictate the following words and sentence. The words in the first set are decodable based on the sounds previously taught. The words in the second set are high-frequency words from this and previous units. Have children write them on a sheet of paper. When they are finished, write the words and sentence on the chalkboard, and have children make any corrections.

- **sight** **my** **kind**
- **very** **does** **want**
- **You must find the light.**

Complete Activity Page Read aloud the directions on page 268. Have children read the poem independently or with partners.

Lesson 169

pages 269–270

Read and Review Words With /ī/igh, y, -ild, -ind

QUICKCHECK ✔

Can children:

✔ **read words with /ī/*igh, y, -ild, -ind*?**

✔ **recognize high-frequency words?**

If YES go to Read and Write.

TEACH

Review Sound-Spellings Review the sound-spelling relationships from the past few lessons, including /ī/*igh, y, -ild, -ind*. Have a volunteer write on the chalkboard the many spellings that can stand for the sound. Then display pictures of objects whose names contain ***-ind*** and ***-ild***. Have children write the letter or letters in each picture's name.

Review High-Frequency Words Review the high-frequency words ***you, first, after, very,*** and ***does***. Write sentences with each word on the chalkboard. Say one word, and have volunteers circle the word in the sentences. Then have children make up sentences for each word.

READ AND WRITE

Build Words Distribute the following letter cards to children: ***c, f, l, r, t, y***. Give children time to build as many words as possible using the letter cards. Children can record their words on paper. Continue building words with these sets of letter cards ***g, h, i, m, r, t; d, f, i, k, n, w;*** and ***c, d, h, i, l, m***.

Build Sentences Write the following words on note cards: ***you, find, may, it, kind, the, have, right, are, hat***. Display the cards, and have children make sentences using the words. Invite children to work in small groups to complete the activity.

Complete Activity Pages Read aloud the directions on pages 269–270. Review the art with children.

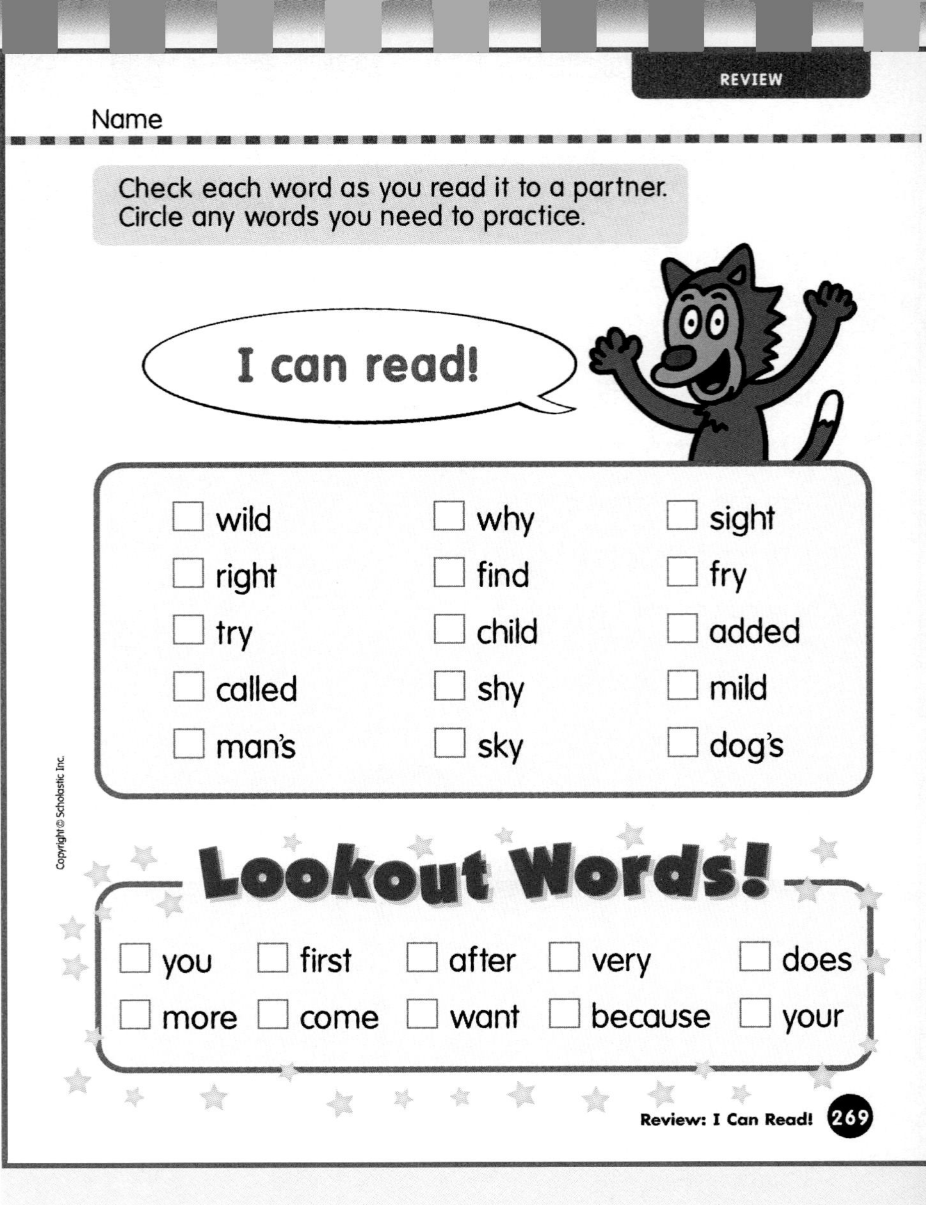

REVIEW

Name

Check each word as you read it to a partner.
Circle any words you need to practice.

☐ wild ☐ why ☐ sight
☐ right ☐ find ☐ fry
☐ try ☐ child ☐ added
☐ called ☐ shy ☐ mild
☐ man's ☐ sky ☐ dog's

Lookout Words!

☐ you ☐ first ☐ after ☐ very ☐ does
☐ more ☐ come ☐ want ☐ because ☐ your

Review: I Can Read! 269

Supporting All Learners

VISUAL LEARNERS

LEARNING CENTER For additional practice with word families, set up Center 4, Activity 3, "Snail's Pace," on pages 31–32 in *Quick-and-Easy Learning Centers: Phonics*. Use word families children have learned recently, such ***-igh, -ild, -ank, -unk, -ink,*** and ***-ing***. In addition to consonants, use blends and digraphs such as ***sp, sn, st, wh,*** and ***ch***.

VISUAL LEARNERS

CHART LETTERS Have children collect and count words with /ī/***igh, y, -ild, -ind***. Children can record the results of their counts in a chart such as the one shown.

6				
5				
4				
3				
2				
1				
	Words With ***-igh***	Words With ***-y***	Words With ***-ild***	Words With ***-ind***

TEST YOURSELF

Fill in the bubble next to the word that best finishes each sentence.

1. We ate lunch ________ the tree.	○ high ○ try ● by
2. The ________ was kicking the ball.	○ mild ● child ○ kind
3. Will you ________ the fish in the pan?	○ my ○ shy ● fry
4. What ________ of dog did you get?	● kind ○ wild ○ blind
5. How ________ is the sunshine today?	○ by ● bright ○ fight
6. I'll ________ you when we play hide-and-seek.	○ fly ● find ○ mild

270 Assess High-Frequency Words and */ī/igh, y, -ild, -ind*

EXTRA HELP

To help children read the I Can Read! word list, model tracking print. Then model using word parts and sound-spelling relationships to decode words. As needed, reteach the use of the apostrophe to show possession.

CHALLENGE

Have children generate a list of words with ***/ī/igh, y, -ild, -ind***. Record these words on the chalkboard. Then have children create a story using as many ***/ī/igh, y, -ild, -ind*** words on the chalkboard as they can. Children can write or dictate the story for group and individual reading. Children can also return to their story, rereading it in subsequent lessons.

Integrated Curriculum

WRITING CONNECTION

ALL ABOUT ME Have children use words with ***/ī/igh, y, -ild, -ind*** as well as high-frequency words to write or dictate sentences about themselves. Encourage children to use as many words with ***/ī/igh, y, -ild, -ind*** and high-frequency words as possible. You may wish to have children illustrate their sentences, too. Then children can share their sentences and point out the words with **/ī/** and the high-frequency words.

SOCIAL STUDIES CONNECTION

WORD SEARCH Have children look through newspapers for words with ***/ī/igh, y, -ild,*** and ***-ind***. Children can list the words they find on a sheet of paper. Then have children use these words and the high-frequency words to write or dictate book titles. You may wish to suggest titles such as *Flight at Night: The Story of Bats,* or *How You Can Find Wild Flowers.* Have children write their book titles on construction paper and illustrate their "book jackets."

I CAN READ! OPTIONS

The I Can Read! page can be used for one or all of the following:

- **paired reading**
- **individual assessment**
- **choral reading**
- **homework practice**
- **program placement**

PORTFOLIO

Invite children to add papers to their portfolios that reflect their progress in writing as well as in reading. Encourage children to select work that shows knowledge of a wide variety of letters, word parts, and lookout words.

Read Words in Context

TEACH

Assemble the Story Ask children to remove pages 271–272. Have them fold the pages in half to form the Take-Home Book.

Preview the Story Preview *The Moon,* a nonfiction story about the moon. Invite children to browse through the first two pages of the story and to comment on anything they notice. Suggest that children point out any unfamiliar words. Read these words aloud as children repeat them. Then have children predict what they think the selection might be about.

READ AND WRITE

Read the Story Have volunteers take turns reading aloud a page at a time. Discuss items of interest on each page. Encourage children to help each other with any blending difficulties. The following prompts may help children while reading:

- **What letter sounds do you know in the word?**
- **Are there any word parts you know?**

Reflect and Respond Have children share their reactions to the story. What did they learn? What else would they like to know about the moon? Encourage children to write another sentence to add to the story.

Develop Fluency Reread the story as a choral reading, or have partners reread the story independently. Provide time for children to reread the story on subsequent days to develop fluency and increase reading rate.

Children can share how they figure out unfamiliar words when reading. Children can also model how to blend words. Encourage children to look for words with ***/ī/igh, y, -ild, -ind*** in other stories and to apply what they learned. Continue to review challenging sound-spelling relationships for children needing additional support.

Supporting All Learners

VISUAL LEARNERS

ADD TO THE STORY This story ends with a question. Invite children to answer it by drawing a picture and labeling it. Children may show themselves as astronauts making a discovery or as explorers searching a new planet.

AUDITORY LEARNERS

RHYME TIME Explain to children that you will play a rhyming game with them. Write the following sentences on the chalkboard: ***The moon looks bright at night; I would like to fly in the sky; What looks like a slice of light at night?; I can look in a book to find pictures of the moon.*** Ask children which words rhyme in each sentence. Have children underline the words that rhyme.

In 1969 a U.S. spaceship went to the moon. The men from the spaceship stood on the moon. They picked up moon rocks to take back with them. The men looked around. What a sight!

3

The moon looks very bright. But it has no light of its own. The light from the sun shines on the moon. That light makes the moon look bright!

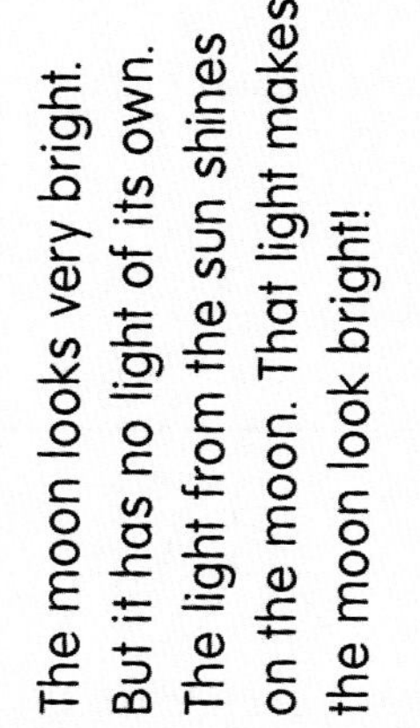

2

Reflect on Reading

ASSESS COMPREHENSION

To assess their understanding of the story, ask children questions such as:

- **Where does the moon get its light?** ***(from the sun)***
- **What did the men who landed on the moon take home?** ***(rocks)***

HOME-SCHOOL CONNECTION

Send home *The Moon*. Encourage children to read the story to a family member.

WRITING CONNECTION

WRITING FACTS Children can write or dictate their own nonfiction story about something they know about. Have children brainstorm ideas that they would like to write about. You may wish to suggest some familiar nonfiction topics such as "pets" or "school." Children can decide on one topic and write what they know about it as a whole group, or children can decide on more than one topic and write about these topics as small groups, one topic per group. Then have children share their nonfiction story or stories.

EXTRA HELP

For additional practice with ***/ī/igh, y, -ild, -ind,*** have children read the following My Books: *There Was Clyde; Right by My Side; Wild Animals;* and *My Family Trip*. For information on using the My Books on the computer, see the WiggleWorks Plus My Books Teaching Plan.

CHALLENGE

Have children write or dictate sentences about the moon. Children can start a moon bulletin board on which they can display their sentences. Children can also draw and cut out pictures of the moon, newspaper and magazine articles about moon-related topics, and anything else related to the moon that may be of interest to place on the bulletin board. Have children take turns explaining the items on the bulletin board to others.

Phonics Connection

Phonics Readers:
#67, *The Week We Cleaned the Park*
#68, *Who Helped Ox?*
#69, *All on a Saturday Night*
#70, *I Spy*
#71, *Wild Lion and the Mice*
#72, *The Kind Child*

Name

Look at each picture. Circle the letter that the picture name begins with. Write the letter on the line.

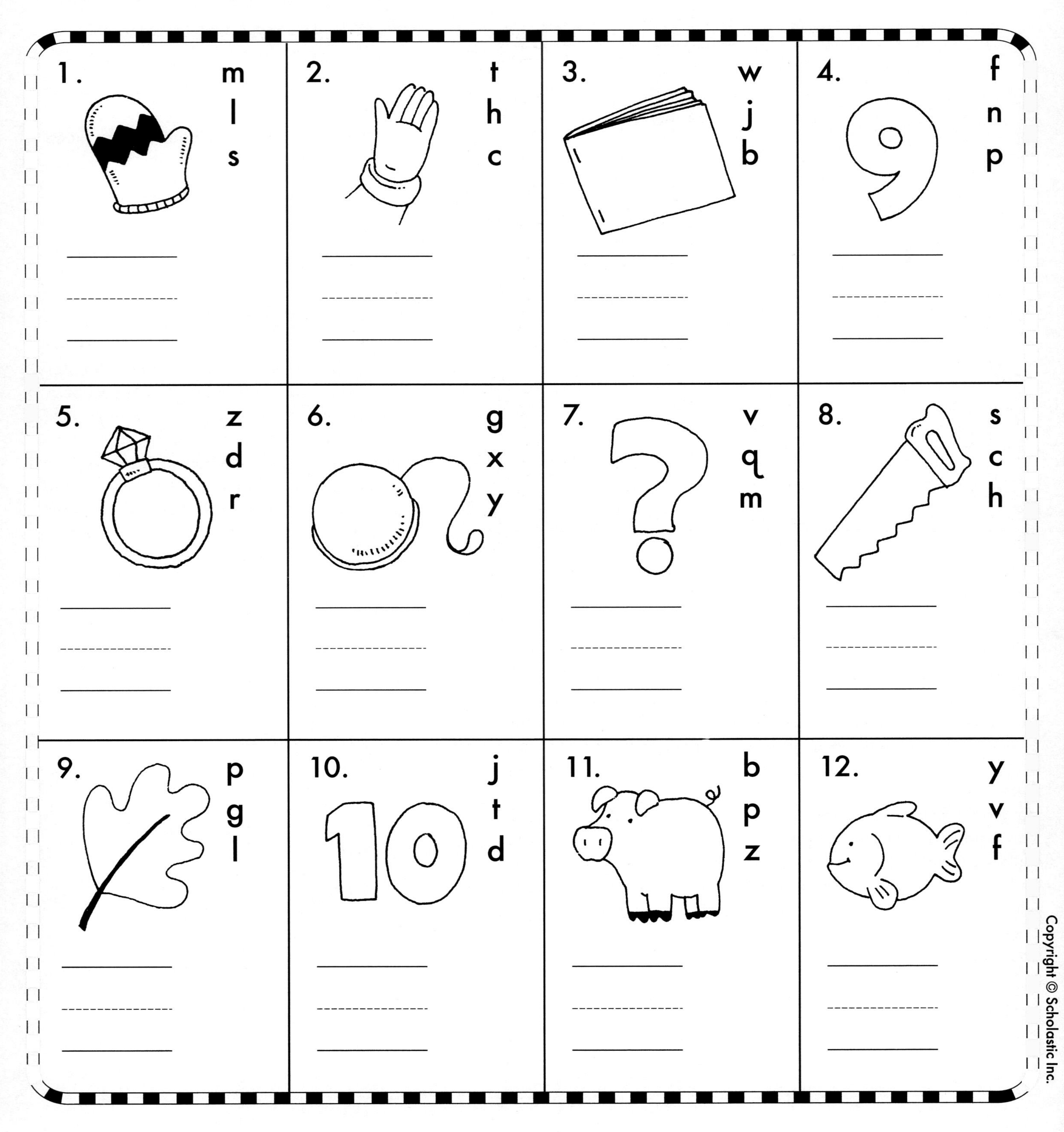

Name

Circle the word that names each picture. Write the word on the line.

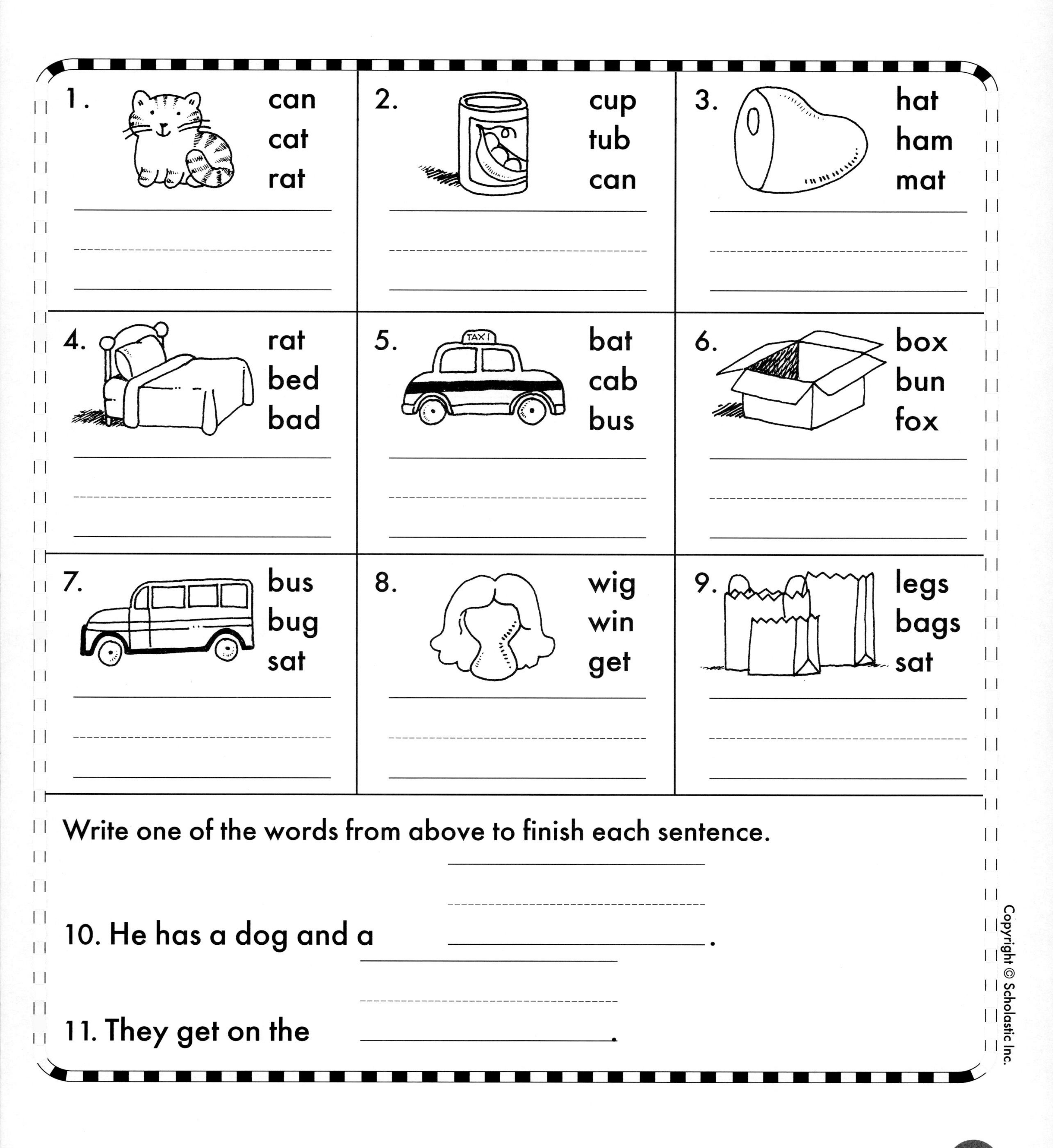

1. can cat rat
2. cup tub can
3. hat ham mat
4. rat bed bad
5. bat cab bus
6. box bun fox
7. bus bug sat
8. wig win get
9. legs bags sat

Write one of the words from above to finish each sentence.

10. He has a dog and a ________.

11. They get on the ________.

Name

Say each picture name. Fill in the bubble next to the letter that stands for the middle sound. Write the letter on the line.

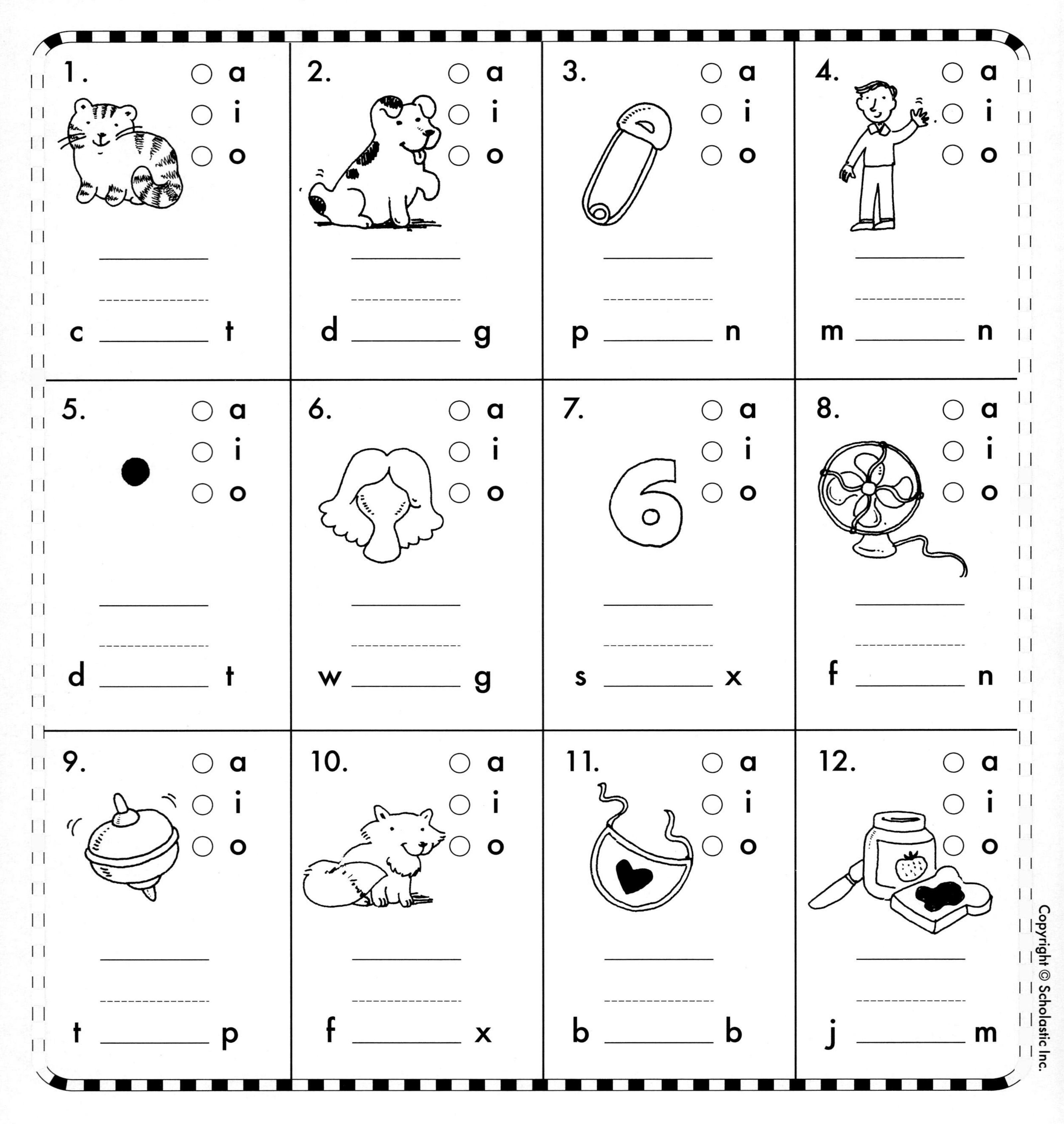

Name ______________________

Say each picture name. Fill in the bubble next to the letter that stands for the middle sound. Write the letter on the line.

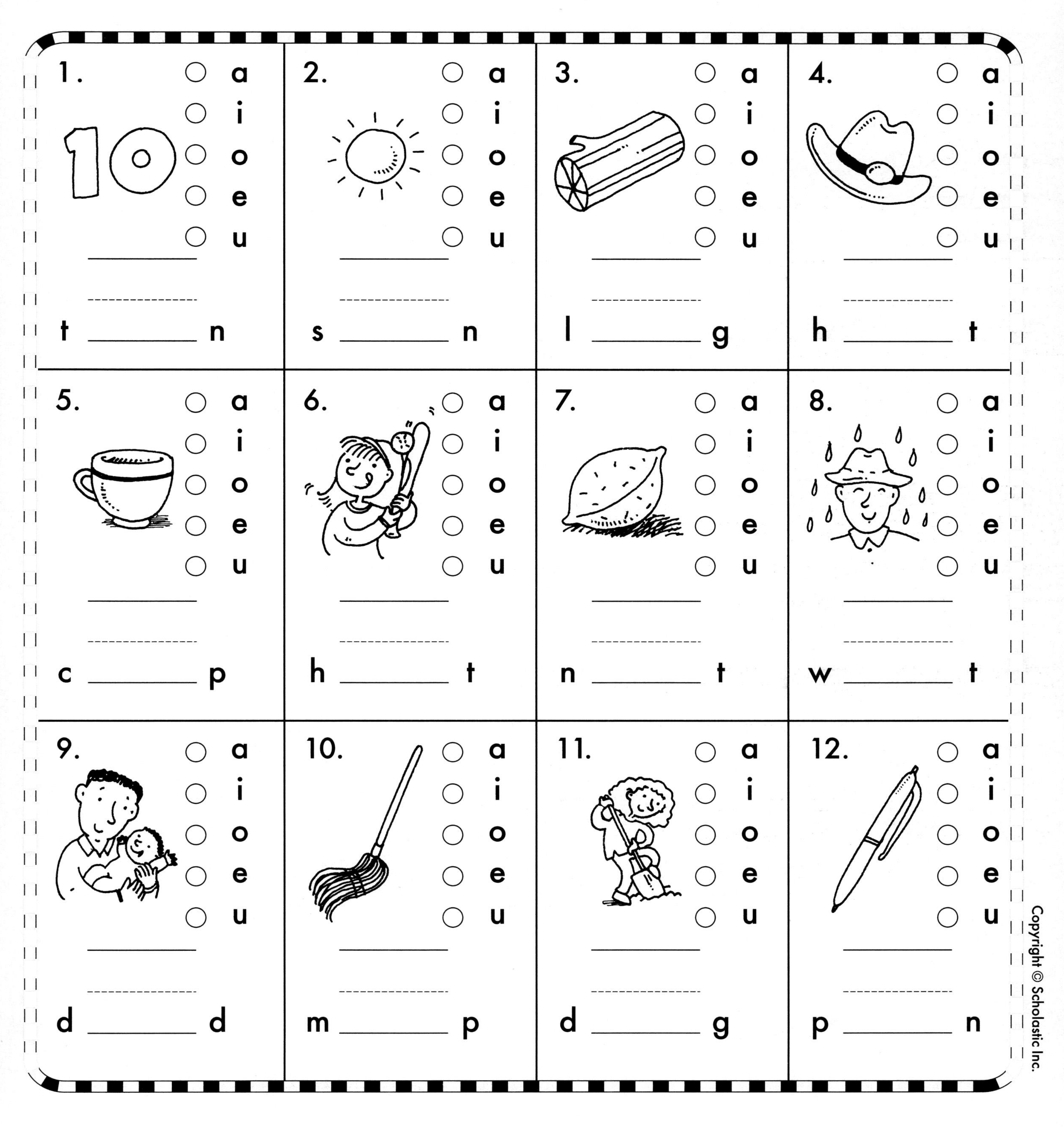

Name ______________________

Fill in the bubble next to the sentence that tells about each picture.

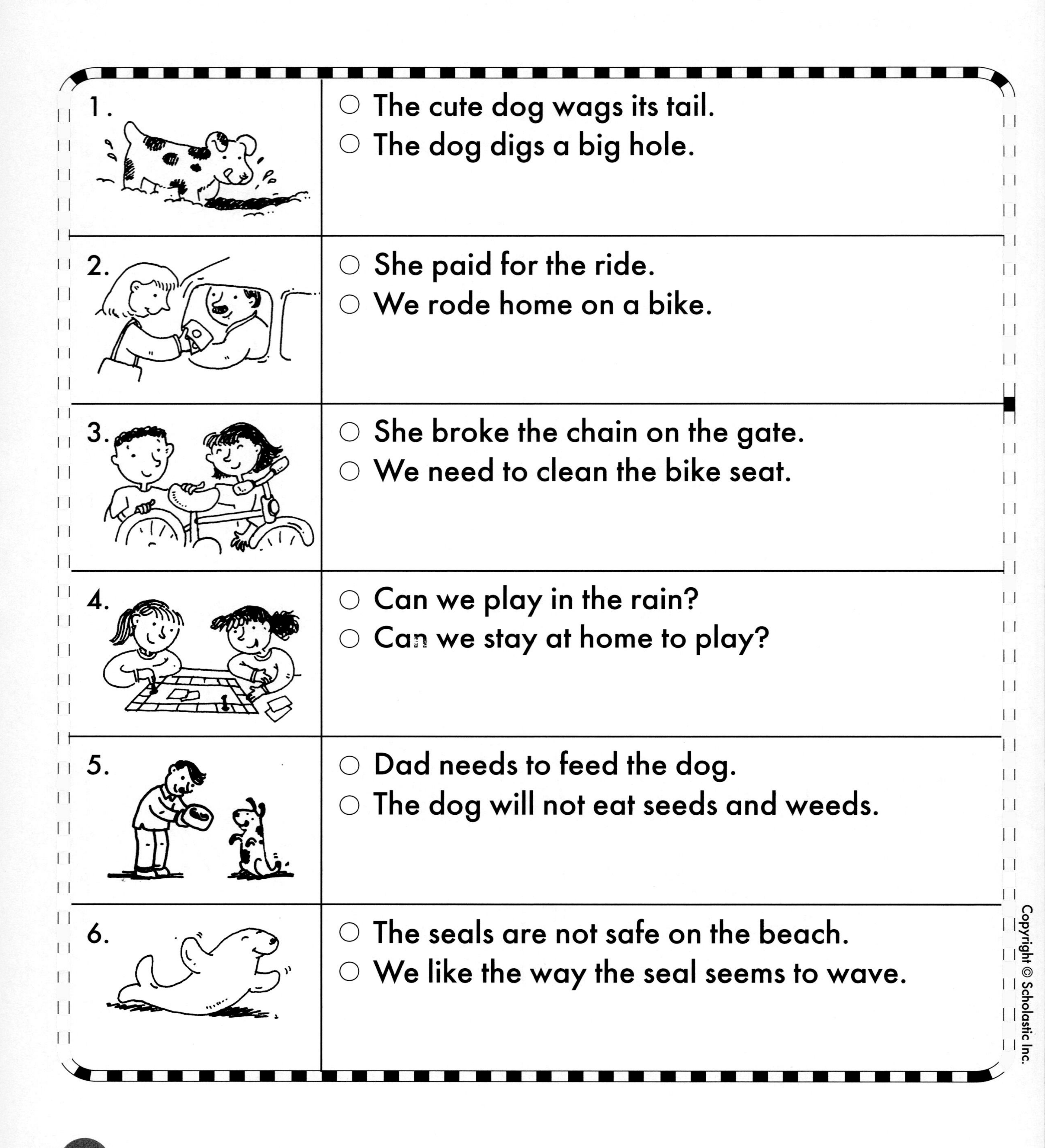

1.
- ○ The cute dog wags its tail.
- ○ The dog digs a big hole.

2.
- ○ She paid for the ride.
- ○ We rode home on a bike.

3.
- ○ She broke the chain on the gate.
- ○ We need to clean the bike seat.

4.
- ○ Can we play in the rain?
- ○ Can we stay at home to play?

5.
- ○ Dad needs to feed the dog.
- ○ The dog will not eat seeds and weeds.

6.
- ○ The seals are not safe on the beach.
- ○ We like the way the seal seems to wave.

Name

Fill in the bubble next to the sentence that tells about each picture.

1.	○ We saw the wild pony and the donkey. ○ Jane rode the pony and the donkey.
2.	○ I try not to cry when I fall. ○ We saw a small toad by the road.
3.	○ All of us might go to the mall. ○ We try to find coats in the shop window.
4.	○ I can't see the sky at night. ○ Each bit of light seems small but bright.
5.	○ The lazy child uses the float. ○ Show me the red rowboat.
6.	○ We smile at the lady and her baby. ○ I call hello to the big hippo.

Name ______________________

Fill in the bubble next to the sentence that tells about each picture.

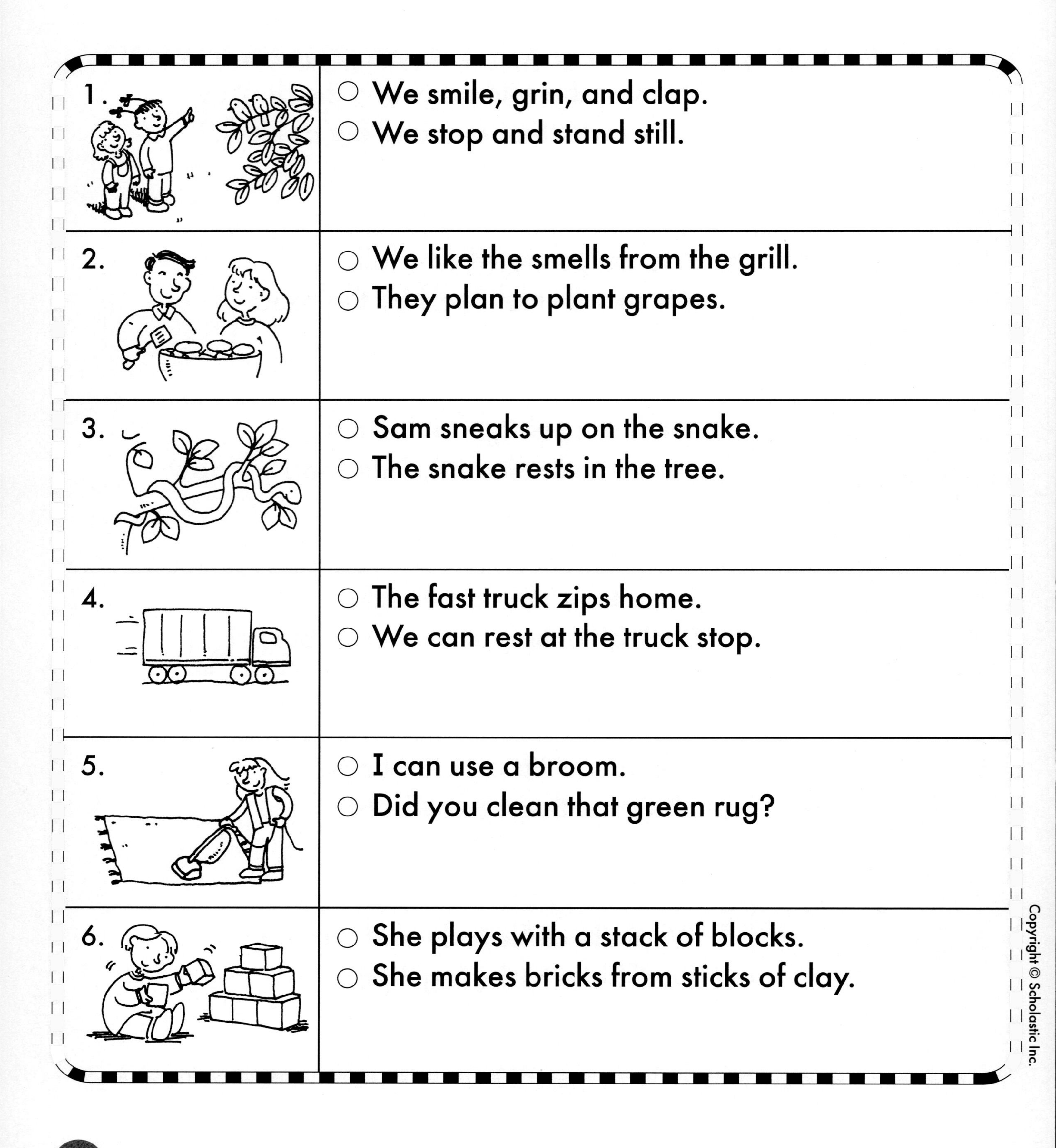

1\.
- ○ We smile, grin, and clap.
- ○ We stop and stand still.

2\.
- ○ We like the smells from the grill.
- ○ They plan to plant grapes.

3\.
- ○ Sam sneaks up on the snake.
- ○ The snake rests in the tree.

4\.
- ○ The fast truck zips home.
- ○ We can rest at the truck stop.

5\.
- ○ I can use a broom.
- ○ Did you clean that green rug?

6\.
- ○ She plays with a stack of blocks.
- ○ She makes bricks from sticks of clay.

Name ______________________

Fill in the bubble next to the sentence that tells about each picture.

1.
- ○ A whale is on the dish.
- ○ That is a big, white whale.

2.
- ○ Chip can't reach the wheel.
- ○ We will fix the wheel in the shed.

3.
- ○ Dad will chop and chill the meat.
- ○ Each of us has seen wheat.

4.
- ○ We see the ship from the beach.
- ○ Do the whales chase the ship?

5.
- ○ I see a cheek, a chin, and a nose.
- ○ I see a long chain.

6.
- ○ They rush to lock up the shop.
- ○ We don't chat much when we shop.

Name ______________________________

Fill in the bubble next to the sentence that tells about each picture.

1.
○ She looks at my loose tooth.
○ She has a black boot on one foot.

2.
○ We took our books to the woods.
○ We found a big nest in the woods.

3.
○ Show me how to use the broom.
○ Dad showed me his tool room.

4.
○ I am too hot when I zip up my hood.
○ Will the clouds go away soon?

5.
○ Mom took me to school.
○ A loud crowd was at the pool.

6.
○ She frowned at the mouse.
○ I shouted at the big brown cow.

Assessment Answer Key

Page T18

1. m	2. h	3. b	4. n	5. r	6. y
7. q	8. s	9. l	10. t	11. p	12. f

Page T19

1. cat
2. can
3. ham
4. bed
5. cab
6. box
7. bus
8. wig
9. bags
10. cat
11. bus

Page T20

1. a	2. o	3. i	4. a	5. o	6. i
7. i	8. a	9. o	10. o	11. i	12. a

Page T21

1. e	2. u	3. o	4. a	5. u	6. i
7. u	8. e	9. a	10. o	11. i	12. e

Page T22

1. The dog digs a big hole.
2. She paid for the ride.
3. We need to clean the bike seat.
4. Can we stay at home to play?
5. Dad needs to feed the dog.
6. We like the way the seal seems to wave.

Page T23

1. We saw the wild pony and the donkey.
2. We saw a small toad by the road.
3. We try to find coats in the shop window.
4. Each bit of light seems small but bright.
5. The lazy child uses the float.
6. I call hello to the big hippo.

Page T24

1. We stop and stand still.
2. We like the smells from the grill.
3. The snake rests in the tree.
4. The fast truck zips home.
5. Did you clean that green rug?
6. She plays with a stack of blocks.

Page T25

1. That is a big, white whale.
2. We will fix the wheel in the shed.
3. Dad will chop and chill the meat.
4. We see the ship from the beach.
5. I see a long chain.
6. They rush to lock up the shop.

Page T26

1. She looks at my loose tooth.
2. We found a big nest in the woods.
3. Dad showed me his tool room.
4. I am too hot when I zip up my hood.
5. Mom took me to school.
6. I shouted at the big brown cow.

Decoding Assessment

Directions: The word list on each I Can Read! page can be used as an assessment. Another way to check children's decoding abilities is to use nonsense words. Using nonsense words prevents children from correctly identifying words based on their sight-word knowledge. Therefore, children must rely solely on their knowledge of sound-spelling relationships.

Have children read aloud the following word list. The words in the list get progressively more complex. Explain to children that each word is a nonsense, or made-up, word. Circle those sound-spelling relationships in each word that children have difficulties with. For example, if children pronounce *lad* for *lat,* circle the letter *t.* If children struggle with several words in sequence, end the assessment at that point.

Name: ______________________________ **Date:** ____________________

1. lat
2. sep
3. fid
4. mot
5. nug
6. shob
7. chun
8. smap
9. flig
10. greff
11. dack
12. nist
13. bame
14. pite
15. foap
16. heen
17. raim
18. teaf
19. vout
20. droin

What do you hear?

Research has determined that children who are not reading on grade level when they are at or just beyond the midpoint of Grade 1 run serious risks of never reading on grade level. That means intervention is crucial at Grade 1 for children who have fallen behind.

WHEN TO INTERVENE

Early and regular evaluation is the key to identifying students in need of remediation. Such assessment should begin with phonemic awareness and progress to phonics and reading skills. It can be based on anecdotal records, formal and informal assessments, or the use of running records. Any of the following difficulties, when displayed consistently and without even subtle signs of progress, may signal the need for intervention:

- **failure to learn sound-spelling relationships**
- **inability to segment or blend words**
- **failure to learn high-frequency words**

STRATEGIES TO USE

Sound-Spelling Intervention

Because children's mastery of sound-spelling relationships is the prerequisite to being able to sound out words, any remediation program must place strong emphasis on these skills.

Letter Cards One approach to sound-spelling intervention is through the use of letter cards. These can be used to review unfamiliar sound-spelling patterns. Such practice should occur frequently. It should also be limited to just a few letters at a time and must start at the point in the sequence of instruction at which difficulties occur. This may require going back to phonemic awareness activities that precede phonics instruction, or to the very first sounds and letters.

It is important to begin at the point where children first experience difficulty and to move slowly through the sequence of instruction from that point.

Wordplay Wordplay can help children become increasingly familiar with the decodable words that they will encounter in print and put to work in their own writing. Once children have mastered a few high-utility vowels and consonants, use magnetic letters or word cards to put the words together letter by letter. Encourage children to say the letters and trace them as you do so. Then have children break apart the words, sound by sound. Scramble the letters, and have children remake the words. Provide multiple opportunities for forming and reforming words.

Blending and Segmenting Intervention

The best way to help children with blending is to continually model blending, to have children do it with you, and then to have children do it on their own. When you model, use the connected form of blending, in which sounds are elongated and run together. This is easier for children who cannot blend isolated sounds.

Similarly, continually model segmentation, and, when possible, do this in tandem with blending. Using manipulatives will help to make this process more concrete. If you have the Phonemic Awareness Kit, you can use the Elkonin boxes and counters that are provided. These can be used for counting the number of sounds in a word. Children can follow along by pointing to the boxes or writing in the boxes. They can also place counters or markers in each box for each sound they hear and then remove the counter and replace it with the letter or letters that represent the sound they hear. With such practice, be sure to move gradually from two- to four-sound words.

High-Frequency Word Intervention

To help children with high-frequency words, review them on a daily basis. Initially, work with very few words at a time, allowing children to master them before moving on. Always introduce the words in context, and have children identify them in that context, by pointing them out or underscoring them. Then have children read the sentences in which they appear.

Other follow-up and reinforcement activities include having children use the targeted high-frequency words in their own sentences, having children write the words, having children find the words in the decodable stories and poems, having children find the words on the Word Wall, and having children build sentences using high-frequency words and other decodable words. In group work, children can read aloud the words and spell them chorally.

Reading Intervention

As children master each cluster of consonants and vowels in a unit of instruction, use the decodable stories and poems to reinforce sound-spelling knowledge and to develop concepts about print. These stories and poems should be read again and again to help children develop fluency. Simple activities, in which children talk about the story or poem, draw or create their responses, or write about what they have read also should be used.

STRUCTURING THE INTERVENTION PROGRAM

The structure of any intervention program will depend on the children's needs, the number of children being served, and the school's resources in terms of both space and staff. Ideally, children can work daily in small groups of three or four for a time span of at least 30 minutes. Personnel resources in the school, such as the curriculum specialists, reading specialists, and trained aides, might all be enlisted to plan, conduct, monitor, or evaluate sessions.

MAKING INTERVENTION WORK

To make sure the intervention strategies are working, continually assess children's progress. Assessment should be a routine part of intervention, and at its heart should be one-on-one assessment procedures. Among the many possibilities for strategies you might use with individuals are the following:

- **Have each child read orally, and listen for the use of decoding strategies.**
- **Record reading progress through the use of running records that note errors, omissions, correction strategies, and pauses.**
- **Ask children to articulate their own progress.**
- **Keep anecdotal records, composing them so that other reading professionals can understand the child's exact degree of progress, as well as short- and long-term instructional needs.**

And, perhaps most of all, reinforce and celebrate any progress children make.

In addition to being vigilant about assessment, be sure to communicate with parents. Explain the strategies you are using, send home evidence of accomplishments, and respond to parents' requests with ways of engaging them in the process of phonics instruction.

RELEASE

The goal of intervention is release. Set this criteria beforehand, combining "hard fact" assessment, such as scores, with your knowledge of the child's increasing use of strategies. When the child is comfortably able to keep up with the rest of your class, you have met the goal of intervention.

Skills Checklist

Phonemic Awareness														
Recognize rhyme														
Recognize words that start and end with the same sound														
Blend words														
Segment words														
Identify beginning, middle, and ending sounds of words														
Manipulate beginning and ending sounds														
Consonants														
/m/*m*														
/l/*l*														
/t/*t*														
/s/*s*														
/h/*h*														
/p/*p*														
/f/*f*														
/n/*n*														
/k/*c*														
/b/*b*														
/w/*w*														
/j/*j*														
/z/*z*														
/d/*d*														
/r/*r*														
/g/*g*														
/ks/*x*														
/k/*k, ck*														
/s/*s*														
/y/*y*														
/v/*v*														
/kw/*qu*														
Vowels														
/a/*a*														
/o/*o*														
/i/*i*														
/e/*e*														
/u/*u*														
final e (*a-e*)														

final e *(i-e)*													
final e *(o-e)*													
final e *(u-e)*													
/ē/*e, ea, ee*													
/ō/*ai, ay*													
/ô/*a(ll), aw*													
/ō/*o, ow*													
/ē/*e, ey*													
/ō/*oa*													
/o͞o/*oo*													
/o͝o/*oo*													
/ou/*ou, ow*													
/oi/*oi, oy*													
/ī/*igh, y*													
Blends													
r-blends: *br, gr, tr*													
l-blends: *bl, cl, pl*													
s-blends: *sp, st, sm, sn*													
Digraphs													
/th/*th*													
/sh/*sh*													
/ch/*ch*													
/hw/*wh*													
Phonograms													
-ad, -at													
-ot, -op													
-id, -ip													
-og, -ig													
-ab, -an													
-en, -et													
-un, -ut													
-ill, -ick													
-ace, -ice													
-ack, -ock													
-ake, -oke													
-ink, -ing													
-ank, -unk													
-ild, -ind													

TO MARKET, TO MARKET

To market, to market, to buy a fat pig;
Home again, home again, jiggety jig.
To market, to market, to buy a fat hog;
Home again, home again, jiggety jog.

OH WHERE, OH WHERE HAS MY LITTLE DOG GONE?

Oh where, oh where has my little dog gone?
Oh where, oh where can he be?
With his ears cut short and his tail cut long,
Oh where, oh where can he be?

THIS LITTLE PIG WENT TO MARKET

This little pig went to market,
This little pig stayed home,
This little pig had roast beef,
This little pig had none,
And this little pig cried, "Wee, wee, wee, wee, wee,"
All the way home.

PETER PIPER

Peter Piper picked a peck of pickled peppers;
A peck of pickled peppers Peter Piper picked.
If Peter Piper picked a peck of pickled peppers,
How many peppers did Peter Piper pick?

LITTLE BOY BLUE

Little Boy Blue
Come blow your horn
The sheep's in the meadow
The cow's in the corn
Where is the boy who looks after the sheep?
He's under the haystack, fast asleep.

WEE WILLIE WINKIE

Wee Willie Winkie
Runs through the town,
Upstairs and downstairs,
In his nightgown,
Rapping at the window,
Crying through the lock,
"Are the children in their beds?
Now it's eight o'clock."

TEDDY BEAR, TEDDY BEAR

Teddy Bear, Teddy Bear, turn around.
Teddy Bear, Teddy Bear, touch the ground.
Teddy Bear, Teddy Bear, show your shoe.
Teddy Bear, Teddy Bear, that will do!
Teddy Bear, Teddy Bear, turn off the light.
Teddy Bear, Teddy Bear, say good night.

SIX LITTLE DUCKS

Six little ducks that I once knew,
Fat ones, skinny ones, fair ones too,
But the one little duck with the feather on his back,
He led the others with a quack, quack, quack!
Quack, quack, quack, quack, quack, quack!
He led the others with a quack, quack, quack!

Down to the river they would go,
Wibble wobble, wibble wobble, to and fro,
But the one little duck with the feather on his back,
He led the others with a quack, quack, quack!
Quack, quack, quack, quack, quack, quack!
He led the others with a quack, quack, quack!
Home from the river they would come,
Wibble, wobble, wibble, wobble, ho-hum-hum!
But the one little duck with the feather on his back,
He led the others with a quack, quack, quack!
Quack, quack, quack, quack, quack, quack!
He led the others with a quack, quack, quack!

TEN IN A BED

There were ten in a bed and the little one said,
"Roll over. Roll over."
So they all rolled over and one fell out.
There were nine in the bed and the little one said,
"Roll over. Roll over."
So they all rolled over and one fell out.
There were eight in the bed and the little one said,
"Roll over. Roll over."
So they all rolled over and one fell out.
There were seven in the bed and the little one said,
"Roll over. Roll over."
So they all rolled over and one fell out.
There were six in the bed and the little one said,
"Roll over. Roll over."
So they all rolled over and one fell out.
There were five in the bed and the little one said,
"Roll over. Roll over."
So they all rolled over and one fell out.
There were four in the bed and the little one said,
"Roll over. Roll over."
So they all rolled over and one fell out.
There were three in the bed and the little one said,
"Roll over. Roll over."
So they all rolled over and one fell out.
There were two in the bed and the little one said,
"Roll over. Roll over."
So they all rolled over and one fell out.
There was one in the bed and the little one said,
"Good night. Sleep tight."

ROW, ROW, ROW YOUR BOAT

Row, row, row your boat,
Gently down the stream,
Merrily, merrily, merrily, merrily,
Life is but a dream.

PEANUT BUTTER

Peanut, peanut butter—and jelly!
Peanut, peanut butter—and jelly!

First you take the peanuts
And you pick 'em, pick 'em,
Pick 'em, pick 'em, pick 'em.
Then you smash 'em, smash 'em,
Smash 'em, smash 'em, smash 'em.
Then you spread 'em, spread 'em,
Spread 'em, spread 'em, spread 'em.

Peanut, peanut butter—and jelly!
Peanut, peanut butter—and jelly!

(continued)

PEANUT BUTTER (continued)

Then you take the berries
And you pick 'em, pick 'em,
Pick 'em, pick 'em, pick 'em.
Then you smash 'em, smash 'em,
Smash 'em, smash 'em, smash 'em.
Then you spread 'em, spread 'em,
spread 'em, spread 'em, spread 'em.

Peanut, peanut butter—and jelly!
Peanut, peanut butter—and jelly!

Then you take the sandwich,
And you bite it, you bite it,
You bite it, bite it, bite it.
Then you chew it, you chew it,
You chew it, chew it, chew it.
Then you swallow it, you swallow it,
You swallow it, swallow it, swallow it.

Mm-mm, mm-mmm mm-mm!
Mm-mm, mm-mmm mm-mm!

MOTHER, MOTHER

Mother, Mother, I am sick!
Call for the doctor
Quick, quick, quick!
In came the doctor,
In came the nurse,
In came the lady with the alligator purse.
"Measles!" said the doctor.
"Mumps!" said the nurse.
"Nothing!" said the lady with the alligator purse.
"Pills!" said the doctor.
"Cough syrup!" said the nurse.
"Ice cream!" said the lady with the alligator purse.

BAA, BAA BLACK SHEEP

Baa, baa, black sheep, have you any wool?
Yes, sir, yes, sir, three bags full;
One for my master, one for my dame,
And one for the little boy who lives down the lane.

OLD MACDONALD

Old MacDonald had a farm,
E-I-E-I-O.
And on that farm he had a cow,
E-I-E-I-O.
With a moo-moo here,
And a moo-moo there,
Here a moo, there a moo,
Everywhere a moo-moo.
Old MacDonald had a farm,
E-I-E-I-O.

[Repeat the song with other farm animal names and noises.]

Scholastic Resources

***Literacy Place* Teacher SourceBooks**
A complete, balanced reading program that offers phonemic awareness activities and a complete program of sequenced phonics instruction for use with specific fiction and nonfiction selections.

***Scholastic Spelling,* Grade 1**
The only research-based program to provide an effective, efficient, and meaningful way to teach spelling.

Phonemic Awareness Kit
A complete phonemic awareness training program for Grades K–2, including:

- A Teacher's Guide with 66 lessons spread out over 13 weeks
- 100 full-color Picture Cards
- Student Activity Book for practice, reinforcement, and assessment
- 100 Counters
- 2 Puppets
- 2 Trade Books
- 35 Smiley Face Response Cards

WiggleWorks Plus Magnet Board
A feature of the WiggleWorks technology system, the Magnet Board allows children to move letters around on the computer screen to form word families, spell words, identify word parts, and hear words read.

Big Books of Rhymes and Rhythms
2 Big Books with 48 rhymes and rhythms that highlight consonants, vowels, digraphs, and blends (also available at K and Grade 2).

Quick-and-Easy Learning Centers: Phonics
A collection of center ideas and related activities for building reading and writing skills.

Quick-and-Easy Learning Games: Phonics A collection of versatile and fun games for building reading and writing skills.

Phonics Readers and Teacher's Guide
72 sequenced readers, each with a two-page lesson that includes a phonemic awareness activity.

Phonics and Word Building Kit
Everything needed to provide concrete contexts for phonics learners and to make phonics accessible to learners of varying abilities and types.

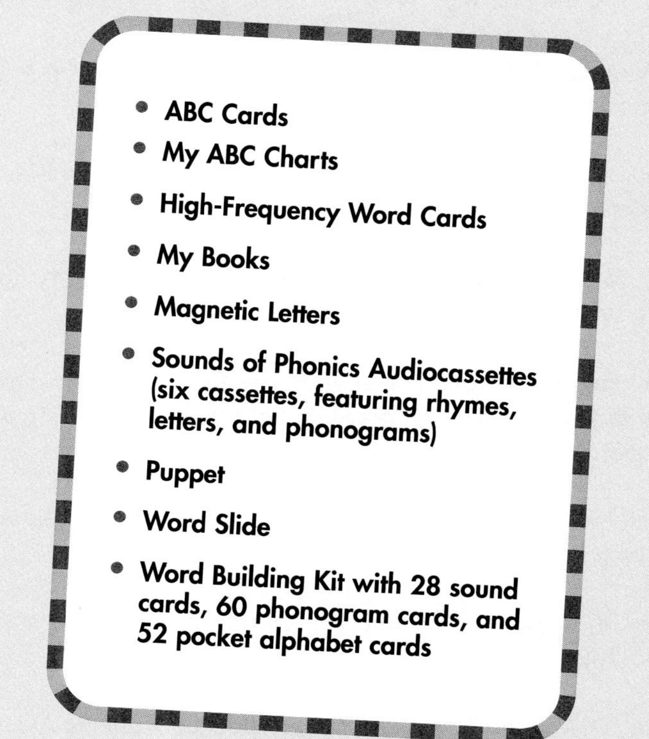

***Scholastic Phonics* Chapter Books**
Full-color beginning readers features decodable text with high-frequency words based on a systematic phonics scope and sequence.

Phonemic Awareness Activities for Early Reading Success
A collection of playful, engaging ideas and activities for developing this critical skill.

Phonics from A to Z: A Practical Guide
Contains word lists, games and activities, information on stages of reading development, assessments, tips for helping struggling readers, and much more.

Professional Bibliography

Adams, M.J. *Beginning to Read: Thinking and Learning About Print.* Cambridge: Massachusetts Institute of Technology, 1990.

Bear, D., et al. *Words Their Way: Word Study for Phonics, Vocabulary, and Spelling.* New York: Macmillan, 1995.

Blevins, W. *Phonics from A to Z: A Practical Guide.* New York: Scholastic, 1998.

Chall, J. *Learning to Read: The Great Debate.* 2nd ed. with new introduction. New York: McGraw-Hill, 1982.

Chall, J. and H. Popp. *Teaching and Assessing Phonics.* Cambridge, Massachusetts: Educators Publishing Service. Inc., 1996.

Cunningham, P. M. *Phonics They Use: Words for Reading and Writing.* 2d ed. New York: HarperCollins, 1995.

Honig, W. *How Should We Teach Our Children to Read?* Center for Systemic School Reform, San Francisco State University, 1995.

Platts, M. E. *Phonics: A Handbook for Teachers of Primary Phonics.* rev. ed. Mansfield, OH: Opportunities for Learning, 1991.

Shefelbine, J. "The Utility of Vowel Spelling-Sound Correspondences in Higher Frequency, Single Syllable Words." A paper presented at the annual meeting of the International Reading Association in Toronto, Canada.

Shefelbine, J. "Learning and Using Phonics in Beginning Reading." New York: Scholastic, *Literacy Research Paper*, Volume 10 (1995)

Snider, V. "A Primer on Phonemic Awareness: What It Is. Why It's Important, and How to Teach It." *School Psychology Review* 24, no. 3 (1995): 443–455.

Tangel, D. M., and B. A. Blackman. "Effect of Phoneme Awareness Instruction on Kindergarten Children's Invented Spelling." *Journal of Reading Behavior* 24 (1992): 233–261.

Treiman, R. T. 1991. "The Role of Intrasyllablic Units in Learning to Read," in L. Rieben & C. A. Perfetti, ed. *Learning to Read: Basic Research and Its Implications,* pp. 149–160. New Jersey: Lawrence Erlbaum Associates.

Wagstaff, J. *Phonics That Works!* New York: Scholastic, 1994.

Credits: Margery Mayer: Executive Vice President, Instructional Publishing Group; **Francie Alexander:** Vice President and Publisher, Curriculum and Development; **Eileen Thompson:** Publisher, Reading and Language Arts; **Alice Dickstein:** Director of Primary Programs; **Wiley Blevins:** Supervising Editor; **Cindy Chapman:** Senior Editor; **Stacie Sinder:** Director of Special Projects; **Janelle Cherrington:** Associate Editor; **Maria Cunningham:** Executive Art Director; **Ellen O'Malley:** Art Director; **Kathy Kelly:** Art Director

Teacher Notes

Teacher Notes

Teacher Notes

Teacher Notes

Teacher Notes

Teacher Notes

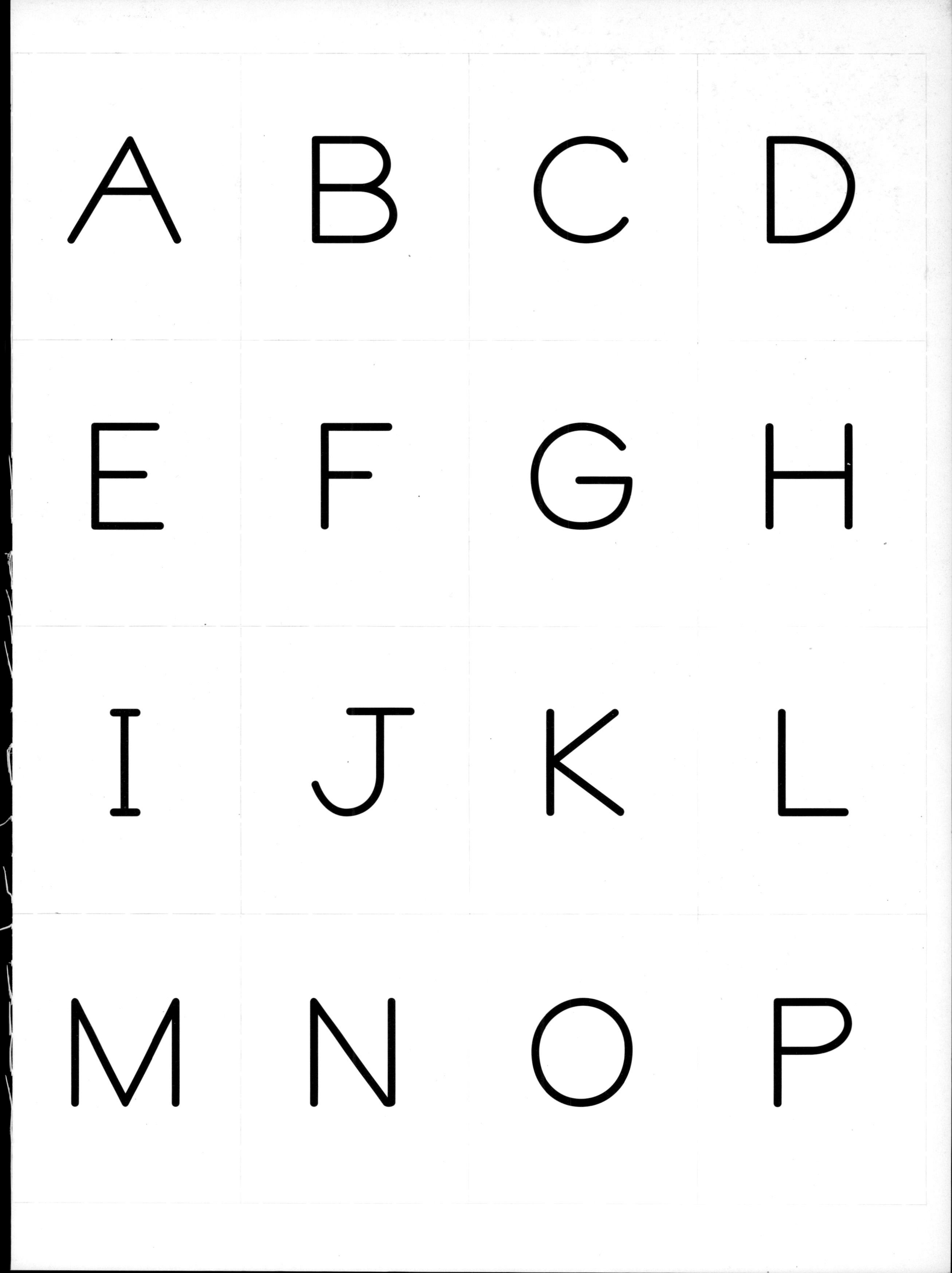
A B C D
E F G H
I J K L
M N O P

NAME
CONTENTS
REG. NO.
MADE IN CHINA

Q
R
S
T
U
V
W
X
Y
Z
a
e
i
o
u
CAPITAL
LETTERS

red
hat
CAPITAL
LETTERS
sun
top
pig

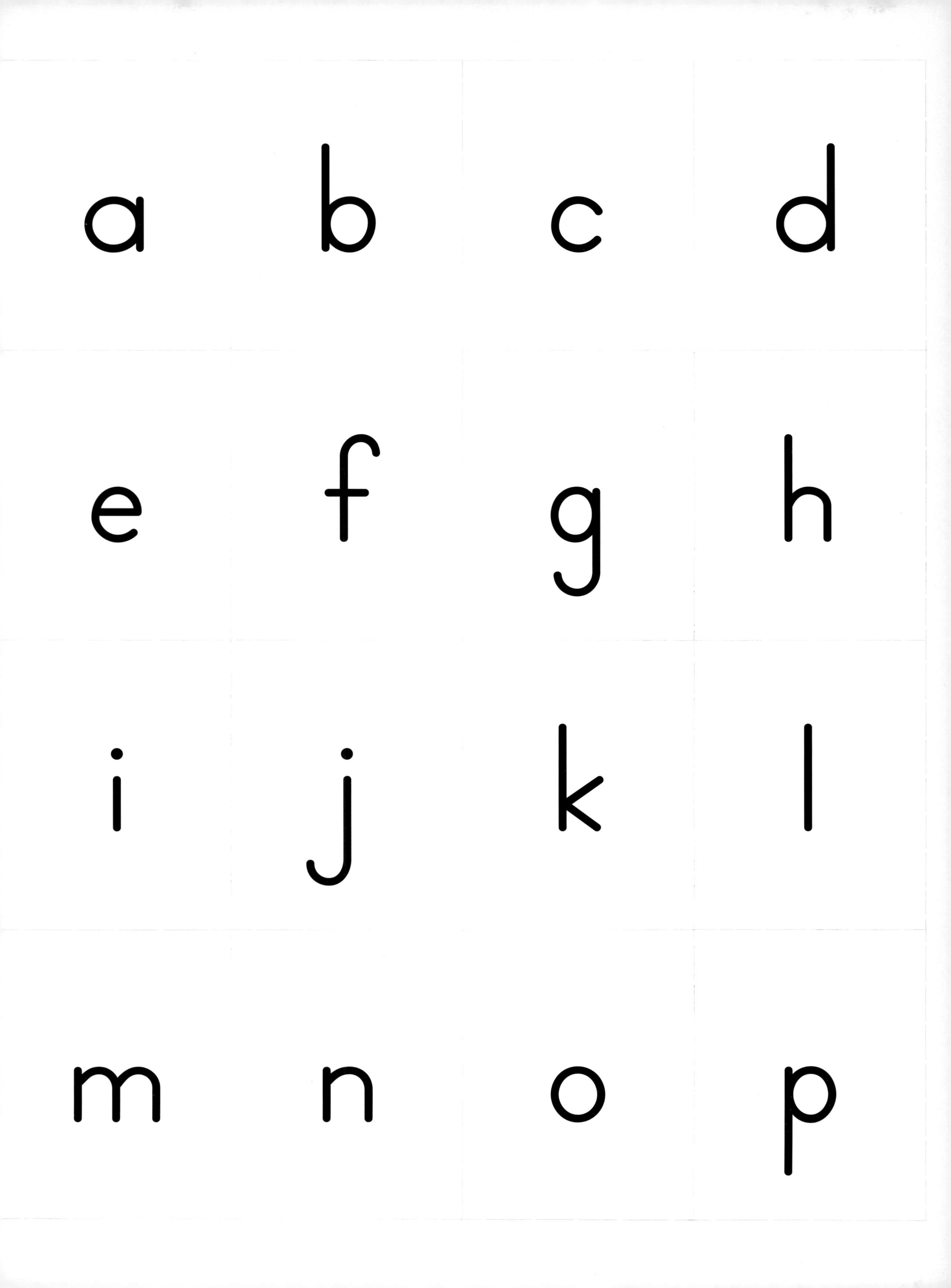
a
b
c
d
e
f
g
h
i
j
k
l
m
n
o
p

did sad
dig fed
dog hid
doll rod
dug mud

cat
can
cap
cot
cut

bat lab
bed web
big rib
box rub
bus tub

bag take
cat made
can day
fan say
hat rain

hand
hat
hen
hit
hop

get bag
gift leg
got pig
grass dog
grab rug

fan if
fell cliff
fish off
fox puff
fun stuff

fed me
hen she
jet bean
let read
men feet

land ball
leg call
let tell
lid sell
luck will

kid kick
king pick
kiss rock
kit sock
kitten luck

jam
jet
job
jog
jump

big bike
did kite
fit light
hill night
hit pie

pan map
pet flip
pick mop
pin top
pop up

box note
dog rope
hot boat
hop go
lot snow

nap ran
neck ten
net chin
not spin
nut fun

man ham
met jam
miss them
mop him
mud gum

q
r
s
t
u
v
w
x
y
z
a
e
i
o
u
small
letters

tag cat ten pet tip mitt top hot tub nut	sad pass sell yes sick miss sock this sun bus	ran our red for rip your rock her run year	queen quick quilt quit quiz
box fox mix six wax	wag wet well win with	van have vet give very leave vest five visit love	bun cute bus huge dug mule fun luck
nest we pen he pet heat red leaf wet seed	man bake map page pan may ran pail sat stain	zag fizz zig quiz zip buzz zoo size zoom prize	yam yak yell yes yet
small letters	mud cube nut fuel run music sun tug	mop broke not home pot coat rock so sock show	lip dime pig ride sit flight win sight zip tie

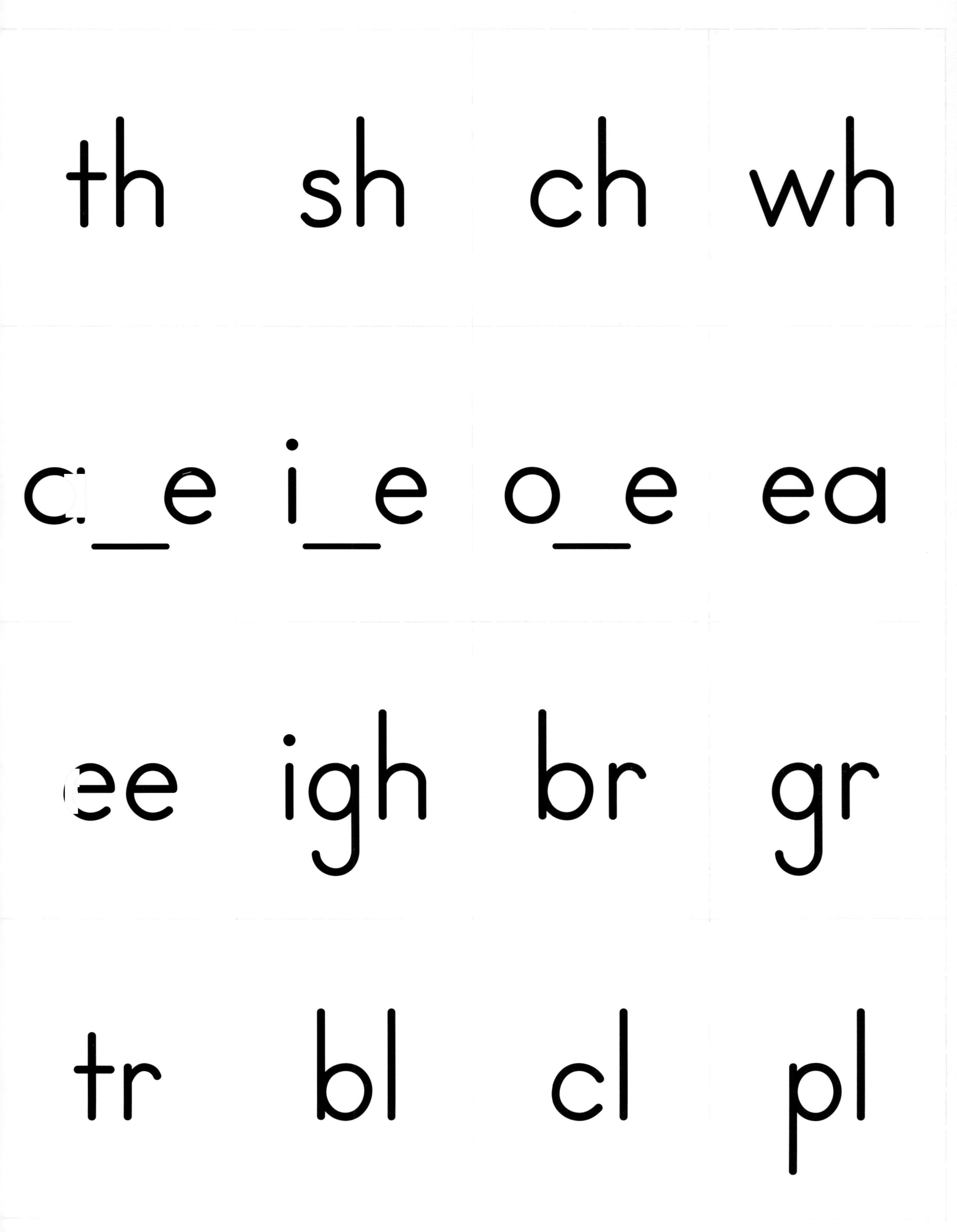

th
sh
ch
wh
a_e
i_e
o_e
ea
ee
igh
br
gr
tr
bl
cl
pl

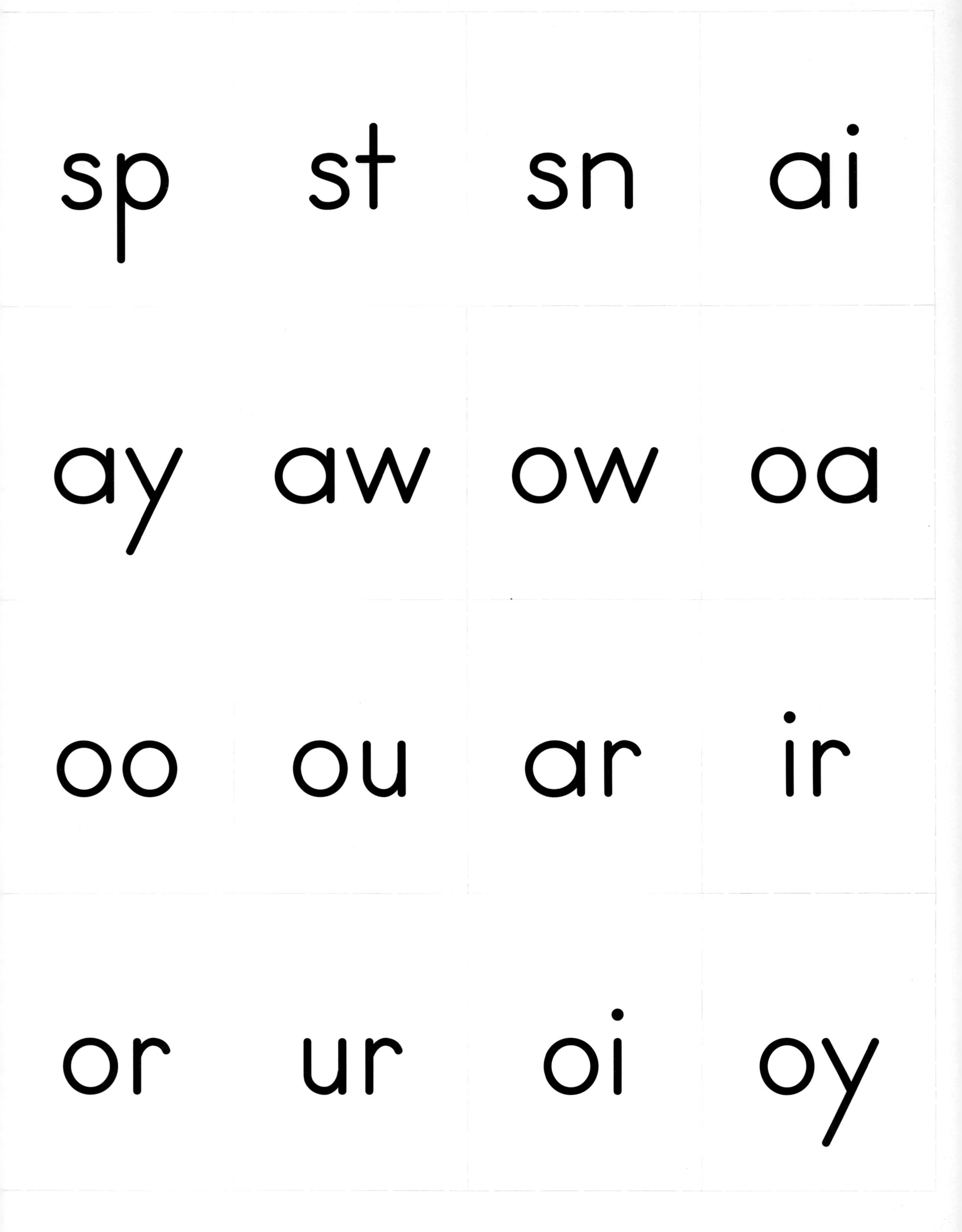
sp
st
sn
ai
ay
aw
ow
oa
oo
ou
ar
ir
or
ur
oi
oy

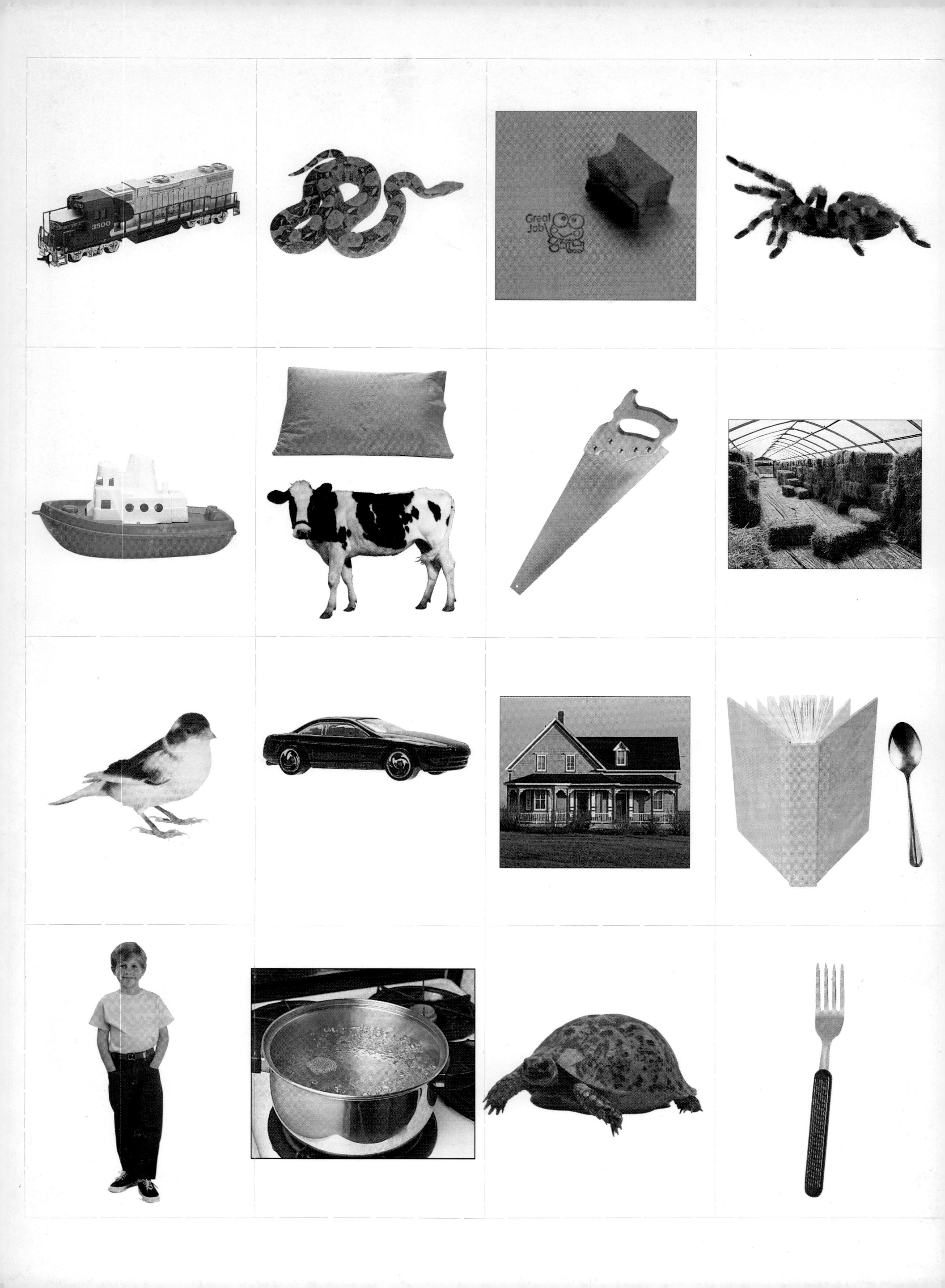
3500
Great
Job